Rocket Design and Construction Fundamentals

Richard Skiba

AFTER MIDNIGHT
PUBLISHING

Skiba, Richard (author)

Rocket Design and Construction Fundamentals

ISBN 978-1-7638046-5-4 (Paperback) 978-1-7638046-6-1 (eBook) 978-1-7638046-7-8 (Hardcover)

Non-fiction

Contents

PART 1

Foundations of Rocketry

Chapter 1
Introduction to Rocketry

The History of Rockets: From Fireworks to Space Exploration

Ancient Beginnings

The origins of rocketry can be traced back to ancient China, where the development of gunpowder in the 9th century led to the creation of early rocket propulsion systems. These primitive devices, known as "fire arrows," were essentially gunpowder-filled tubes attached to arrows, which demonstrated the fundamental principle of rocket propulsion through the expansion of gases [1]. This early innovation laid the groundwork for future advancements in rocketry, illustrating the potential of controlled propulsion systems that would later be refined and expanded upon in various cultures [2].

The history of rockets spans centuries of human ingenuity and technological advancement, beginning with ancient experiments and evolving into modern marvels of engineering. Rockets today are a testament to thousands of years of cumulative experimentation and research, blending principles of physics, chemistry, and engineering that have been refined over time [3].

The foundational principle of rocketry, the action-reaction concept (later formalized as Newton's Third Law), was demonstrated as early as 400 BCE by the Greek philosopher Archytas in Tarentum, southern Italy. Archytas created a wooden bird that flew along suspended wires using escaping steam as propulsion. This experiment amused and mystified his contemporaries, showcasing the potential of controlled propulsion [3].

Three centuries later, Hero of Alexandria, another Greek inventor, developed a device called the aeolipile. It consisted of a sphere mounted on a boiling water kettle. Steam travelled through pipes into the sphere, escaping through two L-shaped nozzles. This caused the sphere to rotate, illustrating the fundamental principles of thrust and propulsion. While these devices were not true rockets, they laid the conceptual groundwork for rocket propulsion [3].

The first true rockets likely emerged in China around the 9th to 10th centuries with the discovery of gunpowder. Initially used in fireworks for religious ceremonies, bamboo tubes filled with gunpowder were thrown into fires to create explosions. Some tubes, instead of bursting, propelled themselves out of the flames, leading to the development of gunpowder-propelled devices [3].

By the 13th century, during the Song dynasty, these fire-arrow rockets were adapted for military purposes. The Battle of Kai-Keng in 1232 marked a pivotal moment when the Chinese used "arrows of flying fire" to repel Mongol invaders. These rockets consisted of a capped tube filled with gunpowder attached to a long guiding stick, providing a simple yet effective weapon [3].

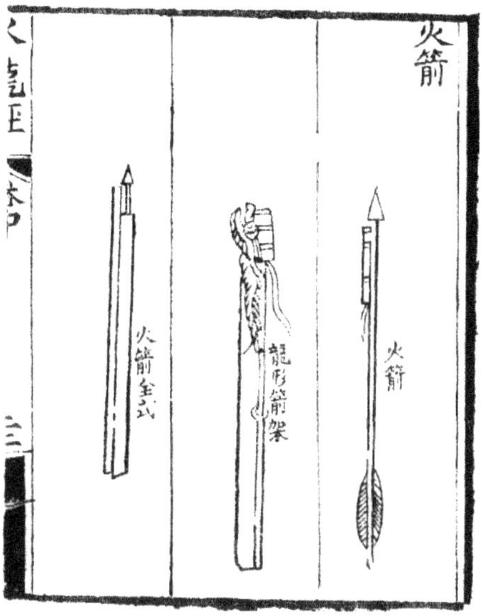

Figure 1: The oldest depiction of rocket arrows. From the Huolongjing. The right arrow reads 'fire arrow,' the middle one is an 'arrow frame in the shape of a dragon,' and the left one is a 'complete fire arrow.' ed.: 焦玉 (Jiāo Yù and 文言 (w:Liu Bowen), CC BY-SA 4.0, via Wikimedia Commons.

The Mongols, impressed by Chinese rocketry, adopted the technology and likely facilitated its spread across Eurasia during their invasions in the 13th century. In the Middle East, scholars such as Hasan al-Rammah documented advanced rocket recipes and techniques, enhancing their effectiveness. By the late medieval period, rockets had reached Europe, where they were adapted for warfare [3].

European innovators contributed significantly to rocketry's evolution. In 14th-century France, Jean Froissart proposed launching rockets through tubes for greater accuracy—a precursor to the modern bazooka. Italian engineer Joanes de Fontana designed rocket-powered torpedoes for naval warfare. Rockets also gained prominence in celebratory displays, with German fireworks maker Johann Schmidlap inventing the "step rocket" in the 16th century, a concept foundational to modern multi-stage rockets [3].

The Kingdom of Mysore in India advanced rocketry in the 18th century with the development of iron-cased rockets, which offered increased range and stability. These rockets were effectively used against British forces during the Anglo-Mysore Wars. The British later adopted and refined the technology, creating the Congreve rocket, which played a role in the Napoleonic Wars and inspired the iconic line in "The Star-Spangled Banner" [3].

The 19th and early 20th centuries marked a shift from military applications to scientific exploration. Russian scientist Konstantin Tsiolkovsky introduced the theoretical foundation of modern rocketry, including the rocket equation and the use of liquid propellants. In the United States, Robert H. Goddard conducted pioneering experiments with liquid-fuelled rockets, launching the first successful model in 1926 [3].

Figure 2: V-2 rocket. Leon Petrosyan, CC BY-SA 3.0, via Wikimedia Commons.

Rocket technology reached unprecedented heights during World War II with the German V-2 rocket, the world's first long-range ballistic missile. Although designed for destruction, the V-2 demonstrated the feasibility of reaching the upper atmosphere, inspiring post-war space exploration efforts.

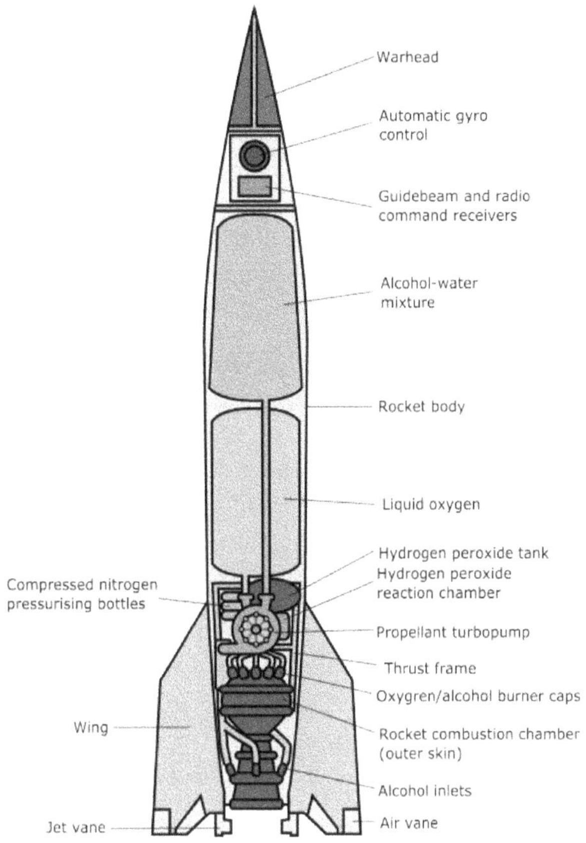

Figure 3: Diagram of the V-2 rocket used in world war 2 (based on a public domain image released by user Fastfission). j4p4n, Public Domain, via Openclipart.

The Birth of Rocket Science in the West

The knowledge of rocketry spread to Europe and the Middle East during the 13th and 14th centuries, primarily through trade and military conflict. The Mongols played a significant role in this transmission, utilizing gunpowder-based weapons and introducing these technologies to distant regions [1]. In Europe, the development of rocketry was sporadic and largely focused on military applications. Notable figures such as William Congreve emerged in the 19th century, refining rocket designs for military use. His Congreve rockets were notably employed by the British during the Napoleonic Wars and the War of 1812, marking a significant advancement in the application of rocketry for warfare [4].

The development of rocketry as a science and practical technology owes much to advancements in understanding physical motion, most notably articulated by Sir Isaac Newton during the 17th century. Newton's three laws of motion laid the scientific foundation for modern rocketry, explaining not only how rockets work but why they function effectively even in the vacuum of space. These principles began to influence practical designs soon after their publication, sparking early experiments that demonstrated the potential of controlled propulsion.

By the early 18th century, engineers and scientists began applying Newtonian mechanics to experimental propulsion systems. Around 1720, Willem Gravesande, a Dutch professor, constructed model vehicles propelled by steam jets. While these were rudimentary in design, they demonstrated the potential of using pressurized gas to generate thrust. Around the same time, rocket experimentation gained momentum in Germany and Russia, where some rockets exceeded 45 kilograms in mass. These powerful prototypes often produced exhaust flames so intense that they scorched the ground prior to launch, showcasing the dramatic potential of early rocket designs [3].

The late 18th and early 19th centuries saw a resurgence of rockets as weapons of war, particularly in response to their success in non-European contexts. Indian forces notably used rockets to repel British advances during the Anglo-Mysore Wars of 1792 and 1799. These iron-cased Mysorean rockets inspired British artillery officer Colonel William Congreve, who developed improved designs for military use. Congreve's rockets gained widespread attention during the War of 1812, notably in the British bombardment of Fort McHenry, an event immortalized in Francis Scott Key's poem, later the U.S. national anthem, "The Star-Spangled Banner" [3].

Despite their effectiveness in large numbers, early rockets were notoriously inaccurate. This limitation led to innovations aimed at improving their stability and precision. Among the most significant was spin stabilization, introduced by Englishman William Hale in the mid-19th century. Hale's design incorporated vanes at the rocket's base, redirecting exhaust gases to induce a stabilizing spin, much like rifling does for bullets. This innovation laid the groundwork for modern rocket guidance systems.

While rockets experienced success in various military engagements throughout Europe, their dominance waned with the advent of more advanced artillery. Rifled, breech-loading cannons with explosive shells proved far more effective in combat, relegating rockets to secondary roles. However, peacetime uses of rocketry, particularly in fireworks and entertainment, preserved the technology and kept experimentation alive. The development of multi-staged rockets during this period, such as those pioneered by Johann Schmidlap, foreshadowed the designs later used for space exploration [3].

The 16th and 17th centuries also saw important theoretical contributions to rocketry. Conrad Haas, a 16th-century military engineer, wrote a manuscript detailing concepts such as multi-staged rockets, delta-shaped fins, and liquid propellants. Rediscovered centuries later, Haas's work demonstrated an advanced understanding of rocketry for his time. Similarly, Kazimierz Siemienowicz's 1650 treatise, *The Great Art of Artillery*, served as a standard reference for artillery and rocket design for over two centuries, offering insights into multi-stage rockets and stabilization mechanisms [3].

William Congreve's work in the early 19th century marked a significant leap in rocket design. Drawing inspiration from Mysorean technologies, he developed iron-cased rockets with improved propellant formulations and aerodynamic designs. These rockets were not only lighter but also capable of delivering payloads over greater distances. Congreve's innovations were adopted across Europe and beyond, and they played prominent roles in battles such as Waterloo. However, their inherent inaccuracies continued to limit their strategic use.

Efforts to address these limitations included Alexander Dmitrievich Zasyadko's work in Russia, where he introduced rocket launch platforms capable of firing salvos. These innovations highlighted the growing interest in maximizing the tactical utility of rockets, even as they competed with emerging artillery technologies [3].

Theoretical Foundations

The late 19th and early 20th centuries marked a pivotal shift from practical applications of rocketry to a more scientific inquiry into its underlying principles. Russian scientist Konstantin Tsiolkovsky is often credited with laying the theoretical groundwork for modern rocketry. His seminal 1903 paper proposed the use of liquid fuels for rockets and introduced the Tsiolkovsky rocket equation, which describes the relationship between a rocket's velocity and its fuel efficiency [5]. Concurrently, Robert H. Goddard, known as the "Father of Modern Rocketry," conducted groundbreaking experiments with liquid-fuelled rockets. In 1926, he successfully launched the world's first liquid-fuelled rocket, demonstrating the feasibility of controlled propulsion systems and establishing principles that continue to underpin modern rocketry [2].

Rockets in War and the Space Race

The development of rocketry accelerated dramatically during World War II, particularly in Germany, where the V-2 rocket became the first long-range guided ballistic missile. Designed by Wernher von Braun, the V-2 not only showcased the destructive capabilities of rocket technology but also hinted at its potential for peaceful exploration [1]. The subsequent Cold War era transformed rocketry from a military tool into an instrument of scientific discovery, culminating in the Space Race between the United States and the Soviet Union. The launch of Sputnik 1 in 1957 marked the beginning of the Space Age, followed by significant milestones such as Yuri Gagarin's historic flight in 1961 and the Apollo 11 mission in 1969, which saw humans land on the Moon [1, 4].

By the mid-19th century, rocketry began transitioning from a primarily military technology to one of scientific inquiry. The refinement of stabilization techniques, advancements in materials, and the increasing availability of detailed theoretical frameworks paved the way for its use in space exploration. Early rocket pioneers such as Robert Anderson proposed metal-cased designs, while others experimented with multi-staged rockets and enhanced propellant systems. These developments set the stage for the transformative breakthroughs of the 20th century [3].

Modern Rockets and Space Exploration

In the post-Apollo era, rocket technology has expanded to encompass a wide range of applications, including satellite deployment and commercial ventures. NASA's Space Shuttle program exemplified a reusable approach to spaceflight, facilitating the construction of the International Space Station (ISS) [1]. Additionally, the rise of private companies such as SpaceX has revolutionized the industry, with innovations like the Falcon 9 and Falcon Heavy rockets significantly reducing the cost of access to space [6]. Other nations, including China and India, have also developed their own space programs, contributing to a global landscape of advancements in rocketry and space exploration [1].

The late 19th and early 20th centuries marked the beginning of modern rocketry, driven by visionaries like Konstantin Tsiolkovsky, Robert H. Goddard, and Hermann Oberth. Their groundbreaking work laid the foundation for space exploration and the development of rocket technology.

In 1898, Russian schoolteacher Konstantin Tsiolkovsky (1857–1935) conceptualized space exploration using rockets. His 1903 report proposed using liquid propellants to achieve greater range and efficiency. Tsiolkovsky's pivotal insight was that a rocket's speed and range were limited only by the exhaust velocity of its escaping gases. This principle became a cornerstone of modern astronautics, earning Tsiolkovsky the title "Father of Modern Astronautics." His theoretical work inspired future generations and provided a scientific basis for achieving space travel [3].

American scientist Robert H. Goddard (1882–1945) turned Tsiolkovsky's theories into reality. Driven by a desire to reach higher altitudes than balloons could achieve, Goddard published *A Method of Reaching Extreme Altitudes* in 1919, exploring the mathematics of meteorological sounding rockets. Initially, Goddard experimented with solid propellants, refining their combustion and measuring exhaust velocities. He soon recognized the superior potential of liquid propellants, a concept that had never been successfully implemented [3].

On March 16, 1926, Goddard launched the world's first liquid-propellant rocket in Auburn, Massachusetts. Powered by liquid oxygen and gasoline, the rocket flew for 2.5 seconds, reached a height of 12.5 meters, and landed 56 meters away. Though modest by modern standards, this flight marked the beginning of a new era in rocketry, comparable to the Wright brothers' first powered flight in aviation history. Goddard's later innovations included gyroscopic flight control, scientific payload compartments, and parachute recovery systems, solidifying his reputation as the "Father of Modern Rocketry" [3].

Hermann Oberth (1894–1989), a German-born physicist, advanced the vision of space exploration. In 1923, he published *Die Rakete zu den Planetenräumen* ("The Rocket into Planetary Space"), a seminal work on using rockets for interplanetary travel. Oberth's writings spurred the formation of rocket societies worldwide, including Germany's Verein für Raumschiffahrt (Society for Space Travel). This organization contributed directly to the development of the V-2 rocket during World War II, a project that brought together German engineers, including Oberth and Wernher von Braun [3].

The V-2 rocket, or A-4, represented the pinnacle of wartime rocketry. Developed under von Braun's leadership at Peenemünde, it was powered by a mixture of liquid oxygen and alcohol, burning at a rate of one ton every seven seconds. Despite its relatively small size, the V-2 could devastate city blocks upon impact. Although introduced too late to change the outcome of World War II, the V-2's technology laid the groundwork for post-war rocket development [3].

After the war, captured V-2 rockets and German scientists became pivotal in shaping the rocket programs of both the United States and the Soviet Union. In the U.S., Wernher von Braun and his team advanced Goddard's early ideas, eventually leading to rockets like the Redstone, Atlas, and Titan, which played key roles in launching astronauts into space [3].

The Soviet Union also capitalized on German technology, developing the R-7 rocket, which launched *Sputnik I*, the first artificial satellite, on October 4, 1957. This achievement ignited the Space Race, culminating in milestones like Yuri Gagarin's historic orbital flight in 1961 and the U.S. Apollo Moon landings later in the decade [3].

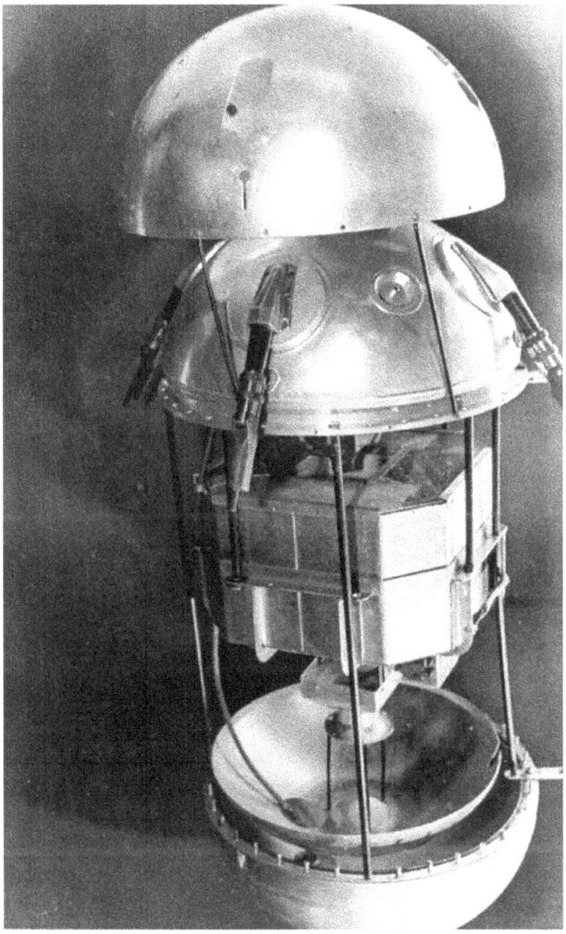

Figure 4: Sputnik 1 with its components visible (Source: NASA). ITU Pictures, CC BY 2.0, via Flickr.

The collective contributions of Tsiolkovsky, Goddard, and Oberth transformed rocketry from a nascent field into the backbone of space exploration. Their vision and ingenuity continue to inspire advancements in aerospace technology, proving that the dreams of the past can fuel the achievements of the future.

The history of rocketry spans centuries. From ancient experiments with gunpowder to modern advancements in space exploration. The following timeline highlights key milestones that have shaped the development of rocket technology.

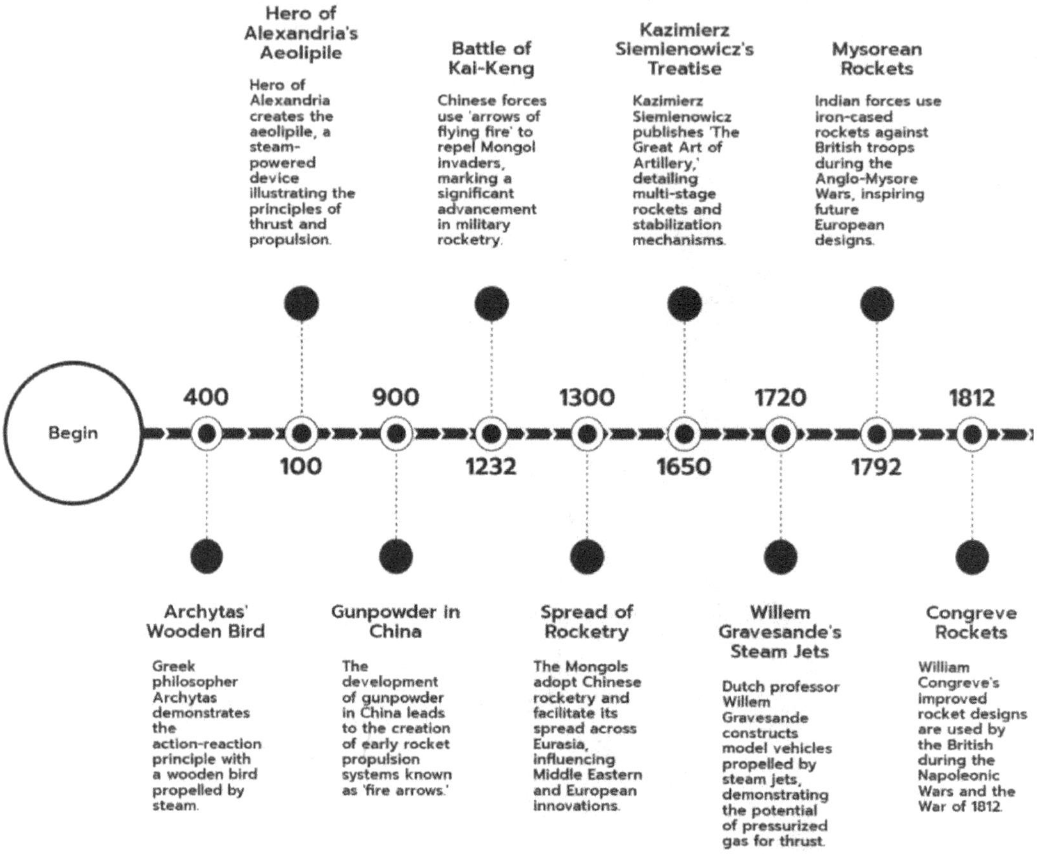
Begin
400
Archytas' Wooden Bird
Greek philosopher Archytas demonstrates the action-reaction principle with a wooden bird propelled by steam.
100
Hero of Alexandria's Aeolipile
Hero of Alexandria creates the aeolipile, a steam-powered device illustrating the principles of thrust and propulsion.
900
Gunpowder in China
The development of gunpowder in China leads to the creation of early rocket propulsion systems known as 'fire arrows.'
1232
Battle of Kai-Keng
Chinese forces use 'arrows of flying fire' to repel Mongol invaders, marking a significant advancement in military rocketry.
1300
Spread of Rocketry
The Mongols adopt Chinese rocketry and facilitate its spread across Eurasia, influencing Middle Eastern and European innovations.
1650
Kazimierz Siemienowicz's Treatise
Kazimierz Siemienowicz publishes 'The Great Art of Artillery,' detailing multi-stage rockets and stabilization mechanisms.
1720
Willem Gravesande's Steam Jets
Dutch professor Willem Gravesande constructs model vehicles propelled by steam jets, demonstrating the potential of pressurized gas for thrust.
1792
Mysorean Rockets
Indian forces use iron-cased rockets against British troops during the Anglo-Mysore Wars, inspiring future European designs.
1812
Congreve Rockets
William Congreve's improved rocket designs are used by the British during the Napoleonic Wars and the War of 1812.

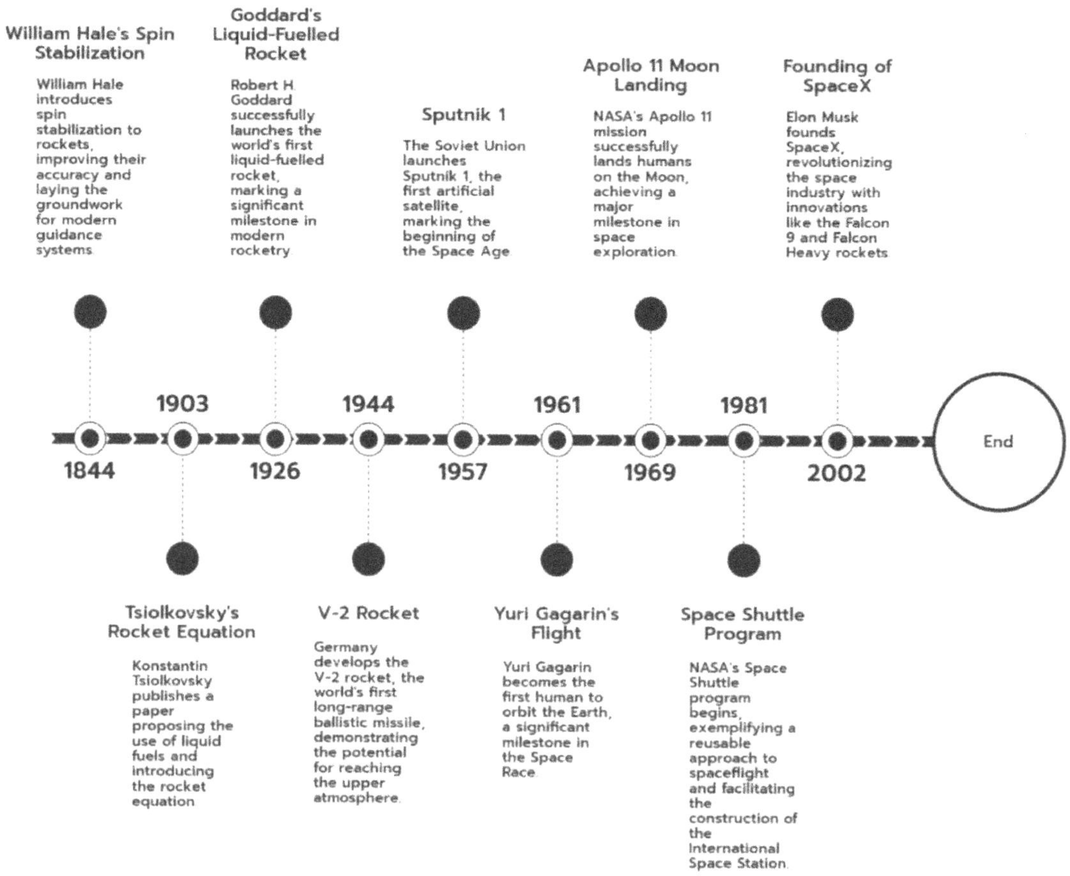
William Hale's Spin Stabilization
William Hale introduces spin stabilization to rockets, improving their accuracy and laying the groundwork for modern guidance systems
Goddard's Liquid-Fuelled Rocket
Robert H. Goddard successfully launches the world's first liquid-fuelled rocket, marking a significant milestone in modern rocketry
Sputnik 1
The Soviet Union launches Sputnik 1, the first artificial satellite, marking the beginning of the Space Age
Apollo 11 Moon Landing
NASA's Apollo 11 mission successfully lands humans on the Moon, achieving a major milestone in space exploration
Founding of SpaceX
Elon Musk founds SpaceX, revolutionizing the space industry with innovations like the Falcon 9 and Falcon Heavy rockets
1844
1903
1926
1944
1957
1961
1969
1981
2002
End
Tsiolkovsky's Rocket Equation
Konstantin Tsiolkovsky publishes a paper proposing the use of liquid fuels and introducing the rocket equation
V-2 Rocket
Germany develops the V-2 rocket, the world's first long-range ballistic missile, demonstrating the potential for reaching the upper atmosphere
Yuri Gagarin's Flight
Yuri Gagarin becomes the first human to orbit the Earth, a significant milestone in the Space Race
Space Shuttle Program
NASA's Space Shuttle program begins, exemplifying a reusable approach to spaceflight and facilitating the construction of the International Space Station

Key achievements in rocket science (Sputnik, Apollo, SpaceX)

Rocket science has seen monumental achievements over the decades, each marking a leap in technology and our understanding of space. Milestones like the launch of *Sputnik*, the Apollo Moon landings, and SpaceX's advancements have redefined what is possible, transforming science fiction into reality.

Sputnik: The Dawn of the Space Age

On October 4, 1957, the Soviet Union launched Sputnik I, the world's first artificial satellite [7, 8]. This beach ball-sized metal sphere, equipped with a simple radio transmitter, orbited Earth and transmitted beeping signals that could be detected by ground stations [7, 8]. Although rudimentary by today's standards, Sputnik demonstrated the feasibility of spaceflight and heralded the start of the Space Age [7, 8].

The implications of Sputnik extended far beyond its technical achievement. It marked the beginning of the Space Race between the United States and the Soviet Union, two superpowers vying for technological and ideological dominance during the Cold War [7, 8]. The satellite's successful launch by the R-7 rocket showcased Soviet expertise in rocketry and spurred the U.S. government to accelerate its own space program [7, 8]. Within months, the U.S. launched Explorer 1 and established NASA in 1958, cementing the institutional framework for future space exploration [7, 8].

Apollo: Humanity's Giant Leap

The Apollo program, initiated by NASA in 1961, remains one of the most significant achievements in human history. President John F. Kennedy's bold challenge to land a man on the Moon and return him safely to Earth galvanized an unprecedented technological and logistical effort [9].

The program culminated in the successful Apollo 11 mission in July 1969, when Neil Armstrong and Buzz Aldrin became the first humans to set foot on the lunar surface [9]. The Saturn V rocket, a towering three-stage launch vehicle, was the backbone of the Apollo program [9]. At 363 feet tall and capable of generating 7.6 million pounds of thrust, it remains one of the most powerful rockets ever built [9].

Figure 5: The Apollo 11 mission, the first manned lunar mission, launched from the Kennedy Space Center, Florida via the Marshall Space Flight Center (MSFC) developed Saturn V launch vehicle on July 16, 1969 and safely returned to Earth on July 24, 1969. Defense Visual Information Distribution Service, Public Domain, via National Archives and Defense Visual Information Distribution Service.

The Apollo missions achieved a series of firsts, including lunar orbit (Apollo 8), precision Moon landings (Apollo 12), and extended lunar exploration (Apollo 17) [10]. These missions yielded not only scientific data but also cultural and political triumphs, solidifying the United States' leadership in space exploration [9].

Figure 6: This is a photograph of the Apollo 8 Capsule being hoisted onto the recovery ship following splashdown on December 27, 1968. The first manned Apollo mission to escape Earth's gravity and travel to the lunar vicinity, the Saturn V, SA-503, Apollo 8 mission liftoff occurred seven days prior, on December 21, 1968. Defense Visual Information Distribution Service, Public Domain, via National Archives and Defense Visual Information Distribution Service.

The Apollo program demonstrated the possibilities of human ingenuity and international collaboration [9]. The Moon rocks it brought back deepened our understanding of planetary formation [11], and its technological advancements laid the groundwork for future space missions, including the International Space Station [12].

The program's scientific legacy is also significant. The Apollo Passive Seismic Experiment, for example, provided valuable seismic data about the Moon's interior structure [13]. Additionally, the Lunar Sourcebook, a comprehensive reference on the Moon's composition and environment, was a key outcome of the program [14].

SpaceX: Revolutionizing Space Access

SpaceX, founded in 2002 by entrepreneur Elon Musk, has transformed the aerospace industry through its innovative and cost-reducing approaches [15]. One of its primary goals is to make space exploration accessible and sustainable, with the ultimate vision of enabling human settlement on Mars [15, 16].

SpaceX achieved a major milestone in 2008 when its Falcon 1 became the first privately developed rocket to reach orbit [15]. This breakthrough was followed by the development of the Falcon 9, a reusable rocket designed to lower the cost of launches significantly [15, 17]. The successful recovery and reuse of Falcon 9's first stage in 2015 represented a paradigm shift in spaceflight economics, proving that rockets could be treated more like airplanes than disposable vehicles [17].

Figure 7: Liftoff of Falcon 1 Flight 4 on September 29, 2008, carrying Ratsat demo. SpaceX, CC BY-SA 3.0, via Wikimedia Commons.

The Dragon spacecraft, another SpaceX innovation, was the first private spacecraft to dock with the International Space Station (ISS) in 2012 [15] . In 2020, SpaceX's Crew Dragon became the first commercial vehicle to carry astronauts to the ISS, heralding a new era of public-private partnerships in space exploration [15].

Figure 8: SPX-8 SpaceX Dragon Spacecraft Grappled by SSRMS. NASA, Public Domain, via Picryl.

SpaceX's most ambitious project is the Starship, a fully reusable spacecraft designed for deep-space missions, including lunar and Martian exploration [15]. When operational, Starship could carry large payloads and crews to distant destinations, significantly expanding humanity's reach into the cosmos [15, 16].

The impact of SpaceX's innovations extends beyond the technical achievements. Researchers have explored the implications of SpaceX's reusable launch vehicles and the resulting pricing competition in the space industry [17, 18]. Additionally, studies have investigated the environmental impacts of space tourism and the need for sustainable practices in the growing space economy [19, 20].

Furthermore, SpaceX's collaborations with international partners and its strategies for managing these partnerships have been analysed, providing valuable insights for future space collaborations [21]. The company's network effects and innovation strategies have also been the subject of academic research [16].

SpaceX has revolutionized space access through its innovative and cost-reducing approaches, making space exploration more accessible and sustainable. The company's achievements, including the successful development of reusable rockets, the Dragon spacecraft, and the ambitious Starship project, have transformed the aerospace industry and paved the way for a new era of space exploration and commercialization.

Legacy and Interconnections

Sputnik, Apollo, and SpaceX represent pivotal moments in the evolution of rocket science [22, 23]. Sputnik broke the barrier of Earth's atmosphere, marking a significant milestone in space exploration [22]. Apollo fulfilled the dream of walking on another celestial body, a major achievement in human spaceflight [23]. SpaceX is redefining how we approach space travel in the 21st century, with its innovative reusable rocket technology [22].

Together, these achievements illustrate the relentless pursuit of knowledge and the human spirit's quest to explore the unknown [22, 23]. They continue to inspire new generations of scientists, engineers, and explorers, ensuring that the legacy of rocket science will propel humanity to even greater heights [22, 23].

The Sputnik launch shocked the United States, triggering the Space Race between the US and the Soviet Union [22]. This event had significant ramifications for science education, leading to increased focus on science, technology, engineering, and mathematics (STEM) fields in the US [24].

Sputnik, Apollo, and SpaceX have also had broader impacts on society, such as advancements in aerospace medicine [23] and the commercialization of space [22]. These milestones continue to inspire new research and technological developments, pushing the boundaries of what is possible in space exploration [22].

Since 2022, significant advancements in rocket design and space exploration have propelled humanity's quest to explore beyond Earth. These developments encompass innovative propulsion systems, reusable rocket technologies, ambitious lunar missions, and groundbreaking space telescopes, collectively enhancing our capabilities in space travel and deepening our understanding of the cosmos.

NASA has validated the Rotating Detonation Rocket Engine (RDRE), a revolutionary propulsion design that utilizes supersonic combustion, or detonation, to generate thrust. This engine produces more power while consuming less fuel compared to traditional propulsion systems, holding promise for future deep space missions to destinations like the Moon and Mars.

SpaceX continues to lead in reusable rocket technology, implementing over a thousand upgrades to its Starship rocket to enhance performance and reusability. These improvements include a major change in stage separation methods and propulsion system enhancements, aiming for rapid turnaround between launches to significantly reduce space travel costs.

Similarly, Europe's Maiaspace, a subsidiary of ArianeGroup, is developing a partially reusable rocket expected to launch by 2026, marking Europe's entry into the reusable rocket market.

NASA's Artemis program has experienced schedule adjustments, with the Artemis II mission, intended to fly astronauts around the Moon, now targeting April 2026. This mission aims to pave the way for a subsequent lunar landing, reinforcing NASA's commitment to returning humans to the Moon and establishing a sustainable presence.

Concurrently, NASA's Space Launch System (SLS) rocket, integral to the Artemis missions, has undergone upgrades to support these ambitious lunar endeavours.

The James Webb Space Telescope (JWST), operational since 2022, has provided unprecedented views of distant cosmic objects and rocky exoplanets within habitable zones. Its observations have brought scientists closer to detecting potential signs of life beyond our solar system, marking a significant leap in astrophysics and space exploration.

Figure 9: James Webb Space Telescope Mirror Seen in Full Bloom. NASA's James Webb Space Telescope, CC BY 2.0, via Flickr.

These advancements reflect a concerted global effort to push the boundaries of space exploration, leveraging cutting-edge technologies and international collaborations to achieve milestones once deemed unattainable.

Overview of rocket design and its role in modern space exploration

Rocket design is the cornerstone of modern space exploration, enabling humanity to venture beyond Earth's atmosphere and into the cosmos. The fundamental principle behind rockets lies in Newton's Third Law of Motion: for every action, there is an equal and opposite reaction. This principle allows rockets to generate thrust by expelling mass (propellant) at high speeds in the opposite direction. Over the decades, advancements in materials, engineering, and propulsion systems have transformed rockets from rudimentary fire arrows into sophisticated machines capable of interplanetary travel.

Modern rockets are highly specialized vehicles composed of several critical components, each playing a vital role in ensuring mission success:

1. **Propulsion Systems**: These include solid, liquid, or hybrid rocket engines that provide the thrust necessary to overcome Earth's gravity. Liquid engines, such as those used in the SpaceX Falcon 9 and NASA's Space Launch System (SLS), allow precise control of thrust and are critical for complex manoeuvres.

2. **Structural Systems**: The rocket's frame must be strong enough to withstand intense forces during launch while remaining lightweight to maximize payload capacity. Advanced composite materials and alloys are commonly used.

3. **Payload Systems**: Rockets are designed to carry satellites, scientific instruments, crew modules, or other cargo to specific destinations. Payloads dictate the rocket's size and design.

4. **Guidance and Control Systems**: These systems include gyroscopes, accelerometers, and computers that ensure the rocket maintains its intended trajectory and adjusts for deviations in real time.

5. **Staging**: Multi-stage rockets discard empty fuel tanks to reduce weight, allowing the upper stages to reach higher velocities and altitudes. This innovation is key to achieving orbit or interplanetary trajectories.

Modern rockets have incorporated reusable technology, significantly lowering the cost of access to space. SpaceX's Falcon 9 and Starship rockets are prime examples, with booster stages capable of landing vertically for refurbishment and reuse. These advancements not only improve efficiency but also enable more frequent and ambitious missions.

Rockets are indispensable for space exploration, performing functions critical to humanity's understanding and utilization of space:

1. **Launching Satellites**: Rockets deploy satellites into orbits around Earth, supporting communication, weather forecasting, navigation, and scientific research.

2. **Interplanetary Missions**: Rockets enable spacecraft to escape Earth's gravity and travel to other planets, moons, and celestial bodies. Missions like NASA's Perseverance rover on Mars and ESA's JUICE mission to Jupiter are made possible by sophisticated launch vehicles.

3. **Human Spaceflight**: Rockets like NASA's SLS and SpaceX's Starship are designed to transport astronauts to destinations such as the International Space Station (ISS), the Moon, and eventually Mars.

4. **Space Commercialization**: Rockets facilitate the deployment of commercial payloads, including space tourism ventures, internet constellations like Starlink, and manufacturing experiments in microgravity.
5. **Scientific Research**: Rockets carry telescopes, probes, and instruments beyond Earth's atmosphere to study phenomena such as cosmic radiation, the origins of the universe, and potential extraterrestrial life.

Despite significant progress, rocket design faces challenges such as reducing costs, minimizing environmental impact, and enhancing payload efficiency. Emerging technologies like electric propulsion, additive manufacturing (3D printing), and new propellant types promise to revolutionize rocket design further. Additionally, the development of fully reusable rockets, such as SpaceX's Starship and Blue Origin's New Glenn, marks a paradigm shift toward sustainable space exploration.

Space launch vehicles are the foundation of modern space exploration, serving as the primary means to lift payloads such as satellites, spacecraft, and crewed missions into orbit. These vehicles are sophisticated rocket systems designed to overcome Earth's gravity and place payloads in specified trajectories or orbits. With the exception of the now-retired U.S. Space Shuttle, the majority of space missions rely on expendable launch vehicles (ELVs), which are single-use systems that are discarded after each mission [25].

The space launch vehicle market is categorized by payload capacity [25]:

1. Small Launch Vehicles:

- Designed for lightweight payloads, these rockets can carry up to a few hundred kilograms into orbit.
- Orbital Sciences Corporation's *Pegasus* is a notable example, capable of air-launching payloads of up to 500 kg into low Earth orbit (LEO). This method increases flexibility and reduces launch costs.

2. Medium Launch Vehicles:

- These are capable of carrying payloads in the range of 800–2,000 kg to LEO. Examples include Orbital Sciences' *Taurus* and Lockheed Martin's *Athena I* and *Athena II*.

3. Heavy Launch Vehicles:

- Used for multiton payloads such as large satellites or interplanetary spacecraft. Leading manufacturers include Lockheed Martin (*Atlas-Centaur* and *Titan*), Boeing (*Delta*), and international counterparts like Russia's *Proton* (Khrunichev), Ukraine's *Zenit* (Yuzhnoye), China's *Long March* (Great Wall Aerospace), and Japan's *H-II*.

In the early 1980s, the United States dominated the commercial space launch market, but by the mid-1990s, its share had declined to approximately 30%. Europe's *Ariane* rockets, developed by the European Space Agency (ESA), captured over half the commercial market, underscoring the competitiveness of international players. These trends highlight the dynamic nature of the industry, where innovation and cost-effectiveness play crucial roles [25].

The U.S. Space Shuttle was an unprecedented system combining the functions of a launch vehicle and a reusable space platform. Capable of carrying up to 30 tons to LEO, the shuttle fleet included four operational orbiters by the early 1990s. However, while components like solid rocket boosters were recoverable and reusable, the external fuel tank was discarded after each launch. The shuttle's versatility allowed for satellite deployment, ISS construction, and scientific research but came at a high operational cost [25].

Spacecraft launched by these vehicles fall into two categories [25]:

1. Uncrewed Spacecraft:

- Satellites in Earth orbit serve purposes ranging from weather observation and communications to remote sensing and navigation. Probes, on the other hand, are sent on interplanetary missions for scientific exploration.
- Commercial satellites have become the backbone of telecommunications, supporting global mobile networks and broadcasting. Companies such as Boeing, Lockheed Martin, and Astrium dominate the manufacturing of these systems.

2. Crewed Spacecraft:

- These systems face greater technical and cost challenges due to the need for life support and safe re-entry mechanisms.
- Examples include Russia's *Soyuz*, the International Space Station (ISS), and the now-retired U.S. Space Shuttle. The ISS, a collaborative project among 15 nations, exemplifies international cooperation in space exploration.

The rocket design and construction industry is a global endeavour, with contributions from countries like the United States, Russia, China, India, and Japan. Government agencies such as NASA, ESA, Roscosmos, and ISRO lead major programs, often collaborating with private aerospace firms. Commercial players like SpaceX and Blue Origin have introduced cost-saving innovations, such as reusable rocket stages, revolutionizing the accessibility and economics of space exploration.

The Rocket Design and Construction Industry

The rocket design and construction industry is a dynamic and rapidly evolving sector that plays a pivotal role in modern space exploration, satellite deployment, and commercial ventures. This industry combines cutting-edge engineering, advanced materials science, and precision manufacturing to create vehicles capable of withstanding the extreme conditions of spaceflight. Over the past decade, the rocket industry has experienced transformative growth, driven by innovations in reusable technologies, increased demand for satellite launches, and a renewed global interest in space exploration.

The global space launch services market is valued at approximately $12 billion as of 2023, with projections to reach $30 billion by 2030, growing at a compound annual growth rate (CAGR) of 12.3% during the forecast period [26]. The rocket design and construction segment represents a significant portion of this market, as launch vehicles are essential for delivering satellites, cargo, and crew into space. The increasing number of small satellites and mega-constellations, such as SpaceX's Starlink and Amazon's Project Kuiper, has significantly boosted the demand for rocket launches. In 2022 alone, SpaceX conducted 61 successful Falcon 9 launches, setting a new record for the company and highlighting the growing commercial demand.

The rocket design and construction industry includes both legacy aerospace companies and innovative private firms:

- **SpaceX**: A market leader, SpaceX revolutionized the industry with reusable rocket technology, significantly reducing launch costs. Its Falcon 9 rocket costs approximately $62 million per launch, compared to hundreds of millions for traditional expendable rockets.
- **Blue Origin**: Focused on reusable rockets like New Shepard and New Glenn, Blue Origin aims to make space more accessible and affordable.
- **United Launch Alliance (ULA)**: A joint venture between Boeing and Lockheed Martin, ULA has developed rockets such as the Atlas V and Delta IV and is working on the Vulcan Centaur for future launches.
- **Arianespace**: This European company provides reliable launch services with its Ariane 5 and Vega rockets, focusing on satellite deployment.
- **Relativity Space**: An emerging player using 3D printing to manufacture rockets like Terran 1, aiming to reduce production time and costs.

The industry has seen groundbreaking advancements in rocket design and construction:

- **Reusable Rockets**: SpaceX's Falcon 9 and Falcon Heavy, as well as Blue Origin's New Shepard, have demonstrated the economic and environmental benefits of reusable stages. Reusability can reduce costs by up to 80% compared to traditional expendable rockets.
- **3D Printing**: Companies like Relativity Space leverage additive manufacturing to produce rocket components in weeks instead of months. This approach reduces material waste and production costs.
- **Composite Materials**: Lightweight and durable materials, such as carbon fibre composites, are increasingly used for rocket structures to enhance performance and payload capacity.
- **Electric Propulsion**: For orbital adjustments and long-duration missions, electric propulsion systems are gaining traction, offering higher efficiency than chemical propulsion.

- **Advanced Propellants**: Research into green propellants and cryogenic fuels aims to make rockets more environmentally friendly and efficient.

The rocket industry is a significant driver of employment and economic growth. In the U.S. alone, the aerospace and defence sector employed over 2.1 million people in 2022, with a substantial portion involved in rocket manufacturing and space exploration. Globally, the industry supports thousands of highly skilled jobs, ranging from aerospace engineers and materials scientists to software developers and technicians. The rise of private space companies has also stimulated the global supply chain, benefiting manufacturers of rocket engines, avionics, and ground support equipment. The economic ripple effects extend to industries such as telecommunications, navigation, and Earth observation, which depend on rocket launches for satellite deployment.

Despite its achievements, the rocket design and construction industry faces several challenges:

- Cost Reduction: While reusable rockets have lowered costs, further innovation is needed to make space access affordable for a wider range of stakeholders.
- Environmental Impact: Rocket launches produce greenhouse gases and other pollutants. Efforts to develop sustainable fuels and mitigate environmental damage are ongoing [27, 28].
- Regulatory Hurdles: Launch licenses, export controls, and space traffic management regulations can complicate operations for rocket companies.
- Technical Complexity: Rockets must perform flawlessly under extreme conditions, leaving no margin for error. Failures can result in catastrophic financial and reputational losses [29].

The rocket design and construction industry is on the brink of unprecedented growth, driven by several transformative trends that promise to redefine the future of space exploration and commercialization. One of the most significant trends is interplanetary exploration, with missions to the Moon, Mars, and other celestial bodies driving the development of advanced rocket technologies. Vehicles such as NASA's Space Launch System (SLS) and SpaceX's Starship are at the forefront, designed to carry heavy payloads and humans on deep space missions, marking a new era in space travel.

Another rapidly growing segment is commercial space tourism, spearheaded by companies like Blue Origin and Virgin Galactic. These pioneers are making suborbital and orbital space travel accessible to private individuals, creating a burgeoning market for recreational and experiential spaceflight. This innovation not only generates new revenue streams but also fosters public interest and investment in the space industry.

The deployment of mega-constellations represents another critical driver of industry growth. Projects like SpaceX's Starlink and Amazon's Project Kuiper aim to launch thousands of satellites to provide global internet coverage. The sheer scale of these constellations necessitates a high frequency of rocket launches, further fuelling demand for cost-effective and reliable launch vehicles.

Finally, global collaboration is playing a pivotal role in the evolution of the rocket industry. International partnerships among space agencies, private companies, and research institutions are fostering the exchange of knowledge, technology, and resources. This collaborative approach accelerates innovation and ensures a more sustainable and inclusive exploration of space, positioning the industry as a cornerstone of global progress. Together, these trends signify a vibrant and transformative future for the rocket design and construction sector.

The design and construction of space launch vehicles is a multidisciplinary effort that involves highly skilled professionals from various fields. Below is an overview of the key roles required:

Aerospace Engineers [30-34]

- **Role**: Aerospace engineers are the backbone of launch vehicle design. They focus on aerodynamics, propulsion systems, structural integrity, and flight dynamics.
- **Specializations**:
  - **Propulsion Engineers**: Design rocket engines, including liquid, solid, and hybrid propulsion systems.
  - **Structural Engineers**: Ensure the rocket's body can withstand high stresses and vibrations during launch.
  - **Systems Engineers**: Coordinate the integration of different subsystems to ensure the vehicle functions as a cohesive system.
- **Skills**: Advanced mathematics, physics, simulation software, and knowledge of materials science.

Mechanical Engineers [30, 35]

- **Role**: Responsible for the design and testing of mechanical systems, including joints, couplings, and moving parts such as gimbals for engine thrust vector control.
- **Tasks**:
  - Analyse thermal loads and mechanical stresses.
  - Develop cryogenic systems for handling liquid fuels like liquid oxygen (LOX) and liquid hydrogen.
- **Skills**: Computer-aided design (CAD), finite element analysis (FEA), and thermal dynamics.

Electrical and Electronics Engineers [30, 31, 33]

- **Role**: Develop and implement electrical systems for the launch vehicle, including avionics and power distribution.
- **Tasks**:
  - Design navigation, guidance, and control systems.
  - Develop telemetry systems for tracking and monitoring vehicle performance.
- **Skills**: Circuit design, embedded systems, and programming languages like C++ and Python.

Software Engineers [30, 31]

- **Role**: Create software that controls the rocket during all phases of the mission.
- **Tasks**:
  - Write algorithms for navigation, propulsion control, and fault detection.
  - Develop ground control software for launch operations.

- **Skills**: Proficiency in real-time programming, machine learning, and simulation tools.

Material Scientists and Engineers [30]

- **Role**: Select and develop materials capable of withstanding extreme conditions such as high temperatures, vibrations, and radiation.
- **Tasks**:
  - Develop lightweight composites for structural components.
  - Test heat-resistant materials for thermal shields.
- **Skills**: Knowledge of advanced materials like carbon composites and metal alloys.

Manufacturing Engineers [30]

- **Role**: Oversee the production and assembly of rocket components.
- **Tasks**:
  - Develop fabrication techniques such as additive manufacturing (3D printing) for engine components.
  - Manage the supply chain for critical materials and parts.
- **Skills**: Process optimization, quality control, and expertise in machining.

Test Engineers [30, 35, 36]

- **Role**: Design and execute tests to validate the performance and safety of the launch vehicle.
- **Tasks**:
  - Conduct static fire tests of engines and stage separation tests.
  - Perform vibration and thermal tests to simulate launch conditions.
- **Skills**: Data acquisition, failure analysis, and instrumentation.

Project Managers [30]

- **Role**: Coordinate teams, budgets, and timelines to ensure the project progresses as planned.
- **Tasks**:
  - Manage cross-functional teams across engineering, testing, and manufacturing.
  - Communicate project milestones to stakeholders.
- **Skills**: Leadership, scheduling software, and risk management.

Quality Assurance (QA) and Safety Specialists [30]

- **Role**: Ensure that all components meet strict safety and quality standards.
- **Tasks**:
  - Perform inspections and audits during manufacturing and assembly.
  - Verify compliance with industry standards and regulatory requirements.
- **Skills**: Attention to detail, documentation, and knowledge of safety protocols.

Propulsion Technicians [30]

- **Role**: Assemble and test propulsion systems, including rocket engines and fuel systems.
- **Tasks**:
  - Handle cryogenic fuels and hazardous materials.
  - Maintain and troubleshoot engine components.
- **Skills**: Mechanical skills, knowledge of safety procedures, and experience in handling high-pressure systems.

Integration Specialists [30]

- **Role**: Assemble and integrate subsystems into a complete launch vehicle.
- **Tasks**:
  - Align and attach stages, payloads, and fairings.
  - Ensure seamless connectivity of electrical and mechanical systems.
- **Skills**: Multidisciplinary understanding and hands-on experience.

Launch Operations Specialists [30]

- **Role**: Prepare and execute the launch of the vehicle.
- **Tasks**:
  - Monitor pre-launch checks, fuelling, and countdown sequences.
  - Respond to anomalies during launch operations.
- **Skills**: Team coordination, real-time problem-solving, and expertise in control systems.

The design and construction of space launch vehicles is a collaborative effort requiring expertise across engineering, science, manufacturing, and operations. As space exploration advances, these roles continue to evolve, offering exciting opportunities in an industry at the forefront of innovation.

Current Trends in the Aerospace Industry

The aerospace industry is experiencing transformative growth, marked by innovations and emerging markets that are reshaping the way humanity interacts with space. From reusable launch systems to space mining, the industry is poised for unprecedented expansion, leveraging advanced technology, private sector involvement, and sustainable practices [37].

Reusable Launch Systems (RLS): Reusable launch systems represent a revolutionary shift in the space industry, challenging the traditional expendable launch vehicle model. By reusing key components of rockets, RLS significantly reduces the cost, environmental impact, and logistical complexity of space missions.

- **Cost Reduction**: The elimination of single-use rockets has led to dramatic cost savings, enabling frequent small satellite launches, affordable space tourism, and cost-effective scientific missions.

- **Increased Launch Frequency**: With faster turnaround times, RLS allows for more frequent launches, facilitating rapid satellite constellation deployments and accelerating advancements in space-based technologies.
- **Environmental Sustainability**: RLS reduces space debris and fuel consumption, contributing to a cleaner and more sustainable future for space exploration.

Pioneered by companies like SpaceX, RLS has achieved significant milestones. The partially reusable Falcon 9 rocket has flown multiple missions, lowering launch costs for satellites and cargo. Starship, SpaceX's fully reusable rocket, aims to extend human presence to Mars and beyond, although its official launch is still pending. According to Fortune Business Insights, the reusable launch vehicle market was valued at US$ 1.61 billion in 2022 and is projected to grow to US$ 5.41 billion by 2030. North America leads the market, holding a 40% share in 2023, followed by Europe and Asia Pacific [37].

Space Tourism Development: Space tourism is transforming the aerospace industry, presenting a lucrative opportunity for commercialization. The global space tourism market, valued at US$ 869.20 million in 2022, is expected to exceed US$ 3.88 billion by 2032, growing at a compound annual growth rate (CAGR) of 16.20% [37].

- **Key Drivers**:
    - Reusable rockets have significantly lowered the cost of accessing space.
    - The emergence of private companies such as SpaceX, Blue Origin, and Virgin Galactic has increased competition and reduced barriers to entry.

Milestones in space tourism include Virgin Galactic's suborbital tourism flights in 2023 and Blue Origin's New Shepard missions. Additionally, organizations like NASA have facilitated public access to the International Space Station (ISS) for research and exploration. Challenges such as high costs, safety concerns, and environmental impacts remain, but advancements in technology and regulatory frameworks are addressing these issues, paving the way for broader participation in space tourism [37].

Emergence of Private Companies: The rise of private companies has revolutionized the aerospace industry, driving rapid technological advancements, increasing global investment, and democratizing space access.

- **Key Achievements** [37]:
    - **SpaceX**: Revolutionized the industry with reusable rockets, frequent launches, and ambitious plans for interplanetary missions.
    - **Blue Origin**: Developed reusable launch systems and suborbital space tourism vehicles.
    - **Rocket Lab**: Provides dedicated small satellite launches and is developing reusable orbital vehicles.

  - **OneWeb**: Deploying large constellations of internet-providing satellites.

Global venture capital funding for space startups reached US$ 16.8 billion in 2021, a 44% increase from 2020, highlighting the growing interest in space exploration. Private companies have fostered public-private partnerships, exemplified by the ISS National Laboratory, where government oversight meets private sector innovation. These collaborations have enhanced research capabilities and reduced financial risks for the public sector, while opening new markets and opportunities in space exploration [37].

Space Mining Initiatives: Space mining is an emerging frontier with the potential to redefine resource acquisition and sustainability. It involves extracting natural resources such as rare earth metals, water, and gases from celestial bodies like asteroids, the Moon, and Mars.

- **Objectives**:
  - Support human sustenance in space by providing essential resources like water for life support and propulsion.
  - Enable sustainable exploration by reducing dependence on Earth's finite resources.
  - Drive economic growth through the creation of a new industrial sector.

Notable initiatives include [37]:

- **Planetary Resources**: Focused on asteroid mining for water and precious metals.
- **iSpace**: Targeting lunar water ice for in-space propulsion.
- **AstroForge**: Aims to extract metals using space-based 3D printing technology.

Although still in its infancy, space mining holds immense potential. Projections estimate its market value could reach US$ 1 trillion by 2040, supported by over US$ 500 million in funding as of 2023. Milestones such as successful asteroid mining simulations underscore its viability.

The aerospace industry is at the forefront of innovation, fuelled by trends like reusable rockets, space tourism, private sector involvement, and space mining. These advancements are not only expanding humanity's reach into the cosmos but also fostering economic growth, technological development, and sustainability. As these trends evolve, they promise to reshape our understanding of space and redefine humanity's place in the universe.

Chapter 2

Basic Principles of Rocket Propulsion

Rocket propulsion is the fundamental mechanism that enables rockets to overcome Earth's gravitational pull and travel into space. At its core, rocket propulsion operates on Newton's Third Law of Motion, which states that "for every action, there is an equal and opposite reaction." By expelling mass (propellant) at high velocity in one direction, a rocket generates an equal and opposite thrust that propels it forward.

The basic principles of rocket propulsion involve several key concepts:

1. **Thrust Generation**: Thrust is the force produced when propellant is burned in a rocket engine, creating high-pressure gases that are expelled through a nozzle. This expulsion accelerates the rocket forward.

2. **Propellant Types**: Rockets rely on chemical energy stored in propellants, which can be solid, liquid, or hybrid. The choice of propellant affects the rocket's performance, efficiency, and application.

3. **Specific Impulse**: A measure of a rocket engine's efficiency, specific impulse represents the amount of thrust produced per unit of propellant consumed over time. Higher specific impulse values indicate better engine efficiency.

4. **Nozzle Design**: The nozzle plays a critical role in directing the exhaust gases and maximizing thrust. De Laval nozzles, for example, are widely used to accelerate gases to supersonic speeds.

5. **Stages of Rocketry**: Multistage rockets discard empty fuel tanks as they ascend, reducing weight and increasing efficiency. This staging process allows rockets to achieve the velocities needed to reach orbit or beyond.

6. **Vacuum Operation**: Rockets are unique in their ability to operate in the vacuum of space. Unlike jet engines that require atmospheric oxygen, rockets carry their own oxidizers, enabling combustion in space.

These principles underpin modern rocketry, from launching satellites into orbit to propelling interplanetary missions. Understanding these basics is essential for advancing rocket design and ensuring successful missions in space exploration and transportation.

Newton's Laws of Motion and their Application to Rockets

Sir Isaac Newton's Three Laws of Motion provide the scientific foundation for understanding how rockets function. These principles explain the forces acting on rockets and the mechanics of their propulsion, enabling them to overcome gravity and navigate through space.

First Law: The Law of Inertia

- **Definition:** An object will remain at rest or in uniform motion in a straight line unless acted upon by an external force.
- **Application to Rockets:** A rocket on a launch pad stays stationary until a force (thrust) is applied. Once in space, where external forces like air resistance are negligible, the rocket continues to move at a constant velocity unless influenced by another force, such as gravity or thrust from a rocket engine.

To fully understand Newton's First Law of Motion—often referred to as the law of inertia—it is essential to grasp the concepts of rest, motion, and unbalanced force [38].

Rest and motion are inherently relative terms. An object is said to be at rest when it is not changing position relative to its surroundings. For example, when you are seated in a chair, you are considered at rest relative to the chair and the objects immediately around you. However, this state of rest is only apparent within your immediate frame of reference. If your chair is part of a moving airplane, your position relative to the airplane remains unchanged, but you are still traveling at high speeds relative to the ground. Even at home, you are technically in motion because you are on a planet spinning on its axis, orbiting a star, which itself moves through a galaxy hurtling through the universe. Hence, absolute rest does not exist in nature. Rest is always relative to a specific frame of reference [38].

Motion, similarly, is relative and defined as a change in position with respect to surroundings. For instance, a ball sitting on the ground is at rest relative to the ground, but it is in motion if it starts rolling. Likewise, a passenger walking down the aisle of an airplane is in motion relative to the plane's interior but still traveling with the plane relative to the ground. In the context of a rocket, rest transitions to motion as the rocket lifts off the launch pad, illustrating a clear change in position relative to the ground [38].

The concept of unbalanced force is critical to understanding how rest and motion interact. When an object is at rest, it is still subject to forces acting upon it. For example, when you hold a ball in your hand, gravity pulls the ball downward, while your hand exerts an equal upward force, balancing the forces and keeping the ball at rest. However, if you release the ball or push it upward, the forces become unbalanced, resulting in motion. Similarly, a rocket on a launch pad remains at rest because the upward force exerted by the pad balances the downward pull of gravity. When the rocket engines ignite, the thrust generated unbalances the forces, causing the rocket to accelerate upward [38].

In rocket flight, the interplay between balanced and unbalanced forces is constant. At launch, the thrust produced by the engines overcomes gravity and any opposing forces, unbalancing the system and propelling the rocket upward. Once the rocket runs out of fuel, gravity reasserts dominance, slowing the rocket until it reaches its peak altitude and eventually falls back to Earth. This constant shift between balanced and unbalanced forces governs the entire trajectory of the rocket [38].

Objects in space, like spacecraft, also respond to forces, though in different conditions. In the vacuum of space, a spacecraft can travel in a straight line indefinitely if all forces acting on it are balanced. However, when the spacecraft approaches a large celestial body, such as Earth or another planet, gravitational forces unbalance the system, altering its path into a curve. This principle is particularly evident in orbital motion. For example, if a satellite is launched parallel to Earth's surface at a high enough speed, gravity will continuously curve its path into an orbit around Earth. Without external forces like atmospheric drag or engine thrust acting to slow it down, the satellite will remain in orbit indefinitely [38].

With a clear understanding of these key terms, Newton's First Law can be restated: An object at rest remains at rest, and an object in motion continues in a straight line at a constant speed, unless acted upon by an unbalanced force. In the context of rockets, this means a force is required to initiate motion, halt motion, change direction, or alter speed. This foundational principle governs the behaviour of all objects, whether on Earth or in space, shaping the mechanics of motion and the science of rocketry [38].

Second Law: The Law of Force and Acceleration

- **Definition:** The acceleration of an object is directly proportional to the net force acting upon it and inversely proportional to its mass. This is expressed mathematically as:

  $F=ma$

  where F is force, m is mass, and a is acceleration.

- **Application to Rockets:** The thrust generated by a rocket engine provides the force that accelerates the rocket. As propellant is burned and expelled, the rocket's mass decreases, allowing for greater acceleration with the same amount of thrust. This is critical during launch, where overcoming Earth's gravity requires significant force.

Newton's Second Law of Motion establishes a fundamental relationship between force, mass, and acceleration, expressed mathematically as $f = ma$. This equation states that the force acting on an object is equal to the product of its mass (m) and its acceleration (a). The equation can also be rearranged to calculate acceleration ($a=f/m$) or mass ($m=f/a$), depending on the situation. This law is key to understanding how objects move and is especially critical in the context of rocket propulsion [38].

To illustrate this law, consider the example of an old-style cannon. When the cannon is fired, an explosion propels a cannonball out of the barrel, sending it flying toward its target. Simultaneously, the cannon itself is pushed backward, albeit by a much smaller distance. This demonstrates action and reaction, as explained by Newton's Third Law, but the specific movements of the cannon and the cannonball are governed by the Second Law [38].

The force of the explosion is the same for both the cannon and the cannonball. This can be expressed as two equations [38]:

$$f = m_{\text{cannon}} \cdot a_{\text{cannon}}$$

$$f = m_{\text{ball}} \cdot a_{\text{ball}}$$

Here, m_{cannon} and a_{cannon} refer to the mass and acceleration of the cannon, while m_{ball} and a_{ball} refer to the mass and acceleration of the cannonball. Since the force is the same, these equations can be combined [38]:

$$m_{\text{cannon}} \cdot a_{\text{cannon}} = m_{\text{ball}} \cdot a_{\text{ball}}$$

Because the cannon has a much larger mass than the cannonball, its acceleration is smaller. Conversely, the smaller mass of the cannonball results in a much greater acceleration [38].

In rockets, the principle is similar. Instead of a cannonball, the gases ejected from the rocket engine take on the role of the smaller mass, while the rocket itself is analogous to the cannon. The force, in this case, is the pressure created by the controlled combustion of propellants within the rocket engine. This force propels the gases out of the engine at high velocity (action) and drives the rocket in the opposite direction (reaction) [38].

Unlike the cannon example, a rocket's thrust continues as long as its engines are firing, creating a sustained force. Additionally, the rocket's mass changes during flight. A rocket's total mass includes its engines, fuel tanks, payload, and control systems, but the largest component is its propellant. As the propellant burns, the rocket's mass decreases. According to Newton's Second Law, if the force remains constant while the mass decreases, the acceleration of the rocket must increase. This is why rockets start off moving slowly and gain speed as they ascend into space [38].

Newton's Second Law is fundamental to designing efficient rockets capable of achieving the extreme speeds required for space exploration. To reach low Earth orbit, a rocket must travel at speeds exceeding 28,000 km/h. To escape Earth's gravity and enter deep space, it must achieve escape velocity, over 40,250 km/h. Achieving these speeds requires the rocket to generate significant thrust in a short period. This is done by burning a large mass of fuel and expelling the resulting gases at high velocities [38].

Newton's Second Law in the context of rocketry can be summarized as follows: the greater the mass of propellant burned and the faster the resulting gases are expelled from the engine, the greater the thrust generated by the rocket. This principle underpins the design of rocket engines, focusing on maximizing the efficiency of fuel combustion and gas expulsion to achieve the required speeds for space travel [38].

Third Law: The Law of Action and Reaction

- **Definition:** For every action, there is an equal and opposite reaction.
- **Application to Rockets:** Rocket propulsion is the clearest example of this law in action. When a rocket expels exhaust gases downward at high speed (the action), an equal and opposite force (thrust) propels the rocket upward. This principle enables rockets to move even in the vacuum of space, where there is no air to push against.

Newton's Third Law of Motion is one of the most intuitive yet profoundly important principles in physics, stating that for every action, there is an equal and opposite reaction. This concept is fundamental to understanding how rockets operate, both on Earth and in the vacuum of space. The principle applies universally, from simple everyday experiences to complex rocket launches [38].

Consider the scenario of stepping off a small boat that is not securely tied to a pier. As you step forward, the boat moves backward. The action is your step off the boat, while the reaction is the boat's movement in the opposite direction. The forces acting on you and the boat are equal in magnitude but opposite in direction. The boat appears to move farther than you, not because the forces are unequal, but because the boat's mass is smaller than yours, making it accelerate more in response to the same force [38].

This principle is the basis for rocket propulsion. A rocket lifts off the ground by expelling exhaust gases out of its engine. The action is the force exerted by the rocket on the escaping gases, and the reaction is the force exerted by the gases on the rocket, propelling it upward. For a rocket to overcome Earth's gravity and achieve liftoff, the reaction force (thrust) must be greater than the gravitational force acting on the rocket's mass [38].

Once in space, even small bursts of thrust can significantly alter the rocket's direction or speed. This is because, in the vacuum of space, there are no counteracting forces like air resistance to impede the rocket's motion [38].

To better understand this, imagine a rider standing still on a skateboard. When the rider jumps off, they exert a force on the skateboard (action), causing the skateboard to move in the opposite direction (reaction). If the rider is much heavier than the skateboard, the skateboard will travel farther due to its smaller mass. This difference in motion highlights the interplay between force, mass, and acceleration, which will be further explained through Newton's Second Law [38].

One of the most common questions about rockets is how they work in the vacuum of space, where there is no air to push against. The answer lies in Newton's Third Law. Rockets do not need air to generate thrust; they rely on the reaction force from expelling exhaust gases [38].

To understand this, think about the skateboard example again. The surrounding air plays no role in the action-reaction forces; it merely creates drag, slowing the skateboard's movement. Similarly, on

Earth, a rocket's exhaust gases must push against the surrounding air, which consumes some of the energy generated by the rocket engine. In the vacuum of space, however, there is no air to impede the exhaust gases. This allows the rocket to use its energy more efficiently, making rockets work even better in space than in an atmosphere [38].

The lack of air resistance in space eliminates energy losses due to drag. As the rocket expels gas at high speed, the exhaust freely escapes, and the full thrust generated by the action-reaction process is directed toward propelling the rocket. This efficiency is critical for achieving the high speeds necessary for space travel, whether reaching orbit, traveling to other planets, or adjusting trajectories for interplanetary missions [38].

Newton's Third Law underpins the entire concept of rocket propulsion. By expelling exhaust gases at high speed (action), rockets generate an equal and opposite reaction that propels them forward. This principle works regardless of whether the rocket is within Earth's atmosphere or in the vacuum of space. Understanding this fundamental law helps explain why rockets can operate in the seemingly empty expanse of space and why their efficiency improves in the absence of atmospheric drag [38].

Combined Application of Newton's Laws

The combined application of Newton's laws is essential to understanding how rockets operate from launch to space travel. These principles guide every stage of a rocket's journey, ensuring efficient propulsion, effective manoeuvring, and optimal design.

At launch, the rocket engines generate a thrust force that must exceed the gravitational pull acting on the rocket's mass. This upward thrust, governed by Newton's Second Law, accelerates the rocket into the sky, overcoming Earth's gravitational force and initiating its ascent.

Efficient propulsion is achieved through the careful design of the rocket nozzle and the high velocity of exhaust gases. According to Newton's Third Law, the downward expulsion of gases creates an equal and opposite reaction force (thrust) that propels the rocket upward. This efficiency is crucial for achieving the required escape velocity or entering orbit.

Figure 10: Forces acting on a rocket at take-off.

In space, Newton's First Law becomes particularly significant. With minimal external forces like drag or friction, a rocket in motion continues at a constant velocity unless acted upon by another force. Adjustments to the rocket's trajectory or speed are made by firing thrusters, applying both Newton's Second and Third Laws to control direction and momentum.

Multistage rockets highlight the practical use of Newton's Second Law. By jettisoning used stages, the rocket's mass decreases, allowing the same amount of thrust to produce greater acceleration. This reduction in mass optimizes fuel efficiency and ensures the rocket can continue its journey to its destination.

Together, these applications of Newton's laws ensure that rockets can successfully launch, travel through space, and reach their intended targets. These principles are foundational to modern rocketry and are at the core of advancements in space exploration.

The Rocket Equation: Thrust, Velocity, and Mass

The rocket equation, also known as the Tsiolkovsky rocket equation, is a fundamental principle in rocket science that describes the relationship between a rocket's velocity change, the mass of the propellant expelled, and the exhaust velocity [39]. Formulated by Konstantin Tsiolkovsky in 1903, this equation provides the mathematical framework to understand how a rocket's velocity changes based on the interplay between thrust, velocity, and mass [40].

The Tsiolkovsky rocket equation highlights the challenges and intricacies of achieving spaceflight [5]. It explains how the expulsion of mass (propellant) at high velocity generates the thrust necessary to accelerate the rocket [41]. This equation is crucial in space mission design, as it allows for the calculation of the delta-v (change in velocity) required for specific manoeuvres, such as orbit transfers or landings [39].

The canonical form of the Tsiolkovsky rocket equation is derived under the assumption that the rocket is moving slowly compared to the speed of light throughout its flight [40]. This equation relates the mass of the propellant (mp), the initial and final mass of the spacecraft, and the effective exhaust velocity (ve) of the propulsion system [39]. The equation illustrates that the fraction of the original vehicle mass that can be accelerated through a given velocity increment, Δv, is a negative exponential in the ratio of that velocity increment to the effective exhaust velocity [5].

The Tsiolkovsky rocket equation has been further developed and extended to account for various scenarios, such as relativistic rockets [42], air-pressurized waterjet rockets [41], and rockets with combustion oscillations [43]. These advancements have helped to refine our understanding of rocket propulsion and its limitations.

At the heart of the rocket equation is the principle of conservation of momentum. When a rocket expels exhaust gases at high speed in one direction, an equal and opposite momentum is imparted to the rocket, propelling it forward. This is a direct application of Newton's Third Law of Motion: every action has an equal and opposite reaction.

The Equation: A Mathematical Representation

The rocket equation is expressed as:

$$\Delta v = v_e \ln\left(\frac{m_0}{m_f}\right)$$

Where:

- Δv = Change in velocity (delta-v) of the rocket.
- v_e = Effective exhaust velocity (speed of the expelled gases relative to the rocket).
- m_0 = Initial mass of the rocket, including propellant.

- m_f = Final mass of the rocket, after the propellant is burned.
- ln = Natural logarithm function.

This equation connects three critical parameters:

1. **Velocity (Δv)**: The total change in the rocket's velocity, which determines its ability to reach orbit or travel through space.
2. **Thrust**: The force generated by the expulsion of exhaust gases.
3. **Mass**: Specifically, the relationship between the initial and final mass of the rocket.

Thrust is the force that propels the rocket forward, created by the high-speed expulsion of exhaust gases from the engine. The magnitude of thrust depends on:

- The mass flow rate of the propellant (amount of mass expelled per second).
- The exhaust velocity (v_e) of the gases.

Thrust is given by the equation:

$$F = \dot{m} v_e$$

Where:

- F = Thrust.
- $\dot{m}$ = Mass flow rate of the propellant.
- v_e = Exhaust velocity.

A higher exhaust velocity or a greater mass flow rate results in more thrust, which directly impacts the rocket's ability to overcome gravity and achieve desired accelerations.

Velocity: Achieving Delta-v

Delta-v (Δv) is a measure of the rocket's ability to change its velocity. It is a critical parameter in space missions, determining whether the rocket can reach orbit, transfer between planets, or escape Earth's gravity. The rocket equation shows that delta-v depends on:

- The effective exhaust velocity (v_e).
- The ratio of the rocket's initial mass (m_0) to its final mass (m_f).

The higher the delta-v, the farther and faster a rocket can travel. However, achieving higher delta-v requires either more efficient propulsion systems or a greater proportion of propellant, leading to design challenges.

Mass: The Key to Efficiency

Mass is a critical factor in the rocket equation, as it directly influences a rocket's performance and efficiency. The total mass of a rocket can be divided into three key components: structural mass, payload mass, and propellant mass. Structural mass refers to the weight of the rocket itself, including essential elements such as the engines, tanks, and other structural components. Payload mass represents the useful cargo the rocket is designed to carry, which can include satellites, scientific instruments, or even a human crew. Propellant mass comprises the fuel required to generate thrust and propel the rocket.

The mass ratio—the ratio of the rocket's initial mass (m_0) to its final mass (m_f) after propellant is consumed—is a crucial metric for determining the rocket's efficiency. A higher initial-to-final mass ratio indicates that a significant portion of the rocket's mass is dedicated to propellant, which can increase delta-v (the change in velocity required for the mission). However, this higher mass ratio often involves trade-offs, such as reduced structural strength or limited payload capacity.

To enhance efficiency and address these challenges, several strategies are employed in rocket design. Multistage rockets are commonly used to shed unnecessary mass during flight, improving the mass ratio as the rocket ascends. Additionally, engineers use lightweight materials for the rocket's structure to minimize structural mass without compromising strength. Advances in propulsion technology also play a vital role, as high-efficiency engines with greater exhaust velocities reduce the amount of propellant required for the mission.

By carefully managing these factors, modern rockets are designed to maximize their performance, enabling them to achieve the demanding speeds and altitudes necessary for space exploration.

Challenges Highlighted by the Equation

The rocket equation reveals some fundamental challenges in rocket design that engineers must overcome to achieve efficient and effective spaceflight. These challenges are inherent to the physics of rocketry and influence the design, cost, and feasibility of missions.

One major challenge is the issue of diminishing returns. Adding more propellant to increase a rocket's delta-v seems like an obvious solution, but it comes with a catch: the added propellant also increases the rocket's mass. This additional mass requires more thrust to overcome, which, in turn, demands even more propellant. This cycle leads to a point where the benefits of adding extra propellant are outweighed by the penalties of its additional weight, making it increasingly inefficient.

Another challenge lies in the payload fraction, which is the proportion of the rocket's total mass that is dedicated to the payload. The rocket equation makes it clear that only a small percentage of the total mass can be allocated to the payload, as the majority is consumed by the propellant and structural components. This limitation means engineers must carefully balance the rocket's mass distribution to maximize its payload capacity while ensuring sufficient thrust and stability.

Exhaust velocity limitations present yet another hurdle. Higher exhaust velocities significantly improve delta-v, allowing the rocket to achieve the necessary velocities for orbital insertion or interplanetary travel. However, achieving higher exhaust velocities requires advanced propulsion technologies, such as cryogenic engines or ion thrusters. These technologies are often complex and expensive to develop and operate, posing financial and engineering challenges.

Addressing these challenges requires innovative solutions, such as multistage rockets, lightweight materials, and advancements in propulsion systems. Each of these approaches plays a critical role in overcoming the inherent difficulties highlighted by the rocket equation, paving the way for more ambitious and cost-effective space missions.

Practical Implications in Modern Rocketry

The rocket equation plays a foundational role in shaping the design and operational strategies of modern space exploration. It serves as the guiding principle behind key engineering decisions, enabling engineers and scientists to tackle the challenges of achieving orbital and interplanetary missions.

The rocket equation encapsulates the fundamental physics of rocket flight, serving as a cornerstone in understanding and designing rocket systems. It applies specifically to rocket-like reaction vehicles when the effective exhaust velocity is constant. For scenarios where the exhaust velocity varies, the equation can be summed or integrated to provide accurate predictions. However, the equation focuses solely on the reaction force produced by the rocket engine, excluding external forces like aerodynamic drag or gravity. When calculating the propellant requirement for missions such as launching from a planet with an atmosphere or landing on one, these external forces must be factored into the delta-v (change in velocity) requirement.

This limitation, often referred to as "the tyranny of the rocket equation," highlights the inherent challenges of rocket design. The equation reveals that the payload fraction—the portion of the rocket dedicated to carrying useful cargo—is constrained by the need for substantial amounts of propellant. Increasing the propellant adds to the rocket's mass, which in turn demands more fuel and greater thrust, creating diminishing returns.

The rocket equation is crucial for calculating the propellant requirements for various space manoeuvres, including orbital changes and interplanetary transfers. For orbital manoeuvres, it assumes impulsive thrust, where propellant burns and velocity changes occur almost instantaneously. This assumption is accurate for short-duration burns, such as mid-course corrections or orbital insertions. However, for low-thrust, long-duration propulsion systems like ion engines, the equation requires integration to account for the gradual changes in the spacecraft's state and trajectory influenced by gravity.

For example, achieving low Earth orbit (LEO) typically requires a delta-v of approximately 9,700 meters per second, which includes overcoming gravity and atmospheric drag. For a single-stage-to-orbit rocket with an exhaust velocity of 4,500 meters per second, 88.4% of the initial mass must be propellant, leaving only 11.6% for the structural mass, engines, and payload. Using a multistage design significantly improves efficiency by discarding used stages and shedding unnecessary mass during flight. In a two-stage rocket scenario, the first stage might require 67.1% of its mass as propellant, with the remaining structure and payload contributing to the second stage.

One of the most practical applications of the rocket equation is in the design of multistage rockets. By dividing a rocket into stages, engineers can shed unnecessary weight as each stage completes its burn and is discarded. This approach optimizes the mass ratio, as the total mass of the rocket decreases during ascent, allowing the remaining stages to achieve higher accelerations with the same amount of thrust. Multistage designs have been pivotal in enabling humanity to reach the Moon, deploy satellites into geostationary orbits, and send probes to distant planets.

The principle of staging allows rockets to overcome the constraints of the rocket equation by sequentially discarding empty propellant tanks and engines. For each stage, the rocket equation applies separately, with the initial mass being the remaining rocket after discarding the previous stage and the final mass being the rocket just before discarding the current stage. Each stage may have a different specific impulse, depending on the engine type and propellant used.

For instance, if the first stage of a rocket comprises 80% propellant and 10% structural mass, with the remaining 10% being the upper stages and payload, the delta-v generated by this stage would be approximately $1.61\times v_e$ (where v_e is the exhaust velocity). Adding subsequent stages with similar mass distributions increases the overall delta-v significantly. For a three-stage rocket with identical configurations, the cumulative delta-v could reach $4.83\times v_e$, enabling the rocket to achieve higher velocities needed for orbital or interplanetary missions.

Another significant implication is the push for reusable rockets, a modern innovation aimed at reducing the high costs associated with space launches. Traditionally, rockets were single-use vehicles, with expensive structural components discarded after a single mission. Companies like SpaceX, with its Falcon 9 rocket, have demonstrated the viability of recovering and reusing first-stage boosters, significantly lowering the cost per launch. This aligns with the rocket equation by rethinking how mass and cost efficiency can be optimized for repeated use.

The rocket equation also underscores the importance of high-efficiency engines in overcoming the limitations of mass and fuel. Engines such as ion thrusters and nuclear thermal propulsion systems are designed to achieve higher exhaust velocities, directly increasing delta-v. Ion thrusters, for instance, use electrically charged particles to produce thrust with exceptional efficiency, making them ideal for long-duration missions in deep space. Nuclear thermal propulsion offers the promise of even greater efficiencies by leveraging the high energy density of nuclear reactions to heat propellant.

Single-stage-to-orbit (SSTO) rockets are less efficient compared to multistage rockets due to their inability to shed mass during ascent. An SSTO rocket with a similar payload fraction might achieve a delta-v of only $2.19\times v_e$, far below the capabilities of a multistage rocket. This limitation underscores the advantage of staging in modern rocketry.

In more complex scenarios, such as when a new stage is ignited before the previous stage is fully discarded (as in the case of solid rocket boosters combined with a liquid-fuel stage), calculating the delta-v involves considering varying specific impulses and overlapping thrust contributions.

Together, these innovations—multistage rockets, reusable systems, and high-efficiency engines—demonstrate how the rocket equation influences every aspect of modern rocketry. By addressing the equation's inherent challenges, these advancements not only make space exploration more feasible and cost-effective but also open the door to new possibilities, such as sustainable lunar bases, Mars colonization, and deep-space exploration missions.

Types of Propulsion Systems (Chemical, Electric, Nuclear)

Space launch vehicles rely on various propulsion systems to achieve their objectives, with the choice of propulsion tailored to mission requirements such as payload, distance, and efficiency. The primary types of propulsion systems include chemical, electric, and nuclear propulsion, each offering distinct advantages and limitations.

Chemical Propulsion Systems

Chemical propulsion systems are the most widely used and well-established propulsion method in space launch vehicles. They involve the combustion of chemical propellants to produce high-energy exhaust gases, which are expelled through a nozzle to generate thrust, as described by Newton's Third Law of Motion [44].

Key Characteristics of Chemical Propulsion:

1. High Thrust: One of the defining advantages of chemical propulsion is its ability to generate immense thrust. This high thrust is essential for overcoming Earth's gravitational pull, especially during the launch phase. The rapid acceleration provided by chemical propulsion systems allows rockets to achieve the velocity needed to escape Earth's atmosphere, making them the primary choice for space launch vehicles.

2. Short Duration: Chemical propulsion systems excel in delivering short, powerful bursts of acceleration. This makes them particularly effective for critical mission phases such as launch, orbital insertion, and rapid trajectory adjustments. Their ability to produce significant thrust in a short time frame ensures the efficient execution of these demanding operations.

3. Diverse Propellant Types: Chemical propulsion systems can be classified based on the type of propellant used. Each type offers unique advantages and applications, allowing engineers to tailor propulsion systems to specific mission requirements.

- **Liquid Propulsion:** Liquid propulsion systems use separate liquid fuels and oxidizers, such as liquid hydrogen (fuel) and liquid oxygen (oxidizer). These components are combined in a combustion chamber, where they ignite to produce thrust. Liquid propulsion offers precise control over thrust, enabling gradual increases or decreases in power. This precision makes it suitable for complex manoeuvres and upper-stage propulsion. Notable examples include SpaceX's Falcon 9 and NASA's Saturn V rockets.

- **Solid Propulsion:** Solid propulsion systems utilize solid propellants that are pre-mixed with fuel and oxidizers. These systems are simpler, highly reliable, and cost-effective, making them a popular choice for rocket boosters. However, they lack the flexibility of liquid systems, as the thrust cannot be controlled or stopped once ignition occurs. Examples include the solid rocket boosters used in NASA's Space Launch System (SLS) and the Space Shuttle.

- **Hybrid Propulsion:** Hybrid propulsion systems combine elements of both liquid and solid propulsion, using a liquid oxidizer and a solid fuel. This design balances the simplicity of solid propulsion with the controllability of liquid systems. Hybrid systems are increasingly being explored for smaller rockets and experimental vehicles.

Liquid propulsion systems are an essential technology in rocketry, offering the ability to generate controlled thrust for a variety of space missions. These systems use liquid propellants, which are either monopropellant (a single substance) or bi-propellant (a combination of fuel and oxidizer). Depending on the design, liquid propulsion systems can be pressure-fed or pump-fed, each with distinct advantages and applications.

Types of Liquid Propulsion Systems [45]:

1. **Monopropellant Systems** Monopropellant systems use a single fluid, such as hydrazine or green propellants, which decomposes to produce thrust when passed over a catalyst. These systems are relatively simple and reliable, making them ideal for attitude control and small orbital adjustments. However, they require heavy catalyst systems for operation, adding to the system's weight and complexity. Green propellants are being developed to replace toxic monopropellants like hydrazine, offering safer handling and improved efficiency.

2. **Hypergolic Bi-Propellant Systems:** Hypergolic propellants are bi-propellants where the fuel and oxidizer ignite spontaneously upon contact. Common hypergolic combinations include hydrazine as fuel and nitrogen tetroxide as the oxidizer. These systems are highly reliable since they do not require separate ignition systems, reducing the risk of ignition failure. Their long-term storability in space makes them suitable for spacecraft like the Orion vehicle's European Service Module. However, hypergolic propellants are toxic and corrosive, posing significant handling challenges and offering lower performance compared to cryogenic propellants.

3. **Cryogenic Bi-Propellant Systems:** Cryogenic systems use propellants stored at extremely low temperatures, such as liquid hydrogen (LH2) as fuel and liquid oxygen (LOX) as the oxidizer. These systems are known for their high performance and relatively benign handling compared to hypergolics. The Space Shuttle and Saturn V rocket utilized cryogenic systems for their main propulsion stages. However, maintaining cryogenic temperatures over long

durations is challenging, as propellants can boil off, requiring advanced insulation and cooling technologies. Cryogenic systems are typically fueled shortly before launch to minimize losses.

4. **Nuclear Thermal Propulsion Systems:** A specialized category of liquid propulsion, nuclear thermal systems utilize a nuclear reactor to heat cryogenic hydrogen to extremely high temperatures. The heated hydrogen is expelled through a rocket nozzle to produce thrust. Nuclear thermal systems offer very high performance, making them suitable for deep-space exploration missions. However, they require heavy nuclear reactors and face regulatory and public perception challenges.

Liquid propulsion systems can deliver propellants to the engine through two main methods [45]:

- **Pressure-fed Systems:** High-pressure tanks push the propellants into the combustion chamber. These systems are simple but require heavier tanks to withstand the high pressures, limiting their efficiency for large rockets.

- **Pump-fed Systems:** Turbopumps transfer the propellants to the engine, enabling the use of lighter tanks and higher performance. Pump-fed systems are commonly used in large launch vehicles due to their efficiency and scalability.

The Advantages of Liquid Propulsion Systems include [45]:

1. **Thrust Control:** Liquid propulsion systems can start, stop, and throttle engines, providing precise control over thrust. This feature is essential for applications such as orbital insertion, docking, and planetary landings.

2. **Wide Thrust Range:** These systems can generate thrust across a broad spectrum, from small corrections in attitude control to powerful liftoffs capable of escaping Earth's gravity.

3. **Versatility:** Liquid propulsion systems are used in a variety of missions, including launch vehicles, orbital adjustments, and deep-space exploration.

In contrast, the disadvantages of Liquid Propulsion Systems include [45]:

1. **Complexity:** The need for multiple components, including pumps, valves, and cooling systems, makes liquid propulsion systems more complex than solid propulsion systems.

2. **Cryogenic Storage Challenges:** Cryogenic propellants require sophisticated insulation and cooling systems to prevent boil-off, limiting their applicability for long-duration missions.

3. **Toxicity:** Hypergolic systems, while reliable, involve handling hazardous materials, requiring extensive safety precautions.

Summary of Advantages and Disadvantages

System Type	Typical Propellants	Advantages	Disadvantages
Monopropellants	Hydrazine, Green Propellants	Simple and reliable	Heavy catalyst systems required
Hypergolic Propellants	Hydrazine, Nitrogen Tetroxide	Reliable, long-term storability	Toxic, corrosive, lower performance
Cryogenic Propellants	Hydrogen, Oxygen, Methane	High performance, safer to handle	Require complex cooling, risk of boil-off
Nuclear Thermal	Hydrogen	Very high performance	Complex, heavy, regulatory challenges

Applications of these propulsion systems include [45]:

- **Launch Vehicles:** Liquid propulsion systems are the backbone of modern rockets, powering primary stages and enabling vehicles to escape Earth's atmosphere. Examples include the Saturn V and Space Shuttle.
- **In-Space Manoeuvres:** Monopropellant and hypergolic systems are commonly used for precise orbital adjustments, station-keeping, and spacecraft docking.
- **Deep-Space Exploration:** Cryogenic and nuclear thermal systems are being developed for interplanetary missions, offering the performance required for long-duration travel to Mars and beyond.

Advancements in material science, green propellant technology, and nuclear thermal propulsion are set to enhance the efficiency and safety of liquid propulsion systems. As space exploration extends to new frontiers, liquid propulsion remains a cornerstone technology, enabling humanity to explore and utilize the cosmos effectively [45].

Solid propellants are one of the oldest and most reliable propulsion technologies in rocketry. Unlike liquid propellants, which require separate storage for fuel and oxidizer, solid propellants combine both components into a single homogeneous mixture. This mixture is cast into a solid material that remains stable and inert until ignited [45].

The solid propellant is a composite material where the oxidizer and fuel are chemically bonded in suspension. The oxidizer supplies the oxygen necessary for combustion, while the fuel provides the energy source. This configuration allows solid propellants to burn efficiently and generate the high-energy gases needed for thrust.

Once ignited, the combustion reaction releases energy that produces hot gases, which expand and are expelled through a nozzle to create thrust. The process adheres to Newton's Third Law of Motion, where the expelled gases generate an equal and opposite reaction, propelling the rocket forward.

One of the primary advantages of solid propellants is their stability and ease of handling. Since the propellant is stored as a solid material, it can be maintained at room temperature without the need for sophisticated cooling systems. This property makes solid rocket motors highly durable and reliable over long storage periods, ensuring they remain "ready-to-go" when needed [45].

Solid propellant rockets are also less complex than liquid propulsion systems, as they do not require pumps, valves, or other intricate mechanisms to deliver the propellant to the combustion chamber. This simplicity contributes to their robustness and reduces the likelihood of mechanical failure.

Solid rocket boosters (SRBs) are a well-known application of solid propulsion technology. For example, the Space Shuttle utilized two massive SRBs mounted on either side of its central fuel tank. These boosters provided the tremendous thrust necessary to lift the shuttle and its payload off the launch pad and into the upper atmosphere.

Solid propellants are also widely used in military applications, such as ballistic missiles, due to their rapid response capabilities and ability to remain in a ready state for extended periods. Additionally, they are employed in smaller rockets for satellite launches, fireworks, and even in emergency ejection systems for aircraft.

Despite their reliability and simplicity, solid propulsion systems have notable limitations [45]:

1. **Lack of Thrust Control:** Once ignited, a solid rocket motor continues to burn until all the propellant is consumed. This inability to throttle, stop, or restart the engine limits its flexibility for missions requiring precise manoeuvring or variable thrust.

2. **Fixed Burn Rate:** The combustion process in solid rockets follows a pre-determined burn rate based on the composition and geometry of the propellant. While this predictability is an advantage for some applications, it restricts the adaptability of the system.

3. **Weight Considerations:** The solid propellant and casing add significant weight to the rocket. Unlike liquid propellant systems, where fuel tanks can be jettisoned after use, the entire solid rocket motor must be carried throughout the mission unless designed to detach as a stage.

4. **Environmental and Safety Concerns:** Solid propellants may release harmful by-products during combustion, and their production often involves hazardous materials. Additionally, once a solid rocket is prepared, it cannot be easily disassembled or modified, presenting safety challenges during handling and transport.

To address some of these limitations, modern advancements in solid propulsion include the development of segmented solid rocket motors and hybrid propulsion systems. Segmented motors allow for staged combustion, while hybrid systems combine the high thrust of solid propellants with the controllability of liquid oxidizers [45].

Hybrid propulsion systems occupy a unique position in rocket propulsion by combining features of both solid and liquid propellant systems. This hybridization addresses limitations inherent in each system while leveraging their respective strengths. The core concept involves using a solid propellant as the fuel and a separate liquid or gaseous oxidizer to sustain combustion, creating a flexible and efficient propulsion system.

In a hybrid propulsion system, the solid propellant serves as the fuel, often made from materials like hydroxyl-terminated polybutadiene (HTPB) or paraffin wax. The oxidizer, stored separately as a liquid such as liquid oxygen or nitrous oxide, or as a gas like gaseous oxygen, is delivered to the combustion chamber where it reacts with the solid fuel to produce thrust. Combustion occurs only when the oxidizer is actively supplied, allowing hybrid systems the flexibility to throttle, stop, or restart thrust by controlling the oxidizer flow. This controllability sharply contrasts with traditional solid propellant systems, which lack the ability to be stopped or restarted once ignited [45].

Hybrid propulsion systems provide significant advantages. The solid propellant delivers robust energy, offering high levels of thrust similar to solid rocket motors, making hybrids suitable for applications requiring substantial lifting power. The ability to throttle, restart, or shut down the engine enhances controllability, beneficial for missions requiring precision manoeuvres or multiple burns during flight. Hybrid systems also provide improved safety by storing the oxidizer separately from the fuel, reducing the risk of accidental ignition. Furthermore, the simpler design eliminates the need for complex turbopumps, reducing system weight, manufacturing, and maintenance costs [45].

However, these systems are not without challenges. Their specific impulse, a measure of efficiency, is generally lower than liquid propulsion systems, making hybrids less competitive for missions demanding extreme performance. Delivering a consistent and controllable flow of oxidizer can be technically complex, particularly under varying pressure and thermal conditions. Combustion uniformity within the chamber is also difficult to achieve due to the nature of solid fuel surface erosion and its interaction with the oxidizer [45].

Hybrid rockets are being increasingly applied in areas where their high thrust and controllability are advantageous. Companies like Virgin Galactic utilize hybrid systems in their SpaceShipTwo spacecraft for suborbital space tourism, delivering the power needed for rapid ascent and controlled descent. The simplicity and safety of hybrid systems make them ideal for educational programs and research experiments, allowing students and researchers to explore rocket science with reduced risk. Hybrid systems are also being explored for planetary landers, reusable launch vehicles, and emergency abort systems, where their safety and controllability features are particularly beneficial [45].

Future innovations in hybrid propulsion include the development of advanced materials for solid fuels and more efficient oxidizer delivery mechanisms to enhance performance and reliability. Research into environmentally friendly oxidizers and fuels aligns hybrid propulsion with the broader goals of sustainable space exploration. Hybrid systems, while not yet rivalling the performance of liquid systems in all aspects, offer a compelling balance of safety, simplicity, and adaptability. This positions them as a valuable technology for a wide range of aerospace applications, from commercial space tourism to advanced scientific missions, as the industry continues to push the boundaries of space exploration [45].

Chemical propulsion is predominantly used in the initial stages of launch vehicles to achieve the high delta-v required for leaving Earth's atmosphere [44]. Gel propulsion systems, which have wide application flexibility for utilization in smart tactical missiles, attitude-control systems, and launch-vehicle boosters, can be throttled similar to liquid rocket engines using semi-liquid propellants [46]. These systems usually have the advantages of conventional liquids and solid propellants [46].

As telecommunication satellites become larger and longer-lived, integrated liquid bipropellant propulsion systems will replace solid propellant/liquid monopropellant thruster propulsion systems for apogee and on-orbit manoeuvres [47].

The propellant tank of a satellite is a key component, as most satellite propulsion systems are primarily based on chemical propulsion, and liquid propellant is the necessary power fuel for the propulsion system [47].

JAXA has been developing a gas-liquid equilibrium propulsion system for small spacecraft, where the propellant is stored as a liquid phase in a storage tank and expelled from a nozzle in gaseous form [48].

Staging of ionic-liquid electrospray thrusters, analogous to launch vehicle staging, has been proposed as a potential solution to the lifetime limitations of individual thrusters, and to increase the overall lifetime of the propulsion system [49].

Ionic liquids have advantages such as low vapor pressure, environmental friendliness, and low toxicity. Multi-component ionic liquids are potential replacements for hydrazine-type liquid propellants by introducing energetic components to improve performance. Ionic liquids can also be used in electric propulsion systems due to their high degree of ionization and easy interaction with electric fields, enabling dual-mode aerospace propulsion technology [50].

Electric Propulsion Systems

Electric propulsion systems are a type of spacecraft propulsion that rely on electricity to accelerate ionized particles (plasma) to generate thrust. These systems are characterized by their high efficiency and low thrust compared to chemical propulsion [51, 52].

Ion propulsion systems use electric power to ionize atoms, creating a charged plasma that is confined within a magnetic bottle. The ions are then expelled through a magnetic nozzle at high speeds, creating thrust without the need for solid matter to interact with the plasma. The process involves two key steps: ionization of the propellant and acceleration of the ions using an electric field. A neutralizer emits electrons to ensure the ion beam does not accumulate charge, maintaining overall electrical neutrality.

The high exhaust velocity characteristic of electric propulsion systems leads to exceptional efficiency, as they consume very little propellant. However, this efficiency comes with trade-offs: these systems require significant electrical energy and generally produce low thrust. As a result, they are best suited for missions where prolonged operation can achieve the necessary impulse, such as station keeping for satellites or primary propulsion for deep-space missions.

Key characteristics of electric propulsion systems include:

- **High Efficiency:** Electric propulsion systems have a much higher specific impulse, meaning they use propellant more efficiently [51, 52].
- **Low Thrust:** They generate only small amounts of thrust, making them suitable for long-duration missions rather than initial launches [51, 52].
- **Propellant Types:** Xenon is a common propellant used in electric propulsion due to its inert nature and high atomic mass [51, 53].

Types of Electric Propulsion:

- **Ion Thrusters:** Use electricity to ionize propellant atoms and accelerate them using electric fields. Examples include NASA's Dawn spacecraft [51, 54].
- **Hall Effect Thrusters:** Create thrust by using a magnetic field to confine electrons, which ionize propellant and produce plasma. Widely used in communication satellites [51, 54].
- **Electrothermal Thrusters:** Heat the propellant electrically before expelling it to generate thrust [55, 56].

Electric propulsion systems require robust power sources. Solar energy is the most common option for missions near the Sun, providing a reliable and lightweight solution. For missions beyond the reach of sufficient solar energy, nuclear power becomes essential. Nuclear electric propulsion systems use nuclear reactors to generate the necessary energy, making them suitable for deep-space missions. Other innovative power generation concepts, such as beamed energy from external sources, are under investigation to overcome the limitations of on-board power generation.

Electric propulsion is ideal for deep-space missions, satellite station-keeping, and interplanetary exploration, where efficiency and prolonged thrust outweigh the need for high initial thrust [51, 57, 58].

The primary limitation of electric propulsion systems lies in their reliance on substantial electrical energy. Current power generation methods, whether solar, nuclear, or chemical, impose constraints on the maximum thrust that can be produced. Additionally, the power source adds significant mass to the spacecraft, impacting overall performance and design. While electric propulsion offers high efficiency, the trade-off is low thrust, which restricts its use to missions where gradual acceleration over time is feasible.

Ongoing advancements in power generation, ionization techniques, and magnetic confinement are expected to enhance the capabilities of electric propulsion systems. As the technology evolves, it holds the potential to revolutionize space exploration by enabling efficient, long-duration missions to distant planets and beyond, while reducing reliance on chemical propellants. With continued innovation, electric propulsion will likely play an integral role in humanity's expansion into the solar system and beyond.

Nuclear Propulsion Systems

Nuclear propulsion harnesses the energy from nuclear reactions, primarily fission, to generate thrust for space exploration [59, 60]. This technology offers significant advantages over chemical propulsion, including higher energy density and specific impulse [61, 62].

There are two main types of nuclear propulsion systems:

1. Nuclear Thermal Propulsion (NTP): NTP uses a nuclear reactor to heat a propellant, typically hydrogen, which is then expelled to generate thrust [61, 62]. This method provides higher specific impulse than chemical propulsion [62].

2. Nuclear Electric Propulsion (NEP): NEP uses a nuclear reactor to generate electricity, which then powers electric thrusters [63-65]. This combines the high energy density of nuclear power with the efficiency of electric propulsion [64].

The central principle of NTP involves using nuclear fission to generate heat. The reactor core is enriched with fissile material, typically Uranium-235. When neutrons are absorbed by the uranium nuclei, they enter an excited state, becoming unstable. This instability leads to the splitting of the nucleus, or fission, which releases significant energy—approximately 190 to 200 megaelectronvolts (MeV) per fission event. Along with this energy, fission also produces gamma rays, neutrons, and fission fragments, which contribute to a sustained nuclear chain reaction within the reactor core [66].

The energy released from the fission process is used to heat hydrogen, the chosen propellant. Hydrogen's low molecular weight is advantageous because it achieves a higher specific impulse (ISP), a measure of engine efficiency. For example, hydrogen in an NTP system can achieve an ISP of approximately 900 seconds at a fuel temperature of around 2800K. By comparison, using water as a propellant in the same system would yield an ISP of only about 375 seconds due to its higher molecular weight [66].

Fusion propulsion is also a potential future option, offering even greater efficiency and thrust, but remains in the research phase [67].

To maintain hydrogen in a liquid state, it must be stored at extremely low temperatures (below 20K), classifying it as a cryogenic propellant. This storage requirement presents challenges, as hydrogen must remain cold to avoid boiling off. Cryogenic systems are distinct from other propulsion types like electric or solid propulsion, where the handling and behaviour of propellants differ significantly.

NTP systems offer several key advantages over conventional propulsion technologies [66]:

- **High Efficiency and Thrust**: NTP systems can achieve almost twice the specific impulse of chemical propulsion systems, with ISPs nearing 900 seconds compared to the ~465 seconds of advanced chemical engines. This efficiency translates to reduced propellant requirements, allowing for lighter spacecraft or heavier payloads.

- **Reduced Transit Times**: NTP enables faster transit times for missions to destinations like Mars. For example, a fast conjunction-class mission using NTP could reduce one-way travel time to approximately 4-6 months. Shorter travel times minimize astronauts' exposure to cosmic radiation and the adverse effects of prolonged microgravity, such as muscle and bone loss.

- **Versatility and Reusability**: NTP systems can support reusable spacecraft designs, reducing development and operational costs. They also allow for artificial gravity operations and provide increased abort capabilities, enhancing mission safety.
- **Broader Mission Applications**: Beyond Mars missions, NTP can be adapted for reusable lunar cargo delivery, crewed asteroid missions, high-energy injection stages for robotic exploration of outer planets, and more. This adaptability makes NTP a "one size fits all" solution for diverse mission profiles.

Nuclear propulsion is seen as a key enabler for ambitious space exploration missions, such as crewed missions to Mars, the outer planets, and potentially even interstellar travel [60, 68]. The high energy density and sustained thrust capabilities of nuclear systems can support these long-duration, high-energy missions [60, 68].

However, the development of nuclear propulsion systems faces several technical and non-technical challenges, including safety, reliability, and public acceptance [69]. Ongoing research and development efforts are focused on addressing these issues and advancing the technology [64, 70].

To address non-proliferation concerns, modern NTP designs prioritize the use of low enriched uranium (LEU) or high-assay low enriched uranium (HALEU) rather than the highly enriched uranium (HEU) used in earlier programs like the NERVA Rover. These advancements maintain fission efficiency while improving safety and security.

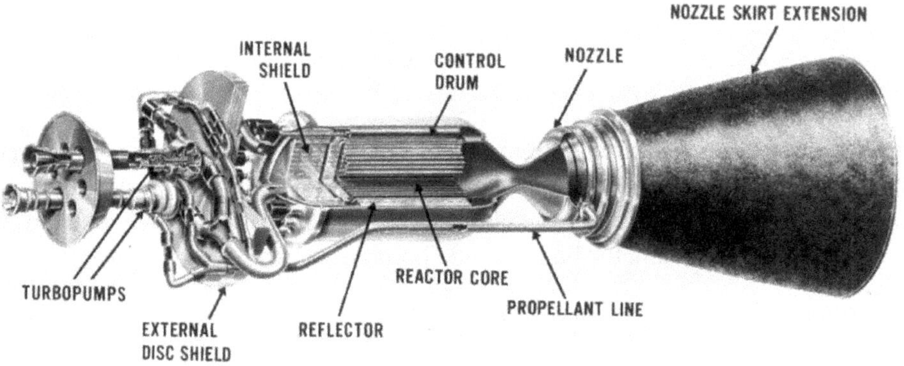

Figure 11: An explanatory drawing of the NERVA (Nuclear Engine for Rocket Vehicle Application)thermodynamic nuclear rocket engine. The main objective of project Rover/NERVA was to develop a flight rated engine with 75,000 pounds of thrust. NASA, Public Domain, via Picryl.

Despite its advantages, NTP faces challenges, including the need for advanced thermal management to sustain cryogenic hydrogen storage and reactor durability during prolonged operation. Additionally, public perception and regulatory hurdles associated with nuclear technology must be carefully managed.

NTP has the potential to revolutionize space exploration. Its high efficiency and thrust capabilities make it a cornerstone for affordable, sustainable missions to Mars and beyond. The reduced initial mass in low Earth orbit (IMLEO) enabled by NTP also decreases the number of launches required, making ambitious missions more feasible [66].

As a reusable and adaptable technology, NTP supports a variety of mission profiles, from lunar landings to robotic exploration of outer planets. With ongoing research and development, NTP is poised to play a central role in the next generation of space exploration, combining efficiency, versatility, and reliability to achieve humanity's interplanetary aspirations.

Comparison and Use Cases

Feature	Chemical	Electric	Nuclear
Thrust	High	Low	Moderate to High
Specific Impulse	Low to Moderate	High	High
Efficiency	Moderate	High	High
Complexity	Low (solid) to Moderate (liquid)	High	Very High
Primary Use	Launch and orbital insertion	Deep-space missions, station-keeping	Long-distance and crewed exploration

Advancements in materials, engineering, and computational modelling are driving innovation across all three propulsion types. Reusable chemical rockets, high-efficiency electric systems, and experimental nuclear propulsion technologies are likely to coexist, each optimized for specific phases of missions. The integration of these propulsion systems will continue to shape the future of space exploration and enable humanity to venture deeper into the cosmos.

Chapter 3
Materials and Structures in Rocket Design

Anatomy of a Rocket

A rocket is a highly engineered vehicle designed to overcome Earth's gravity and reach space. Rockets are used for a variety of purposes, such as transporting satellites, scientific instruments, supplies, or humans into orbit or beyond. Each component of a rocket is carefully designed to perform specific functions essential to achieving a successful mission. Below, we break down the critical elements that make up a rocket:

In a multi-stage system, the first stage provides the initial thrust to lift the rocket off the ground and push it through the dense layers of Earth's atmosphere. Once its fuel is exhausted, the first stage is jettisoned to reduce weight, and the second stage takes over. This stage propels the vehicle to higher altitudes where air resistance is minimal, making it easier to achieve greater speeds with less thrust. When the second stage's fuel is depleted, it too is jettisoned. The third stage then accelerates the payload to the required orbital velocity and altitude. After the payload is deployed, the third stage is discarded, completing its function [71].

The primary advantage of staging is the significant reduction in the rocket's weight as each stage is jettisoned after fuel depletion. This weight reduction allows the rocket to achieve higher final velocities with less total propellant. This staging process maximizes the payload ratio, enabling the launch vehicle to carry a larger payload relative to its overall mass. For missions targeting low Earth orbit, two-stage rockets are often sufficient, while three-stage rockets are typically required for higher altitudes or deep-space missions, such as those involving geostationary orbits or interplanetary trajectories [71].

Achieving the necessary thrust to reach orbit requires high propellant flow rates through the rocket engines. In a liquid-fuelled engine, turbopumps regulate and supply the fuel and oxidizer to the combustion chamber at the required rates. A technique known as regenerative cooling is employed to maintain the nozzle's structural integrity by circulating fuel around the nozzle walls. This process not only cools the nozzle but also preheats the fuel, enhancing combustion efficiency and increasing thrust [71].

Another advantage of multi-stage rockets is the ability to optimize each stage's engine for specific operating conditions. First-stage engines are designed for performance in the atmosphere, where aerodynamic forces are significant. In contrast, upper-stage engines are optimized for operation in the vacuum of space, where no air resistance is present. Different fuels may also be used for each stage to maximize efficiency based on the mission's requirements [71].

However, multi-stage rockets have disadvantages, including the loss of stages after they are jettisoned. These discarded components typically burn up upon re-entry into Earth's atmosphere, making them unrecoverable. Recent advancements in reusable launch technology, like SpaceX's Falcon 9, have begun addressing this issue. The first stage of the Falcon 9 is equipped with aerodynamic control surfaces and thrusters for precise guidance during descent. Using its own propulsion systems, it performs a controlled landing, either on a launch pad or an offshore barge, making it reusable for future missions. While this innovation reduces costs significantly, it requires additional fuel and engineering complexity to ensure safe recovery [71].

Solid rocket boosters (SRBs) are often used to augment the main engines during launch, providing additional thrust to lift heavy payloads. These boosters are typically jettisoned after their fuel is exhausted. In some cases, SRBs can be recovered and refurbished for reuse if jettisoned at lower altitudes and equipped with parachutes. Their relatively low cost and modularity make SRBs a flexible option for enhancing the capabilities of primary launch vehicles [71].

Historically, the Space Shuttle program represented a significant step forward in reusable spacecraft technology. The Orbiter, which housed the payload, was designed to return to Earth and land like an aircraft after completing its mission. Its two SRBs were recovered from the ocean and refurbished, while the external fuel tank was discarded. This approach demonstrated the feasibility of partially reusable launch systems, which laid the groundwork for modern advancements in rocket reusability [71].

Payloads launched by multi-stage rockets encompass a wide range of missions, from placing satellites into geostationary orbit to deploying deep-space probes and crewed capsules. For example, the Voyager probes carried scientific instruments and communication systems to explore the outer planets and interstellar space. The Hubble Space Telescope (HST), launched into low Earth orbit by the Space Shuttle, has revolutionized our understanding of the cosmos. The HST's modular design, with dedicated sections for optical assemblies, control electronics, and scientific instruments, demonstrates the precision engineering required for space payloads [71].

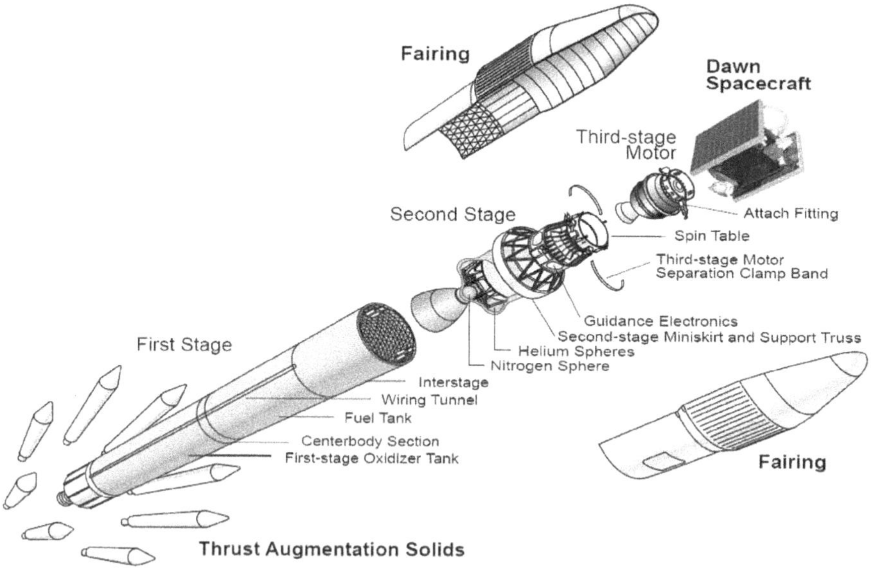

Figure 12: Delta II Heavy 2925H-9.5 including Star 48 upper stage. NASA, CC0, via Picryl.

Payload: The Mission Objective

The payload is the primary cargo the rocket is built to deliver to space. This can include:

- **Satellites** for communication, weather monitoring, or Earth observation.
- **Telescopes** like the Hubble Space Telescope, designed to study distant galaxies.
- **Supplies** for the International Space Station (ISS), such as food, equipment, and experiments.
- **Crewed Missions**, including astronauts traveling to the ISS or other destinations.

The payload's weight and dimensions dictate the design and capabilities of the rocket. Ensuring the payload remains intact and functional during the intense forces of launch is a crucial engineering challenge.

The payload system of a rocket is designed to carry and protect the cargo that serves the rocket's mission. Over the years, rocket payloads have evolved from simple explosive devices to complex scientific instruments, satellites, and even human space farers. This evolution reflects the advancements in rocket technology and the expanding scope of space exploration and utilization, showcasing the versatility of payload systems in meeting diverse objectives.

The earliest rocket payloads were rudimentary, such as the fireworks used for celebrations in ancient China. The use of rockets as military weapons marked a significant shift in their purpose. During

World War II, the German V2 rocket became a pivotal development in rocket technology, capable of delivering several thousand pounds of explosives over long distances. The V2 was the first rocket to breach the boundary of space, demonstrating the potential for rockets to carry larger and more advanced payloads, a milestone that paved the way for future developments.

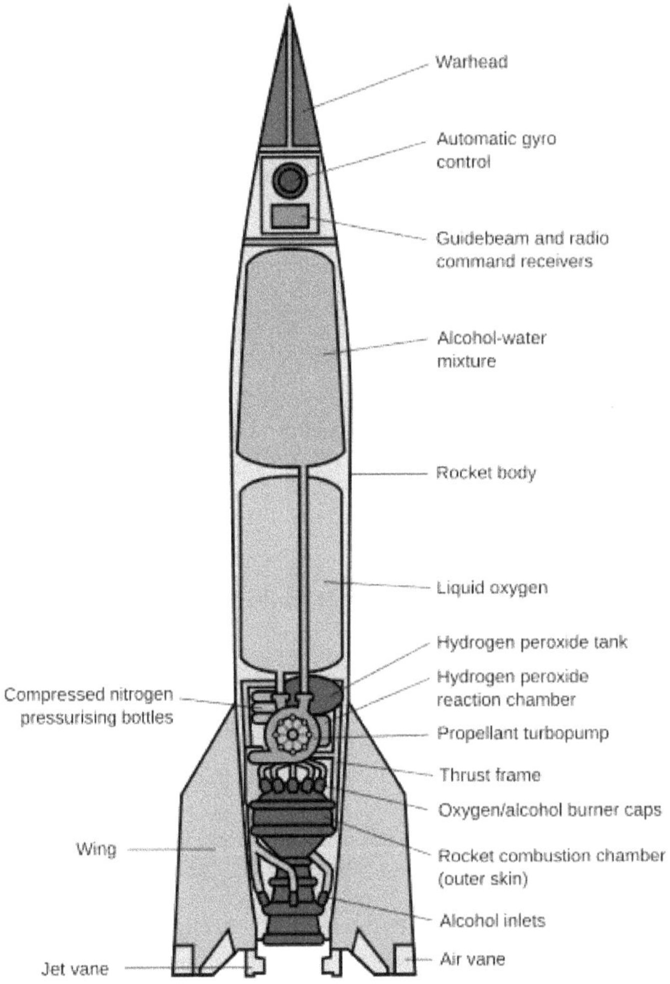

Figure 13: Schematic diagram of a V-2 rocket design. Fastfission, Public Domain, via Wikimedia Commons.

Following World War II, rocket development accelerated during the Cold War, driven by geopolitical tensions. Many nations, particularly the United States and the Soviet Union, developed guided ballistic missiles equipped with nuclear warheads. These rockets were designed to be highly accurate and powerful, focusing primarily on destructive military payloads rather than scientific or exploratory purposes. This era highlighted the dual-use nature of rocket technology, laying the groundwork for its adaptation to other applications.

As rockets transitioned from military to civilian and scientific applications, the launch of Sputnik 1 by the Soviet Union in 1957 marked the dawn of space exploration. Satellites became a common payload, serving a variety of missions including communications, weather monitoring, and Earth observation. Communication satellites revolutionized global connectivity, while weather satellites provided critical forecasting data. Reconnaissance satellites further expanded the applications of payload systems, gathering vital information for both civilian and military purposes. The payload systems of rockets adapted to these missions by incorporating advanced shielding, structural modifications, and optimized delivery mechanisms to ensure the success of increasingly complex objectives.

With the expansion of space exploration, rockets began to carry scientific payloads aimed at understanding the cosmos. Observatories such as the Hubble Space Telescope were launched to study distant galaxies, while space probes like Voyager and Cassini were sent to explore other planets and moons. These payloads required precise engineering to protect delicate instruments from the harsh conditions of launch and the vacuum of space. This era underscored the importance of payload systems in enabling groundbreaking discoveries that advanced humanity's knowledge of the universe.

The ability of rockets to carry humans into space marked a significant milestone in payload system design. Special rockets, such as those used in the Apollo program, were built to launch astronauts to the Moon. These rockets included crew modules that provided life support and safety for astronauts, along with landing modules for exploring the lunar surface. Today, modern rockets like SpaceX's Falcon 9 and NASA's Space Launch System are used for a range of missions, including resupplying the International Space Station and advancing plans for future crewed missions to Mars. These advancements demonstrate the adaptability of payload systems to accommodate human exploration.

The payload system of rockets continues to evolve to support new and diverse missions. Planetary exploration payloads, such as Mars rovers and landers, are designed to analyse extraterrestrial surfaces, pushing the boundaries of our understanding of the solar system. Commercial satellites, launched by companies like SpaceX and OneWeb, form the backbone of global communication networks, enabling new technologies and services. Additionally, the rise of space tourism has led to the adaptation of rockets to carry private citizens into space, opening a new frontier in human exploration and commercial opportunity.

The payload system of a rocket is a versatile and essential component, reflecting the mission it serves. From early explosive warheads to sophisticated scientific instruments, satellites, and crewed modules, payload systems have driven the evolution of rocket technology. These advancements are central to humanity's exploration of space, enabling both scientific discovery and commercial applications. As technology progresses, payload systems will continue to play a pivotal role in shaping the future of space exploration and utilization, unlocking new possibilities and expanding our presence beyond Earth.

Stage 1: The First Stage

The first stage is the foundational section of the rocket, containing the primary engines and fuel tanks. It provides the initial thrust needed to lift the rocket off the ground and begin its ascent through Earth's dense atmosphere.

Once its fuel is depleted, the first stage is discarded to lighten the load for subsequent stages.

Figure 14: Installation the H-1 engines into the S-IB stage - Saturn Apollo Program. NASA, CC0, via Picryl.

The first stage of a rocket is the critical foundational section responsible for providing the initial thrust required to lift the entire vehicle off the ground and propel it through the dense lower layers of Earth's atmosphere. This stage is the largest and most powerful part of the rocket, as it must overcome the forces of gravity and atmospheric drag to begin the rocket's journey into space. Its design and operation are optimized for maximum efficiency and power during this initial phase of flight.

The first stage consists of two main components: the engines and the fuel tanks.

Engines: The engines in the first stage are some of the most powerful on the rocket, designed to generate immense amounts of thrust. They work by igniting fuel and oxidizer in a combustion chamber, creating a high-pressure, high-temperature reaction that produces exhaust gases expelled at high velocity through the rocket nozzle. This action generates the thrust needed to lift the rocket. First-stage engines are typically liquid-fuelled or solid-fuelled, depending on the rocket's design. Liquid-fuelled engines offer the advantage of throttling and control, while solid-fuelled engines are simpler and more robust.

Fuel Tanks: The fuel tanks in the first stage store the propellants needed for the engines to operate. These tanks are massive, as the first stage requires a significant amount of fuel to generate the thrust necessary to overcome gravity. In liquid-fuelled rockets, the tanks hold both the fuel (such as liquid hydrogen or RP-1 kerosene) and the oxidizer (such as liquid oxygen). These propellants are kept under controlled conditions to ensure stability and are fed into the engines via turbopumps during operation.

The first stage has distinct features that define its role and performance:

High Power: The first stage engines are designed to generate immense thrust, often producing millions of pounds of force. This power is necessary to overcome the combined forces of gravity and atmospheric drag acting on the rocket. The thrust-to-weight ratio of the first stage is a critical metric, ensuring the rocket can lift off the ground and accelerate effectively.

Limited Range: While the first stage is powerful, its range is limited by the fuel it carries. The stage is designed to burn through its propellant rapidly to achieve maximum thrust during its brief operational phase. Typically, the first stage operates for a few minutes before its fuel is depleted.

The first stage ignites at launch and burns continuously until its fuel is exhausted. Once the first stage has completed its role, it is jettisoned from the rest of the rocket to reduce weight. This staging process is crucial for the rocket's efficiency and overall performance. By discarding the empty first stage, the rocket sheds unnecessary mass, allowing the subsequent stages to operate more effectively with less energy expenditure.

In many modern rockets, first-stage separation is a carefully controlled process. For instance, SpaceX's Falcon 9 uses a pneumatic or explosive release mechanism to detach the first stage, followed by a brief ignition of the second stage engines to ensure a smooth transition. Some rockets, like Falcon 9, even employ technologies to recover and reuse the first stage, further enhancing cost-efficiency and sustainability.

The design and capabilities of the first stage vary among different rockets based on their missions and payload requirements. For example:

- **Saturn V**: The first stage, known as the S-IC, was powered by five F-1 engines and produced over 7.5 million pounds of thrust. It operated for approximately 2.5 minutes before separation, propelling the rocket to an altitude of about 68 kilometres.
- **Falcon 9**: The Falcon 9's first stage is powered by nine Merlin engines and produces about 1.7 million pounds of thrust at liftoff. Its recoverable design includes landing legs and grid fins for controlled descent and reuse.

- **Delta IV Heavy**: The first stage features a common booster core with a single RS-68 engine, supplemented by two additional boosters to generate extra thrust.

The first stage is the workhorse of any rocket, providing the raw power needed to initiate flight and clear Earth's atmosphere's densest layers. Its high-power engines and large fuel tanks are optimized for short, intense operation. The staging process ensures efficiency by jettisoning the depleted first stage, reducing weight for subsequent stages to complete the mission. Advances in design, such as reusability, continue to enhance the performance and cost-effectiveness of first-stage systems, making them an essential component of modern space exploration and commercial launches.

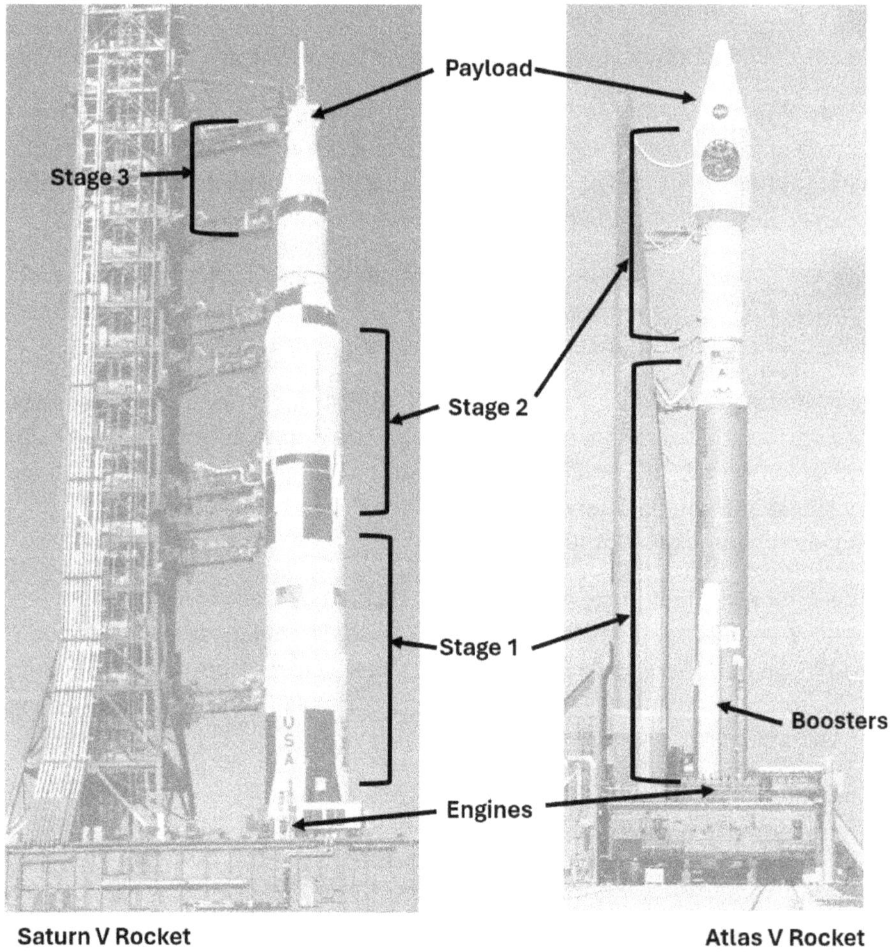

Figure 15: Basic anatomy of a rocket.

Stage 2: Taking Over Mid-Flight

The second stage takes over after the first stage is jettisoned. Positioned atop the first stage, it is smaller and less powerful because:

- **Reduced Atmospheric Resistance**: At higher altitudes, thinner air minimizes drag, so the engines require less power.
- **Efficiency**: The empty first stage is discarded to conserve energy, allowing the second stage to focus entirely on propelling the payload further toward orbit.

The second stage engines are typically optimized for operation in the vacuum of space, where aerodynamic considerations are minimal.

Figure 16: The second stage of the Delta II rocket is raised off the transporter for its lift up the launch tower on Pad 17-A, Cape Canaveral Air Force Station. NASA, CC0, via Picryl.

The second stage of a rocket plays a critical role in continuing the mission after the first stage has fulfilled its purpose and been jettisoned. Positioned directly above the first stage, the second stage is smaller and less powerful but highly efficient, tailored for the conditions encountered at higher altitudes. Its primary function is to accelerate the payload further and achieve the orbital velocity or trajectory required for the mission.

The second stage consists of the following essential elements:

- **Engine**: The engine(s) of the second stage are specifically designed for operation in the vacuum of space. Unlike first-stage engines, which must account for atmospheric drag and pressure, second-stage engines are optimized for the near absence of atmospheric resistance. This design often includes a larger nozzle to maximize efficiency by better expanding the exhaust gases in low-pressure environments. These engines are typically liquid-fuelled, as this allows for precise throttling, ignition, and shutoff capabilities, critical for placing payloads in exact orbits or trajectories.
- **Fuel Tanks**: The second stage carries its own fuel and oxidizer in smaller tanks compared to the first stage. The fuel is usually a combination of lightweight, high-energy propellants such as liquid hydrogen (LH2) and liquid oxygen (LOX) or RP-1 (refined kerosene) and LOX. The tanks are pressurized to ensure a steady supply of propellant to the engine during operation, even in the microgravity of space.
- **Structural Framework**: The second stage must be strong enough to withstand the stresses of launch and stage separation while being as lightweight as possible to maximize efficiency. Advanced materials such as aluminium-lithium alloys or carbon composites are often used to achieve this balance.
- **Payload Mounting and Interface**: The second stage includes a secure mounting system to hold the payload or third stage in place during ascent. This interface may also include mechanisms for separating the payload or transferring forces evenly during acceleration.

The second stage has distinct features that enable it to perform its role effectively:

- **Reduced Atmospheric Resistance**: By the time the second stage is activated, the rocket has ascended to an altitude where the atmosphere is significantly thinner. At these altitudes, drag is minimal, and the second stage engines require less power to accelerate the payload. This reduced resistance allows for more efficient fuel usage and higher speeds.
- **Vacuum-Optimized Engines**: The second stage engines are tailored for space conditions, focusing on high specific impulse (a measure of engine efficiency). These engines often feature larger nozzles, which improve performance by allowing the exhaust gases to expand more fully in the vacuum of space, maximizing thrust.
- **Energy Efficiency**: With the first stage jettisoned, the rocket's weight is significantly reduced, allowing the second stage to operate more efficiently. By focusing solely on the remaining payload and structure, the second stage achieves greater acceleration with less propellant.

Once the first stage separates, the second stage ignites to continue the rocket's ascent. This process typically involves a carefully timed ignition to ensure a smooth transition and avoid destabilizing the rocket. The second stage operates for several minutes, during which it increases the rocket's velocity and altitude to reach the desired orbit or trajectory. Once its fuel is depleted, the second stage is either jettisoned or left in orbit, depending on the mission requirements.

Advantages of the Second Stage

The second stage brings several advantages to multi-stage rockets:

1. **Optimized Performance**: By focusing its engines and design on vacuum conditions, the second stage operates with much higher efficiency than the first stage could in the same conditions.
2. **Flexibility**: Second stages can be precisely controlled to place payloads in specific orbits, making them ideal for missions involving satellites, space probes, or human transport.
3. **Payload Protection**: The second stage often includes structural support for the payload or additional fairings, ensuring that the cargo remains secure and undamaged during flight.

Examples of Second Stages

Different rockets have uniquely designed second stages tailored to their missions:

- **Saturn V (S-II Stage)**: The second stage of the Saturn V rocket was powered by five J-2 engines fuelled by liquid hydrogen and liquid oxygen. It played a crucial role in propelling the Apollo spacecraft toward lunar orbit.
- **Falcon 9**: SpaceX's Falcon 9 second stage uses a single vacuum-optimized Merlin engine powered by RP-1 and LOX. This stage is highly precise, capable of multiple ignitions to place payloads in diverse orbits.
- **Delta IV**: The Delta IV second stage uses the RL10 engine, which is renowned for its efficiency and high specific impulse, making it ideal for geostationary transfer orbits and interplanetary missions.

While second stages are highly efficient, they pose unique challenges, such as managing thermal loads during prolonged operation in space and ensuring reliable stage separation. Recent innovations include designing second stages for partial reuse, reducing space debris, and integrating advanced control systems for greater orbital precision.

Stage 3: For Massive Payloads

When carrying particularly heavy payloads or targeting high orbits, a third stage may be necessary. This stage provides the additional thrust required to achieve the mission's objective. Third stages are rare in modern rockets unless dealing with exceptionally challenging missions, as advancements in engineering often make two-stage rockets sufficient.

Figure 17: Saturn V Rocket, 3rd Stage. Alan Wilson from Stilton, Peterborough, Cambs, UK, CC BY-SA 2.0, via Wikimedia Commons.

The third stage of a rocket is essential for missions involving exceptionally heavy payloads or those targeting higher orbits and deep-space trajectories. Acting as the final propulsion system, it ensures the payload reaches its intended orbital velocity or escapes Earth's gravitational pull for interplanetary travel. Although advancements in engineering have made two-stage rockets sufficient for many missions, the third stage remains vital for specific, demanding objectives where additional thrust and precision are required.

The engine in the third stage is designed for optimal performance in the vacuum of space. Unlike earlier-stage engines that prioritize raw thrust to overcome gravity and atmospheric resistance, third-stage engines focus on controlled, sustained acceleration. These engines are smaller and more efficient, achieving high specific impulse to maximize fuel usage and minimize waste. The third stage also includes compact fuel tanks, typically filled with high-energy propellants like liquid hydrogen and liquid oxygen. These tanks are constructed from lightweight materials such as aluminium-lithium alloys or carbon composites to reduce weight without compromising structural integrity.

The third stage directly connects to the payload through a secure interface that ensures stability during ascent and precise detachment upon deployment. This interface prevents any unintended forces from altering the payload's trajectory. Additionally, advanced guidance and control systems in the third stage monitor its position and velocity, making real-time adjustments to maintain an accurate trajectory. These systems rely on sensors, gyroscopes, and onboard computers to deliver the payload to its designated orbit or path with precision.

Operating entirely in the vacuum of space, the third stage has unique functions and characteristics. It provides the final thrust needed to achieve orbital velocity or escape velocity for interplanetary missions. For example, low Earth orbit requires speeds exceeding 28,000 kilometres per hour, while interplanetary missions demand even greater velocities. The third stage's reduced mass, following the jettisoning of the first and second stages, allows it to operate more efficiently, requiring less fuel to achieve its objectives. Furthermore, its design supports flexibility for various mission profiles, enabling it to place payloads in high orbits, geostationary orbits, or deep-space trajectories. The engine's ability to throttle, reignite, and control direction adds to this versatility.

Third stages are indispensable for missions involving heavy payloads or complex orbital insertions. For example, they are used to propel large satellites or cargo into geostationary orbit, ensuring sufficient velocity to overcome Earth's gravity. In interplanetary missions, the third stage provides the critical push needed to achieve escape velocity, as seen with the Saturn V rocket's S-IVB stage, which propelled Apollo missions toward the Moon. Other examples include the Ariane 5's third stage,

designed for precise placement of commercial satellites, and the Vulcan Centaur rocket, which features a third stage option for high-energy missions, demonstrating its importance in contemporary space exploration.

The third stage offers several advantages, including extended capability to carry larger payloads to higher orbits, precision in orbital manoeuvres, and flexibility for diverse mission profiles. However, it also presents challenges, such as increased design complexity, higher costs, and the risk of contributing to space debris if not managed carefully. Innovations in propulsion technology and materials science are addressing these challenges by improving engine efficiency, enabling multiple restarts, and reducing structural mass through advanced materials. These advancements ensure that third stages remain relevant for modern and future space missions.

The third stage of a rocket is a critical component for achieving mission objectives that are beyond the capabilities of simpler two-stage designs. Its precision, efficiency, and adaptability make it indispensable for ambitious space exploration efforts. Although less common in modern rockets, the third stage remains a cornerstone of complex missions, enabling humanity to push the boundaries of what is possible in space exploration and technology.

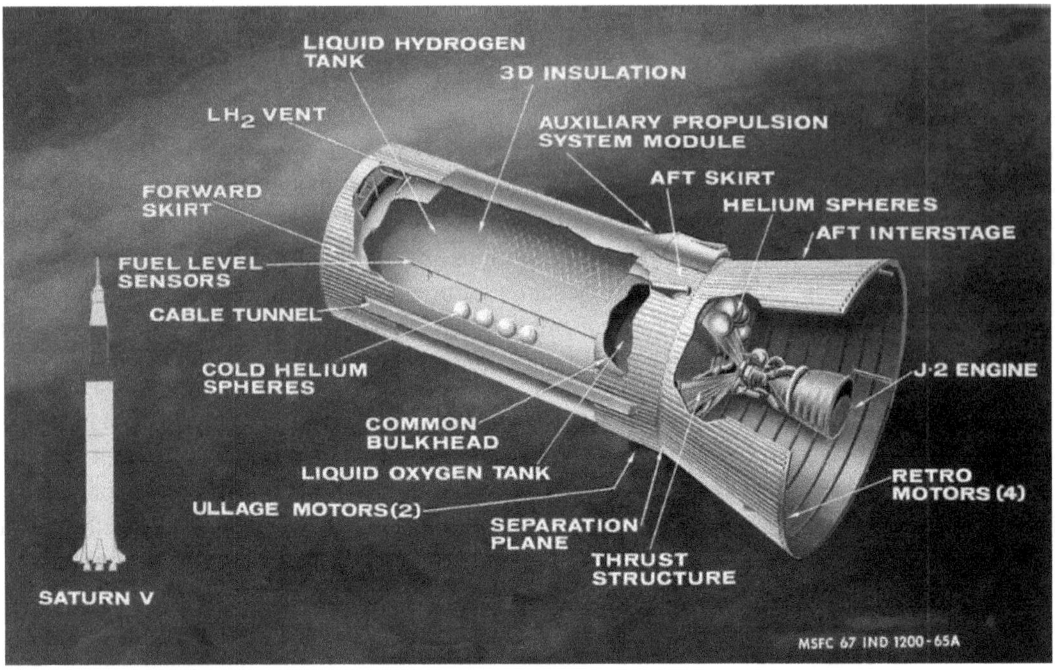

Figure 18: This cutaway illustration shows the Saturn V S-IVB (third) stage with the callouts of its major components. NASA, CC0, via Picryl.

Boosters: Supplemental Thrust

Boosters are smaller, detachable rockets attached to the first stage to provide extra thrust during liftoff. They are particularly useful for heavy payloads or missions requiring more power to achieve orbit. Once their fuel is exhausted, boosters are jettisoned to reduce weight. For example:

- **Space Shuttle Solid Rocket Boosters** provided significant thrust during launch and were reusable.
- **Falcon Heavy** employs side boosters that return to Earth for recovery and reuse.

Boosters are an integral part of many rocket designs, providing additional thrust during the crucial moments of liftoff and ascent. These smaller, detachable rockets are mounted alongside the main first stage and serve to increase the overall thrust of the vehicle, enabling it to carry heavier payloads or achieve higher altitudes. Boosters are particularly valuable in missions where the main engines of the rocket alone would not generate sufficient power to overcome gravity and atmospheric resistance during the initial phase of launch.

The primary function of boosters is to supply extra thrust during the rocket's initial ascent through Earth's dense lower atmosphere. This phase requires the most power because the rocket must overcome gravity and drag while accelerating to reach escape velocity. Boosters provide this additional power by igniting alongside the main engines of the first stage. They burn their fuel rapidly to deliver a significant surge of thrust, effectively lightening the load on the main engines and allowing the rocket to lift off with greater efficiency.

Once the fuel in the boosters is depleted, they are jettisoned to reduce the rocket's overall weight. This staging process is essential for maintaining efficiency, as it allows the remaining stages of the rocket to continue their ascent without carrying the dead weight of spent fuel tanks and engines. Booster separation is a carefully controlled process, often involving explosive bolts or pyrotechnic devices to ensure a clean and safe detachment from the main vehicle. After separation, boosters typically fall back to Earth and are either discarded or, in some cases, recovered for reuse.

Boosters come in various designs and configurations, with two main types being solid-fuel and liquid-fuel boosters. Solid-fuel boosters, such as those used on the Space Shuttle, are simple and reliable, containing a pre-mixed solid propellant that burns consistently throughout their operation. These boosters are robust and require minimal maintenance, making them an economical choice for many missions. On the other hand, liquid-fuel boosters, such as those used on SpaceX's Falcon Heavy, offer greater control and flexibility, allowing throttling and multiple ignitions. This makes them suitable for more complex missions that require precise adjustments during ascent.

The Space Shuttle's Solid Rocket Boosters (SRBs) are a classic example of booster technology. These SRBs provided most of the thrust required during the Shuttle's liftoff and initial climb, each generating approximately 3.3 million pounds of thrust. They were also one of the first examples of reusable boosters, designed to be recovered after launch. Equipped with parachutes, the SRBs would splash down in the ocean and be retrieved for refurbishment and reuse in subsequent missions. This reusability significantly reduced the overall cost of the Space Shuttle program.

In contrast, SpaceX's Falcon Heavy employs liquid-fuel side boosters derived from the first stage of its Falcon 9 rocket. These boosters use RP-1 kerosene and liquid oxygen as propellants and are equipped with advanced landing systems for recovery. After providing supplemental thrust during liftoff, the Falcon Heavy's side boosters separate and return to Earth, landing either on designated landing pads or autonomous drone ships at sea. This innovation allows SpaceX to reuse its boosters multiple times, dramatically lowering the cost of space launches and making commercial spaceflight more accessible.

Boosters also play a crucial role in enabling modularity and flexibility in rocket designs. By adding or removing boosters, engineers can customize a rocket's thrust capacity to suit the specific needs of a mission. For example, a launch vehicle may be equipped with additional boosters for a heavy payload or high-altitude mission, while fewer boosters may be used for lighter loads. This adaptability makes boosters a cost-effective and efficient solution for a wide range of launch requirements.

Figure 19: Atlas V Booster Arrival and Offload. NASA, CC0, via Picryl.

Boosters are an essential component of modern rocket design, providing the supplemental thrust needed to lift heavy payloads into space. Whether solid-fuel or liquid-fuel, boosters enhance a rocket's performance during liftoff and early ascent, allowing it to achieve higher velocities and altitudes. By detaching when their fuel is spent, boosters improve the overall efficiency of the rocket. Innovations in booster recovery and reuse, as seen in the Space Shuttle and Falcon Heavy, continue to revolutionize the economics of space exploration, paving the way for more ambitious missions in the future.

Engines: The Heart of Propulsion

Rocket engines are the core components that generate thrust by igniting fuel in a combustion chamber. The energy released propels the rocket in the opposite direction. Engine designs vary:

- **Single-Burn Engines**: Fire once and burn continuously until the fuel is depleted.
- **Restartable Engines**: Can be shut off and reignited as needed, offering flexibility for complex missions, such as placing multiple payloads into different orbits.

Modern rocket engines, such as SpaceX's Merlin engines, are designed to be reusable, significantly reducing mission costs.

Figure 20: Blue Origin BE-4 rocket engine, sn 103. N2e, CC BY-SA 4.0, via Wikimedia Commons.

Rocket engines are the most vital components of any launch vehicle, serving as the primary means of generating the thrust needed to propel the rocket into space. By igniting fuel in a combustion chamber, rocket engines convert chemical energy into kinetic energy, expelling high-velocity exhaust gases to create thrust in the opposite direction, in accordance with Newton's Third Law of Motion. Their design and functionality determine a rocket's performance, reliability, and overall mission success.

The basic structure of a rocket engine consists of several critical elements, including the combustion chamber, propellant feed system, and nozzle. The combustion chamber is where fuel and oxidizer combine and ignite, producing high-pressure, high-temperature gases. These gases are then expelled through the nozzle, which is shaped to accelerate the exhaust flow to supersonic speeds, maximizing thrust. The propellant feed system, often powered by turbopumps, ensures a steady and controlled supply of fuel and oxidizer to the combustion chamber.

Rocket engines are broadly categorized into two main operational types: single-burn engines and restartable engines. Single-burn engines are designed to fire once and burn continuously until their fuel is depleted. These engines are typically used in the first stages of rockets, where maximum and sustained thrust is required to overcome Earth's gravity and atmospheric drag. Once the fuel is exhausted, these engines are no longer operational, and the spent stage is jettisoned to reduce weight.

In contrast, restartable engines offer greater flexibility and precision, as they can be shut down and reignited multiple times during a mission. These engines are commonly used in upper stages of rockets, where precise orbital adjustments or multiple payload deployments are required. Restartable engines are particularly advantageous for interplanetary missions, as they allow spacecraft to perform mid-course corrections or transfer to higher orbits after initial insertion.

Modern rocket engines incorporate advanced technologies to enhance their performance and reusability. For example, SpaceX's Merlin engines are a class of liquid-fuelled engines that use RP-1 kerosene and liquid oxygen as propellants. These engines are designed for high efficiency, with a specific focus on reusability. By incorporating robust materials and innovative cooling systems, Merlin engines can withstand the intense heat and pressure of repeated launches, significantly reducing costs and turnaround times between missions. SpaceX has successfully demonstrated the reusability of its engines through numerous launches and recoveries, marking a transformative shift in the economics of space exploration.

Figure 21: A SpaceX Merlin 1C Vacuum engine built at the company's Hawthorne, California facility. SpaceX, CC0, via Wikimedia Commons.

Rocket engine design also varies depending on the stage in which the engine is used. Engines on the first stage prioritize raw power and high thrust to lift the rocket off the ground and push it through the dense lower atmosphere. These engines are optimized for operation at sea-level pressure, with smaller nozzles tailored to the dense air conditions. On the other hand, upper-stage engines are optimized for performance in the vacuum of space, where the absence of atmospheric pressure allows for the use of larger nozzles. These vacuum-optimized engines achieve higher specific impulse, a measure of fuel efficiency, making them ideal for long-duration burns.

Thermal management is another critical aspect of rocket engine design. During operation, the combustion chamber and nozzle experience extreme temperatures that could compromise structural integrity. To counter this, many engines employ regenerative cooling, where the propellant is circulated around the engine walls before being injected into the combustion chamber. This system not only cools the engine but also preheats the propellant, enhancing combustion efficiency.

The versatility of rocket engines extends beyond traditional chemical propulsion systems. In recent years, alternative propulsion technologies, such as ion thrusters and nuclear thermal propulsion, have been explored for deep-space missions. While not as powerful as chemical engines, these systems offer exceptional fuel efficiency, making them suitable for prolonged missions in the vacuum of space.

Fins: Atmospheric Steering

Fins are found on some rockets to assist with steering while the rocket travels through the lower atmosphere. Their roles include:

- **Aerodynamic Stability**: Helping the rocket maintain a steady trajectory.
- **Emergency Functionality**: On the Saturn V, fixed tail fins were a contingency measure to stabilize the rocket in emergencies, giving astronauts time to activate escape systems.

However, many modern rockets forego fins and instead rely on engine gimbaling (swivelling the engines) for precise directional control.

Fins are aerodynamic structures attached to the lower sections of some rockets, primarily serving as stabilizing and steering mechanisms during the early phase of flight through the dense lower atmosphere. Their design and placement are critical for maintaining a rocket's stability, reducing aerodynamic drag, and, in certain cases, providing emergency stabilization. Although modern advances in engine technology have rendered fins less common, they remain a vital feature in certain rocket designs.

One of the primary functions of fins is to provide aerodynamic stability. As a rocket ascends, it must maintain a steady trajectory to ensure it reaches the desired altitude and orbit. Fins act as control surfaces that counteract any unwanted lateral or rotational movements caused by atmospheric disturbances or imbalances in thrust. Positioned symmetrically around the rocket's base, fins create a restoring force that keeps the rocket aligned along its intended flight path. This stability is particularly crucial during the initial stages of launch when the rocket is traveling through dense atmospheric layers and is more susceptible to aerodynamic forces.

Fins can also play a role in emergency functionality, as demonstrated in the design of the Saturn V rocket. The Saturn V featured large, fixed tail fins that were not intended for routine steering but were included as a contingency measure. In the event of a malfunction that disrupted the rocket's stability, the fins would help stabilize the vehicle long enough for the crew to activate escape systems. Fortunately, such scenarios never occurred during the Saturn V missions, but the inclusion of fins highlights their potential role as a backup safety feature in critical launch systems.

The design and effectiveness of fins depend on several factors, including their shape, size, and placement. Most rocket fins are designed to minimize drag while providing sufficient surface area for stabilization. In some cases, fins are fixed, meaning they do not move and serve only to passively stabilize the rocket. In other designs, fins may be movable, functioning as active control surfaces that can adjust their angle to steer the rocket. This adjustability is achieved through hydraulic or electronic actuators that respond to commands from the rocket's guidance system.

While fins are effective for stabilization, they become less necessary as rockets ascend into thinner atmospheric layers. Once the rocket reaches altitudes where air density is minimal, aerodynamic forces have little impact on its trajectory. At this point, steering and stabilization are achieved through other mechanisms, such as engine gimbaling, where the rocket's engines swivel to direct the thrust vector. Engine gimbaling provides precise control over the rocket's pitch, yaw, and roll, making it the preferred method for modern rockets that operate predominantly in vacuum conditions.

The shift away from fins in many contemporary rockets is also influenced by advancements in materials and propulsion technology. Modern rockets, such as SpaceX's Falcon 9, rely entirely on engine gimbaling for steering, eliminating the need for fins. This approach reduces weight and drag, improving overall efficiency. However, fins are still used in specific applications, such as in smaller rockets, sounding rockets, or vehicles that require additional stabilization during the atmospheric phase of flight.

Atlas V rocket

Figure 22 shows a detailed cutaway schematic of the Atlas V rocket, highlighting its structural components, propulsion systems, and payload housing. The diagram is divided into key sections, each labelled with annotations to describe the specific components and their roles within the rocket's architecture.

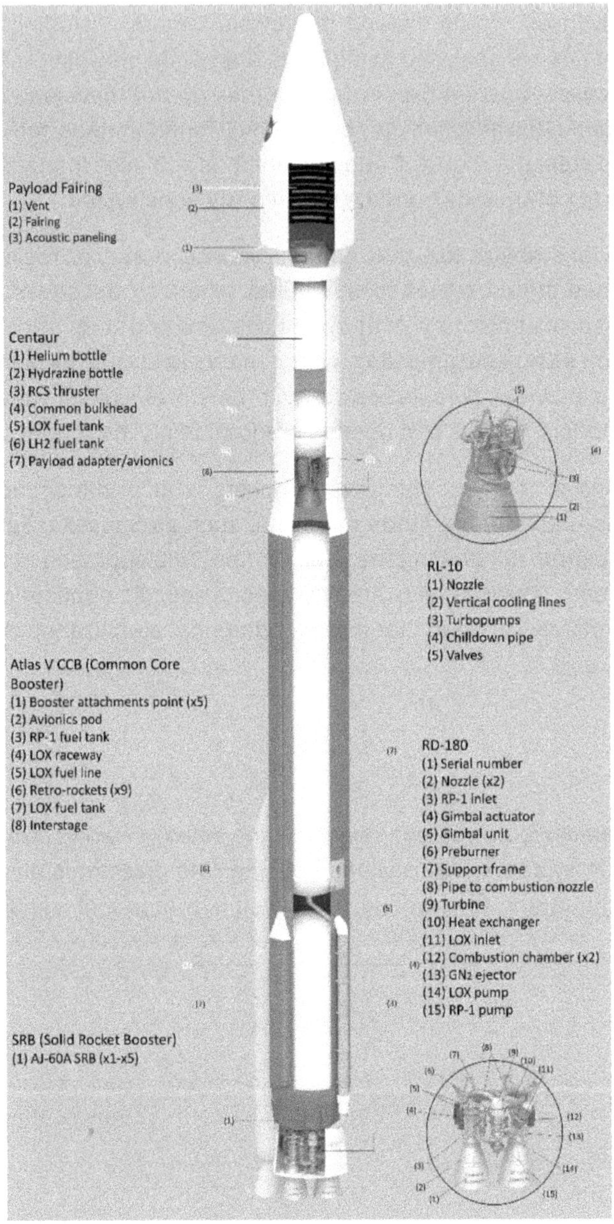

Figure 22: Atlas V 4 Meter Fairing Cutaway. Fac-tory-o, CC BY-SA 4.0, via Wikimedia Commons.

At the top of the rocket is the payload fairing, designed to house and protect the payload during ascent through Earth's atmosphere. This nose cone serves multiple functions. Vents allow for pressure equalization inside the fairing during ascent, preventing structural stress. The external shell of the fairing shields the payload from aerodynamic forces and heat, while internal acoustic panelling reduces intense vibrations and sound waves generated during launch. This ensures that sensitive payloads, such as satellites or scientific instruments, remain intact and functional.

The Centaur stage, located below the payload fairing, is the upper stage of the Atlas V rocket, responsible for placing the payload into its final orbit. It features several components critical to its functionality. Helium bottles store gas for pressurizing fuel tanks and controlling valves, while hydrazine bottles supply fuel for the Reaction Control System (RCS) thrusters, which provide fine attitude adjustments and stability in space. The Centaur's LOX and LH2 fuel tanks, separated by a common bulkhead to reduce weight, store the cryogenic propellants required to power its RL-10 engine. The stage also includes a payload adapter and avionics systems, securing the payload and housing the electronics needed for stage guidance and communication.

The RL-10 engine, which powers the Centaur stage, is designed for high efficiency and optimized performance in a vacuum. Its nozzle expels exhaust gases at high velocity to generate thrust, while vertical cooling lines prevent overheating by circulating coolant around the engine. Turbopumps deliver liquid hydrogen and liquid oxygen to the combustion chamber at high pressure, with chill-down pipes pre-cooling the fuel lines to avoid thermal shock when cryogenic propellants flow. Valves control the precise flow of fuel and oxidizer, ensuring reliable and efficient operation.

The Common Core Booster (CCB) serves as the rocket's first stage, providing the primary thrust during liftoff. This stage includes booster attachment points for securing solid rocket boosters when additional thrust is needed. The avionics pod houses critical guidance, navigation, and control systems. The RP-1 fuel tank stores highly refined kerosene, while the LOX fuel tank contains liquid oxygen, the oxidizer required for combustion. The LOX raceway, an insulated pathway, allows liquid oxygen to flow along the core's exterior. Retro-rockets assist in separating the first stage from the Centaur, and the interstage physically connects the two stages.

Powering the first stage is the RD-180 engine, a highly efficient propulsion system designed for immense thrust at liftoff. This engine features dual nozzles for expelling exhaust gases, RP-1 inlets for kerosene fuel intake, and gimbal actuators for swivelling the engine nozzles to steer the rocket. The pre-burner ignites a small amount of fuel and oxidizer to drive the turbopumps, which deliver propellants to the combustion chamber. Inside the combustion chamber, the fuel and oxidizer mix and burn, creating the high-pressure exhaust gases that propel the rocket upward.

The Atlas V can also be equipped with one to five AJ-60A solid rocket boosters (SRBs) for additional thrust during liftoff. These boosters, powered by solid propellant fuel, provide reliable high-thrust performance. They attach securely to the Common Core Booster and include separation systems that eject them once their fuel is expended, reducing weight and allowing the rocket to continue its ascent efficiently.

Overall, the Atlas V rocket is a versatile and modular launch vehicle capable of delivering payloads to low Earth orbit, geostationary transfer orbit, or even interplanetary trajectories. Its design, which allows for optional solid rocket boosters and an efficient Centaur upper stage, makes it adaptable to a wide range of mission requirements. Unlike some modern rockets, the Atlas V is not designed for reusability; components such as the CCB and boosters are not recovered after flight. This detailed cutaway offers a comprehensive understanding of how the rocket's components work in unison to achieve successful payload delivery, demonstrating the complexity and precision of modern rocket design.

NASA's Space Launch System (SLS)

Figure 23 illustrates the components and systems of NASA's Space Launch System (SLS) rocket, which is designed to support missions under the Artemis program. The NASA Space Launch System (SLS) is a powerful heavy-lift launch vehicle that is being developed to enable human exploration of deep space, including missions to the Moon and Mars [72, 73]. The SLS is designed to be an evolvable architecture that can be upgraded over time to increase its capabilities [72, 73].

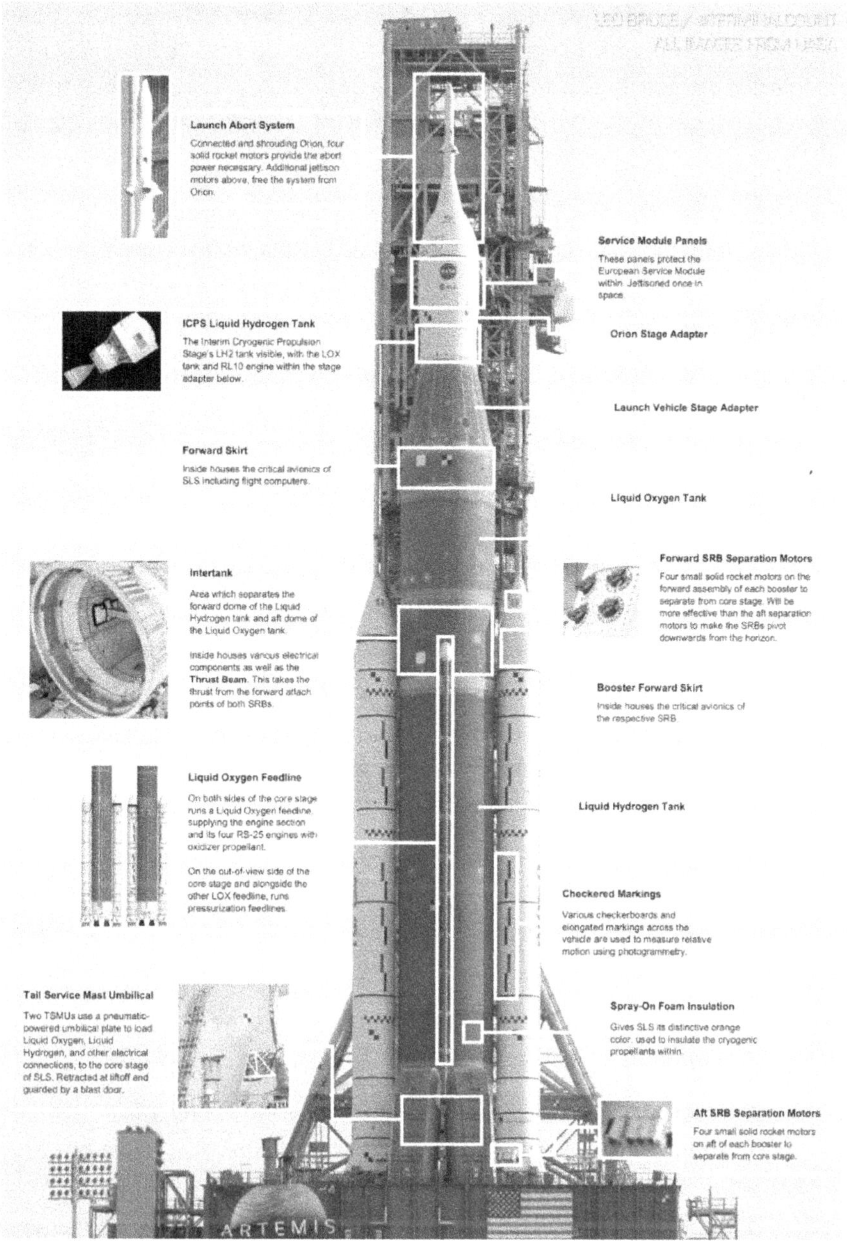

Figure 23: Technical diagram of NASA's Space Launch System for Artemis I. Leo Bruce, CC BY 2.0, via Flickr.

One key feature of the SLS is its use of friction stir welding (FSW) technology to manufacture the large liquid hydrogen tank, which is 39 meters long [74]. This was made possible by the construction of a custom, high-performance FSW system that was specifically designed for the SLS [74].

The SLS will be used to launch the Orion Multi-Purpose Crew Vehicle (MPCV), as well as other important payloads, on missions beyond Earth orbit [72, 73]. The SLS is designed to simplify spacecraft complexity, provide improved mass margins and radiation mitigation, and reduce mission durations, which offers attractive advantages for ambitious missions such as a crewed mission to Mars [73].

In addition to its primary role of launching the Orion MPCV, the SLS will also provide opportunities for secondary payloads, such as CubeSats, to be deployed on deep space missions [75, 76]. This will enable a wide range of science and exploration capabilities to be carried out alongside the primary mission objectives [75, 76].

The development of the SLS has faced a number of challenges, including logistics challenges [77] and the need to develop new technologies such as the adaptive augmenting control (AAC) system for the flight control system [78, 79]. However, NASA has been working to address these challenges through a range of risk reduction activities and demonstrations [80, 81].

The Launch Abort System (LAS) is positioned at the very top of the rocket, connected to the Orion spacecraft. It is equipped with four solid rocket motors that provide the thrust needed to pull Orion away from the rocket in the event of an emergency during launch. Additional jettison motors are used to separate the LAS from Orion once it is no longer needed.

The Orion Stage Adapter connects the spacecraft to the rocket and ensures structural integrity during launch. The Service Module Panels enclose and protect the European Service Module, which houses the propulsion and life-support systems for the Orion spacecraft. These panels are jettisoned once the rocket reaches space, exposing the spacecraft for operation.

The Interim Cryogenic Propulsion Stage (ICPS) contains a Liquid Hydrogen (LH2) tank, paired with a Liquid Oxygen (LOX) tank and the RL-10 engine within the stage adapter. This stage provides additional propulsion to deliver the spacecraft into its intended orbit or trajectory after separation from the core stage.

The forward skirt is part of the core stage and houses the critical avionics of the SLS, including its flight computers. These systems provide guidance, navigation, and control throughout the launch.

The intertank is the structural section separating the liquid hydrogen tank and the liquid oxygen tank. It houses electrical components and the thrust beam, which transfers the thrust generated by the Solid Rocket Boosters (SRBs) to the rocket's core. This region plays a vital role in maintaining the structural integrity of the rocket.

The liquid oxygen tank stores the oxidizer required for combustion in the rocket's RS-25 engines. Feedlines on either side of the core stage deliver liquid oxygen from the tank to the engines. These feedlines are insulated and pressurized to maintain the stability of the cryogenic propellant.

The SLS includes two powerful Solid Rocket Boosters mounted on either side of the core stage. Each booster includes the Booster Forward Skirt, which houses avionics systems for the booster's operation. The Forward SRB Separation Motors help detach the boosters from the core stage after they have expended their fuel. Similarly, the Aft SRB Separation Motors at the lower end of the boosters ensure a clean detachment.

This massive tank stores cryogenic liquid hydrogen, the primary fuel for the RS-25 engines. It is insulated with spray-on foam to maintain the low temperatures required for the propellant and prevent heat transfer that could lead to evaporation or instability.

The Tail Service Mast Umbilicals (TSMUs) are located at the base of the rocket. These are pneumatically powered systems that transfer liquid oxygen, liquid hydrogen, and other critical supplies, such as electrical power and data, into the core stage during pre-launch operations. The umbilicals retract shortly before liftoff.

The orange colour of the core stage comes from its spray-on foam insulation, which prevents heat transfer to the cryogenic propellants inside the tanks. The checkered and elongated markings across the rocket are used for photogrammetry, enabling engineers to measure relative motion during flight.

Spacecraft Platforms

The SmallSat market has rapidly evolved to provide a diverse range of mission-enabling components, including both individual systems and fully integrated spacecraft bus solutions. A spacecraft bus refers to the segment of a satellite mission that provides essential services, such as power, propulsion, and data handling, to the payload. This field has expanded significantly with offerings that cater to various mission profiles, from academic research to commercial ventures. The state-of-the-art in SmallSat platforms offers insights into the diverse options available and the programmatic considerations for mission development [82].

The SmallSat market represents a rapidly growing industry focused on the development, production, and deployment of small satellites for various applications. SmallSats are compact satellites typically classified by their mass, ranging from less than one kilogram to a few hundred kilograms. This market has experienced significant expansion in recent years, driven by advancements in miniaturized technology, reduced launch costs, and increasing demand for satellite-based services.

The SmallSat market offers two primary types of spacecraft platform options: hosted payloads and dedicated spacecraft buses. These options are not inherently superior to one another, as the selection largely depends on the specific needs of the mission [82].

Hosted payloads, often referred to as "satellite-as-a-service," enable multiple independent payloads from different customers to share a single platform. This configuration emphasizes resource sharing, including cost, operational autonomy, and data handling. Hosted payloads can take two forms: (1) platforms that integrate multiple independent customer payloads without a primary mission goal and (2) platforms designed for a provider's primary mission that allow secondary payloads to utilize excess resources. Hosted payloads have gained popularity due to their cost-efficiency and scalability, particularly for academic and government scientific missions [82].

In contrast, dedicated spacecraft buses provide the entire platform exclusively to a single customer or mission, offering complete control over the bus's resources. These platforms are further categorized into PocketQubes, CubeSats, and ESPA-Class buses, each with distinct capabilities tailored to different mission scales. PocketQubes, for example, are ultra-small platforms with strict mass and volume constraints, whereas CubeSats follow a modular 10 cm cube standard, supporting missions from 1U to 27U. ESPA-Class buses cater to larger payloads and often serve as secondary payloads on launch vehicles [82].

SmallSats are categorized into several types based on their size and functionality. CubeSats, which are modular satellites with a standard unit size of 10 cm^3 (1U), can be scaled up to sizes like 3U, 6U, or larger. They are widely used in academic, commercial, and government projects. PocketQubes, smaller than CubeSats with 5 cm^3 units, provide ultra-low-cost access to space. MicroSats and NanoSats, with masses ranging from a few kilograms to hundreds of kilograms, support more advanced payloads and missions. MiniSats, which weigh between 100 and 500 kilograms, are designed for more complex operations and are frequently utilized for commercial purposes.

Figure 24: 1U cubesat structure without outer skin. Svobodat, CC BY-SA 3.0, via Wikimedia Commons.

The SmallSat market serves a variety of industries. Telecommunications is a major application, providing low-latency broadband and global internet coverage through constellations like SpaceX's Starlink and OneWeb. In Earth observation, SmallSats monitor agriculture, urban development, natural disasters, and climate change with high-resolution imaging. Scientific research benefits from these satellites by enabling experiments in space, such as astrophysics research, biological studies, and atmospheric analysis. SmallSats are also used in navigation and positioning to augment existing GPS services with enhanced navigation systems. Defense and security applications include supporting military operations through reconnaissance and secure communications.

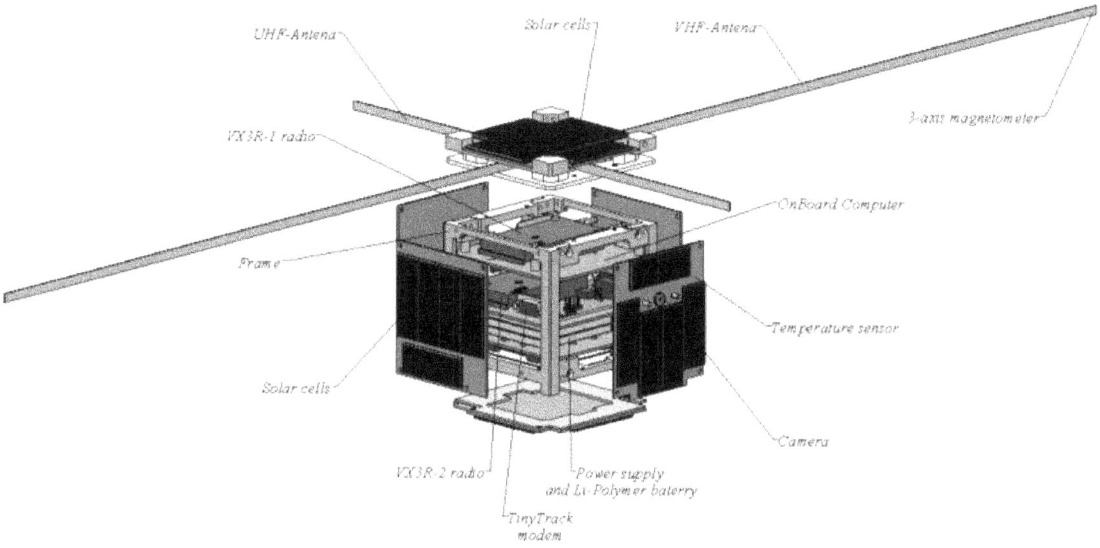

Figure 25: F-1 CubeSat exploded view. Thuvt, CC BY-SA 3.0, via Wikimedia Commons.

Several factors drive the SmallSat market. SmallSats offer cost-effective alternatives to traditional large satellites, with miniaturized technology enabling affordable production and launch. Technological advancements in sensors, processors, and propulsion systems have significantly enhanced SmallSat capabilities. Commercial opportunities abound as companies leverage SmallSats to provide satellite-based services, including global connectivity and remote sensing. Additionally, SmallSats can be launched as secondary payloads on larger rockets or as part of dedicated launch missions, with companies like Rocket Lab and SpaceX developing vehicles specifically for this market.

The development of these platforms involves significant logistical and engineering challenges, including integrating payloads, ensuring mission compatibility, and adhering to strict size and weight constraints. Additionally, programmatic considerations such as risk assessment, system reliability, and production timelines heavily influence design and implementation choices [82].

Emerging technologies are poised to revolutionize the SmallSat industry. As launch services become more cost-effective and frequent, the accessibility of SmallSat platforms will expand, enabling universities, research institutions, and smaller enterprises to undertake missions that were previously infeasible. Advances in key subsystems—such as In-space propulsion, navigation control, optical communications, and radiation tolerance—are expected to enhance performance and reliability. These improvements will likely result in platforms with greater capabilities, fostering innovation and attracting new vendors to the market [82].

To aid mission developers in navigating this complex landscape, resources like NASA's CubeSat 101 Book, the NASA Systems Engineering Handbook, and the Small Spacecraft Technology Guidebook provide comprehensive guidance on design, selection, and implementation processes. These tools emphasize understanding mission requirements, evaluating system performance, and conducting thorough risk-based trade studies [82].

Looking ahead, SmallSat platforms will continue to play a pivotal role in democratizing access to space. The combination of modular designs, reduced costs, and turnkey solutions offered by vendors positions the industry for further growth. Emerging trends, such as turnkey platforms and customizable solutions, highlight the industry's adaptability to diverse customer needs. As the market matures, statistical reliability data and subsystem improvements will refine design criteria, enabling more robust and sustainable space missions. This evolution underscores the transformative potential of SmallSat technologies in addressing global challenges and advancing space exploration [82].

Structural Requirements for Each Stage of a Rocket

The structural design of each stage of a rocket is a critical aspect of its functionality and success. Each stage is tailored to specific operational conditions, dictated by its position in the launch vehicle, the environmental stresses it encounters, and the performance requirements it must meet. These requirements ensure that the rocket remains structurally sound and efficient throughout its journey from the ground to orbit.

First Stage: Withstanding Launch Stresses

The first stage of a rocket is responsible for providing the initial thrust needed to lift the entire vehicle off the launch pad and through the densest layers of Earth's atmosphere. This stage faces the highest structural demands due to the immense forces exerted during liftoff, including gravitational forces, aerodynamic drag, and vibrations from the engines.

The structural design of the first stage must prioritize strength and rigidity to handle these stresses. Materials such as high-strength aluminium alloys, stainless steel, or carbon composites are commonly used for the outer shell and internal supports. The structure must also securely house large fuel tanks, which contain cryogenic or other high-energy propellants, and prevent deformation under high internal pressures. Additionally, the stage must include mounting points for solid rocket boosters or auxiliary engines, if used, ensuring these components remain securely attached during the high-thrust phase of launch.

The structural integrity of the first stage is further tested by acoustic vibrations generated by the engines, which can reach destructive levels. Engineers design the stage to dampen these vibrations, protecting sensitive components and maintaining overall stability. The first stage must also incorporate aerodynamic features, such as fairings or fins, to stabilize the rocket during ascent through the atmosphere.

Second Stage: Transitioning to Higher Altitudes

The second stage operates at higher altitudes where the atmospheric density is significantly reduced. Its primary role is to continue accelerating the payload after the first stage is jettisoned. The structural requirements for this stage are different from those of the first stage because it encounters lower aerodynamic forces but must endure high velocities and sustain its own propulsion.

The second stage structure must be lightweight to maximize efficiency while maintaining sufficient strength to house fuel tanks, engines, and other critical systems. Aluminium-lithium alloys are frequently used for this stage because of their high strength-to-weight ratio and resistance to thermal stress. The structure must also include secure mounting mechanisms for attaching the payload or the next stage of the rocket.

A key consideration for the second stage is thermal protection. During ascent, the rocket experiences heating from residual atmospheric friction and the intense energy generated by the engines. The second stage often includes lightweight thermal shielding to prevent overheating and maintain the integrity of its structural components. Additionally, precise structural alignment is crucial to ensure the stage can deliver the payload or upper stage to its intended trajectory without deviation.

Third Stage: Delivering Orbital Velocity

The third stage, if present, is tasked with delivering the payload to its orbital velocity or transferring it to a trajectory for deep-space missions. Operating in the vacuum of space, the third stage does not encounter aerodynamic forces but must be structurally optimized to withstand high acceleration, intense vibrations during ignition, and the challenges of prolonged space exposure.

The structural design of the third stage prioritizes lightweight construction, as any excess weight directly reduces payload capacity. Advanced materials such as carbon fibre composites are often employed to minimize mass while ensuring durability. The structure must securely hold the payload, which may include satellites, scientific instruments, or crewed modules. The mounting mechanism must be robust enough to handle the forces of acceleration while allowing for precise deployment of the payload.

Thermal and radiation shielding is another critical aspect of the third stage's structure. In space, the stage is exposed to extreme temperature fluctuations and cosmic radiation. Insulating materials and reflective coatings are used to protect sensitive components and prevent thermal stress on the structure. Additionally, the third stage may include mechanisms for attitude control, such as reaction wheels or thrusters, which must be integrated into the structural design.

Structural Integration Across Stages

The connection points between rocket stages must also meet stringent structural requirements. These interstage adapters transfer forces between stages during launch and ensure clean separation when a stage is jettisoned. They must be strong enough to withstand the combined weight and thrust of all upper stages while being lightweight to avoid unnecessary mass. The design of separation mechanisms, such as explosive bolts or pneumatic systems, must allow for a smooth and reliable detachment without imparting excessive forces on the remaining stages.

The structural requirements for each stage of a rocket are highly specific, reflecting the unique challenges and roles of that stage during the mission. The first stage must handle immense forces at liftoff and through the atmosphere, the second stage must balance efficiency with strength for high-altitude acceleration, and the third stage must endure the vacuum of space while delivering the payload to its final destination. Each stage is engineered to work seamlessly with the others, creating a launch vehicle capable of achieving the precise demands of modern space exploration. Through advanced materials, precise engineering, and innovative designs, rockets continue to meet these structural challenges, enabling increasingly ambitious missions.

Materials Used in Rocket Construction (Alloys, Composites, Ceramics)

Rocket construction involves a complex interplay of materials designed to endure extreme conditions, such as high temperatures, intense mechanical stresses, and corrosive environments. To meet these demands, engineers rely on advanced materials like alloys, composites, and ceramics, each offering unique advantages based on their properties and applications.

Alloys are among the most widely used materials in rocket construction due to their strength, durability, and resistance to high temperatures. Aluminium alloys, such as 7075 and 6061, are extensively used in the fuselage and fuel tanks of rockets. These alloys offer an excellent strength-to-weight ratio, which is crucial for reducing the overall mass of the rocket while maintaining structural integrity.

Aluminium alloys, particularly aluminium-magnesium and aluminium-lithium alloys, are also commonly used in rockets. Aluminium-magnesium alloys are valued for their ease of deformation and welding, while aluminium-lithium alloys provide a strong and lightweight combination [83].

Titanium alloys are another critical material, often used in areas subject to higher temperatures and stresses, such as the rocket's engine components. Titanium's corrosion resistance and ability to withstand high temperatures without deformation make it ideal for applications in oxidizing environments. Similarly, nickel-based superalloys, such as Inconel, are utilized in turbine blades and combustion chambers, where temperatures can exceed 1,000°C. These materials maintain their strength under extreme thermal and mechanical loads, ensuring reliable performance during launch and flight.

Titanium alloys are widely used in rocket construction due to their high strength-to-weight ratio and excellent corrosion resistance [84-86]. Titanium alloys such as Ti-6Al-4V and Ti-17 are commonly used for components like rocket shells, pressure vessels, and engine parts [85, 86].

Lightweight composite materials, such as ceramic matrix composites (CMCs) and carbon-carbon composites, are becoming increasingly popular in the space industry. These materials offer high strength and low density compared to traditional materials [87-90].

Composites are gaining prominence in rocket construction due to their exceptional strength-to-weight ratio and versatility. Carbon fibre-reinforced polymers (CFRPs) are among the most popular composites, offering high tensile strength, stiffness, and reduced weight compared to traditional metallic materials. CFRPs are often used in the construction of rocket fairings, payload shrouds, and structural supports, where weight reduction is critical.

Glass fibre-reinforced polymers (GFRPs) and Kevlar composites are also utilized in areas requiring additional flexibility and impact resistance. Advanced composites, such as ceramic matrix composites (CMCs), combine the advantages of ceramics and polymers, making them suitable for high-temperature environments like rocket nozzles and heat shields.

Ceramics play a vital role in rocket construction, particularly in components exposed to extreme thermal conditions. High-performance ceramics, such as silicon carbide (SiC) and aluminium oxide (Al2O3), are used in thermal protection systems, rocket nozzles, and ablative heat shields. These materials can withstand temperatures well above 2,000°C, making them essential for protecting the rocket during re-entry or in sustained high-temperature environments.

Ceramic tiles, as seen in space shuttle heat shields, offer lightweight insulation and thermal resistance, protecting the spacecraft's underlying structures from extreme heat. In addition to their thermal properties, ceramics are also highly resistant to corrosion and wear, ensuring long-term durability in critical applications.

In modern rocket construction, the integration of alloys, composites, and ceramics creates a synergistic effect, leveraging the strengths of each material. For instance, a rocket's fuselage may combine aluminium alloys for structural support with carbon composites for weight reduction, while ceramic coatings protect engine components from high-temperature corrosion. This multi-material approach ensures rockets are not only robust but also optimized for efficiency and performance.

Materials Used in an Orbital Rocket's Structural System

The structural system of an orbital rocket is an intricate assembly of advanced materials tailored to meet the rigorous demands of space travel. These materials are chosen based on their ability to withstand extreme stresses, temperatures, and other harsh conditions while optimizing the rocket's performance and weight. The following materials are integral to modern orbital rockets [91]:

Aluminium Alloys: Lightweight and Versatile - Aluminium alloys are a cornerstone of rocket construction due to their high strength-to-weight ratio, corrosion resistance, and workability. The aerospace-grade 6061 aluminium is widely used in various rocket structures, including the fuselage and support frameworks.

6061 aluminium is a widely used alloy known for its versatility, strength, and excellent corrosion resistance [92-95]. It is part of the 6xxx series of aluminium alloys, which are primarily composed of aluminium, magnesium, and silicon [92, 96, 97].

The composition of 6061 aluminium typically includes [92, 96, 97]:

- **Aluminium**: 95.8–98.6%
- **Magnesium**: 0.8–1.2%
- **Silicon**: 0.4–0.8%
- **Copper**: 0.15–0.4%
- **Chromium**: 0.04–0.35%
- **Iron**: Up to 0.7%
- **Zinc**: Up to 0.25%
- **Titanium**: Up to 0.15%
- **Other elements**: 0.05% max each, 0.15% total

This combination of elements contributes to the alloy's strength, corrosion resistance, and machinability.

Key properties of 6061 aluminium include [92-95, 97]:

- **Strength**: 6061 aluminium is medium to high in strength compared to other aluminium alloys. It offers a good balance between structural integrity and weight.
- **Corrosion Resistance**: It exhibits excellent resistance to corrosion in most environmental conditions, including marine environments, due to the presence of magnesium and silicon.
- **Machinability**: This alloy is easy to machine, weld, and form, making it a preferred choice for intricate designs and applications.
- **Heat Treatment**: 6061 aluminium can be heat-treated to increase its strength and hardness. It is typically used in its T6 temper (solution heat-treated and artificially aged) for maximum performance.
- **Weight**: As an aluminium alloy, it is lightweight, making it ideal for applications where weight reduction is crucial.
- **Thermal and Electrical Conductivity**: While not as conductive as pure aluminium, it still offers decent thermal and electrical conductivity.

Different types of aluminium alloys are selected based on their specific advantages [91]:

- **Duralumin** (aluminium, copper, and manganese) offers exceptional strength and hardness but is challenging to weld, requiring riveting or bolting for assembly.
- **Aluminium-magnesium alloys** are preferred for components requiring ease of deformation and welding.
- **Aluminium-lithium alloys** (e.g., AlLi 2198) are both lightweight and strong, making them ideal for reducing overall rocket mass. Notable examples include the Saturn V, which used 2014T6 aluminium, and the SpaceX Falcon 9, which employs aluminium-lithium alloys in its structural systems.

Stainless Steel: Durable and Cost-Effective - Although heavier than aluminium, stainless steel has emerged as a crucial material for rockets like SpaceX's Starship, primarily due to its durability, cost-effectiveness, and ability to withstand significant temperature fluctuations. Stainless steel resists cracking and offers excellent thermal properties, enabling it to endure the intense conditions of space travel. SpaceX's innovative approach to using stainless steel instead of carbon fibre highlights advancements in manufacturing that make it a viable and robust choice for large launch vehicles [91].

Titanium: High Strength for Critical Components - Titanium is a high-strength, corrosion-resistant material used in components requiring additional protection and durability. While less common than aluminium or stainless steel in large structural elements, titanium finds critical applications in high-stress areas. For example, in the Juno spacecraft, titanium was used for radiation shielding to protect electronics from Jupiter's intense radiation belts [91].

Silica (Ceramic) Fibers and Carbon Composites: Advanced Thermal Protection - Silica fibres and carbon composites are used primarily in thermal protection systems, such as heat shields, which protect the vehicle during re-entry. Silica-based ceramics can withstand re-entry temperatures of up to 1600°C (3000°F), ensuring structural integrity in the most extreme conditions. Reinforced carbon-carbon (RCC), with its exceptional heat resistance, is employed in rocket nozzles and external surfaces exposed to intense heat [91].

Carbon Composites: Lightweight and Strong - Carbon composites are increasingly used in rocket structures, particularly for small launch vehicles like Rocket Lab's Electron, whose main structure is almost entirely carbon fibre. Carbon composites are not only lightweight but also possess superior stiffness and strength, making them ideal for structural components where weight reduction is a priority. Rocket Lab's innovative approach includes 3D printing carbon composite parts in just 12 hours, demonstrating the material's adaptability and efficiency [91].

Materials Used in an Orbital Rocket's Propulsion System

The propulsion system of a rocket requires materials capable of withstanding extreme temperature variations—from cryogenic conditions to the scorching heat generated during combustion. Key materials include:

Stainless Steel and Aluminium-Lithium Alloys: Propellant Tanks - Stainless steel is the primary material for propellant tanks due to its rigidity and suitability for cryogenic fuels like liquid hydrogen and liquid oxygen. These fuels need to be stored at temperatures as low as -253°C (-423°F). However, aluminium-lithium alloys are also used for their lightweight properties and higher tensile strength compared to conventional aluminium. The Space Shuttle's external fuel tank and SpaceX Falcon 9's propellant tanks exemplify the use of aluminium-lithium alloys to reduce weight without compromising strength [91].

Copper Alloys and Inconel: Combustion Chambers and Nozzles - The extreme heat generated in rocket engines requires materials with exceptional thermal resistance and conductivity. Copper alloys, particularly chromium copper, are commonly used in combustion chambers for their ability to conduct and dissipate heat effectively. Inconel, a nickel-based superalloy, is often combined with copper to create the chamber walls, which are regeneratively cooled by pumping cryogenic fuel through internal channels [91].

Niobium: Nozzle Extensions - Niobium, a metal with a high melting point and excellent heat conductivity, is used in nozzle extensions, particularly for vacuum-optimized engines. It was employed in the Apollo Service Module and is currently used in SpaceX Falcon 9's Merlin engine nozzle, highlighting its effectiveness in handling high temperatures in space [91].

Materials Used in an Orbital Rocket's Payload System

The payload system of an orbital rocket is one of the most critical components, as it houses the mission's primary objective—whether it be satellites, resupply cargo, exploratory spacecraft, or even astronauts. This system must meet stringent requirements for strength, durability, and protection while being lightweight to maximize the rocket's efficiency. It comprises the payload fairing, forward adapter, and payload itself, each requiring specialized materials to endure the challenges of launch and space environments [91].

Payload Fairing: Lightweight Protection - The payload fairing is a protective covering that encloses the payload during launch and ascent. Positioned at the front of the rocket, it serves two primary purposes: to shield the payload from aerodynamic forces and environmental conditions during launch, and to reduce drag by providing a streamlined shape.

Carbon fibre composites are the material of choice for payload fairings due to their exceptional strength-to-weight ratio. These composites are made from carbon fibres embedded in a polymer matrix, offering high tensile strength, rigidity, and resistance to deformation. During launch, the fairing experiences immense aerodynamic pressures and vibrations, making these properties vital for maintaining payload integrity [91].

Modern rockets like the SpaceX Falcon 9, Atlas V, and Ariane 5 all use carbon fibre composites for their payload fairings. These fairings are designed to split into halves and jettison once the rocket exits the dense part of the atmosphere, ensuring the payload is exposed to space at the appropriate time [91].

Forward Adapter: Structural Connection - The forward adapter is a universal component in most payload systems, serving as the structural link between the rocket's upper stage and the payload. It must withstand the mechanical stresses of launch while securely holding the payload in place.

For this purpose, aluminium-lithium alloys are commonly used. These alloys combine aluminium's lightweight properties with lithium's ability to enhance strength and stiffness. Aluminium-lithium alloys also exhibit excellent corrosion resistance and thermal stability, making them well-suited for the high-stress and varying thermal conditions experienced during ascent [91].

The use of aluminium-lithium in forward adapters is particularly advantageous in reducing the rocket's overall weight, which improves fuel efficiency and payload capacity. This material is an essential component in many modern orbital rockets, contributing to their structural reliability and performance.

Payload Materials: Mission-Specific Adaptations - The payload itself varies significantly depending on the mission. It can include [91]:

- **Satellites**: Built with materials such as aluminium and titanium alloys for structural components, and composites for lightweight panels.
- **Space Station Resupplies**: Packed in containers made of durable polymers or lightweight alloys.
- **Exploratory Spacecraft**: Equipped with specialized materials like high-performance ceramics, radiation-resistant metals, and heat shields for planetary exploration.
- **Astronaut Capsules**: Constructed with a combination of high-strength alloys and thermal protection materials to ensure safety during re-entry.

Each payload component must endure not only the stresses of launch but also the vacuum, radiation, and extreme temperatures of space. This requires a careful selection of materials tailored to the specific demands of the mission [91].

Performance Criteria for Payload Materials

1. **Strength-to-Weight Ratio**: Materials must be lightweight to maximize payload capacity while being strong enough to withstand launch forces.
2. **Thermal Stability**: The ability to endure temperature extremes, from the heat of re-entry to the cold of space.
3. **Durability**: Resistance to vibrations, acoustic stresses, and radiation.
4. **Precision**: For mission-critical payloads like satellites, materials must support precise configurations and maintain structural integrity.

Summary List of Materials Used to Construct a Rocket

Rockets are constructed using a wide range of materials, each chosen to meet specific structural, thermal, and functional requirements. These materials are selected based on their properties, such as strength, weight, thermal resistance, corrosion resistance, and cost-effectiveness. Below is an exhaustive list of materials commonly used in rocket construction, categorized by their applications:

1. Structural Materials

These materials form the primary body, frames, and load-bearing components of a rocket.

- **Aluminium Alloys**: Widely used for lightweight and strong structural elements, such as fuel tanks, outer shells, and payload adapters. Examples:
    - 6061 Aluminium (aircraft-grade alloy)
    - 7075 Aluminium
    - Aluminium-Lithium Alloys (e.g., AlLi 2198)
    - Duralumin (aluminium-copper-manganese alloy)
- **Stainless Steel**: Used for components requiring higher strength, rigidity, and thermal resistance. Examples:
    - 301 Stainless Steel (used in SpaceX's Starship)
    - 321 Stainless Steel (for high-temperature applications)
- **Titanium Alloys**: Ideal for high-strength, corrosion-resistant parts. Common in critical areas such as fasteners, tanks, and structural components.
- **Carbon Fiber Composites**: Lightweight and strong, used for payload fairings, structural supports, and fuselage sections. Examples:
    - Carbon-fibre-reinforced polymers (CFRP)
    - Carbon-carbon composites (RCC)
- **Magnesium Alloys**: Occasionally used for lightweight structural components, though less common due to lower strength compared to aluminium.

2. Propellant Tank Materials

Materials used to construct fuel and oxidizer tanks must withstand extreme pressures and cryogenic temperatures.

- **Aluminium-Lithium Alloys**: High strength-to-weight ratio and corrosion resistance. Common for cryogenic tanks (e.g., Space Shuttle external fuel tank).
- **Stainless Steel**: Durable and resistant to cracking, used in high-pressure cryogenic tanks.
- **Titanium Alloys**: Lightweight and strong, suitable for smaller high-performance tanks.
- **Carbon Composites**: Emerging material for ultra-lightweight and pressure-resistant tanks.

3. Thermal Protection Materials

These materials protect the rocket from extreme heat during launch, ascent, and re-entry.

- **Ceramics**:
  - Silicon Carbide (SiC): High-temperature resistance, used in heat shields.
  - Aluminium Oxide (Al2O3): Used in thermal tiles and coatings.
  - Reinforced Carbon-Carbon (RCC): Heat-resistant, used in leading edges and nozzles.
  - Silica Tiles: Found in re-entry systems (e.g., Space Shuttle heat shields).
- **Ablative Materials**: Designed to burn away and dissipate heat during re-entry. Examples:
  - Phenolic resin composites
  - Carbon phenolic
- **Insulating Foams**: Polyurethane foams used for thermal insulation on cryogenic fuel tanks.

4. Engine Components

Rocket engines require materials capable of withstanding extreme heat, pressure, and corrosive environments.

- **Copper Alloys**: Used in combustion chambers and nozzles for thermal conductivity and heat resistance.
- **Nickel-Based Superalloys**: High strength and temperature resistance, used in turbines and combustion chambers. Examples:
  - Inconel (e.g., Inconel 718)
  - Hastelloy
- **Niobium**: High melting point, used in vacuum nozzles (e.g., SpaceX Falcon 9 Merlin Vacuum engine).
- **Molybdenum**: Occasionally used in nozzles and heat exchangers.
- **Tungsten**: Used in heat-intensive applications like nozzle throats.
- **Columbium (Niobium)**: Applied in upper-stage engine nozzles for its resistance to heat and oxidation.

5. Payload Fairing Materials

Payload fairings protect the cargo during launch and are often made from lightweight materials.

- **Carbon Fiber Composites**: Commonly used due to their high strength-to-weight ratio.
- **Aluminium Honeycomb Structures**: Often combined with composite face sheets for lightweight but strong fairing construction.
- **Glass Fiber Composites**: Occasionally used for less demanding applications.

6. Fasteners and Joints

These materials secure various rocket components together.

- **Titanium Fasteners**: Lightweight and corrosion-resistant, used in critical areas.
- **Inconel Fasteners**: High-strength fasteners for high-temperature environments.
- **Steel Bolts**: Used in non-critical joints.

7. Electronics and Avionics Materials

Electronics require protective casings and high-performance materials for functionality in space.

- **Aluminium and Titanium Casings**: For electronic housing to provide strength and shielding.
- **Gold and Silver Coatings**: Used in electronic circuits and wiring for conductivity.
- **Kapton**: A high-performance polyimide film used for insulation in wiring and circuits.
- **Silicon and Gallium Arsenide**: Semiconductor materials used in solar cells and sensors.

8. Coatings and Surface Treatments

Rockets require specialized coatings to reduce heat absorption, prevent corrosion, and improve durability.

- **Thermal Barrier Coatings**: Ceramic-based coatings applied to engine components.
- **Anodized Aluminium**: Protects aluminium surfaces from corrosion.
- **Reflective Coatings**: Thin films of gold or silver for radiation protection in space.
- **Paints and Foils**: Special heat-resistant paints or Mylar foils for thermal regulation.

9. Solid Rocket Booster Materials

Solid rocket boosters use materials designed to contain and burn solid propellants.

- **Steel Casings**: For structural support and containment of propellant.
- **Graphite Nozzle Liners**: High-temperature resistance for exhaust flow.
- **Rubber-Based Insulators**: Prevent heat transfer to the casing.

10. Specialized Components

Some components use unique materials for specific applications.

- **Pyrotechnic Materials**: Used in separation mechanisms (e.g., explosive bolts).
- **Kevlar**: High-strength fibres used for straps, parachutes, and composite reinforcements.
- **Beryllium**: Lightweight metal used in mirrors and structural supports for scientific payloads.

Rocket construction relies on a diverse array of materials, each chosen to meet specific requirements for strength, weight, heat resistance, and durability. As materials science continues to advance, new materials and composites are being developed to make rockets lighter, more efficient, and capable of meeting the demands of increasingly complex missions. This exhaustive list underscores the complexity and precision involved in building these engineering marvels.

Structural Design Considerations (Strength, Weight, and Durability)

When designing the structure of a rocket, engineers must consider several key factors, including material selection, mass, strength, accuracy, and cost. The references provided offer insights into these important considerations.

The material used for the rocket's structure should be strong, lightweight, and have good mechanical properties such as modulus, toughness, and strength [98]. Common materials used include aluminium, honeycomb, composite materials, high-strength aluminium and titanium alloys, and high-resistance steel [98]. The selection of the appropriate material is crucial, as it can impact the rocket's performance, durability, and cost [99, 100].

The rocket's body, including its diameter and wall thickness, can make up a large portion of its total mass. Doubling the diameter or wall thickness will increase the mass significantly, but the exact relationship can vary based on design specifics [98]. Therefore, engineers must carefully consider the mass of the structure to optimize the rocket's overall performance [101].

The rocket body will experience compression and bending loads during flight, and the structure must be designed to withstand these loads [98]. The factor of safety approach and the load and resistance factor design (LRFD) approach are two common methods used to ensure the structural integrity of the rocket [98].

The structure may need to maintain its geometry with great accuracy, whether it's fixed or deployed in space [98]. This requirement can impact the material selection and manufacturing processes used.

The cost of the materials should be considered, as it can significantly impact the overall cost of the rocket [102]. Factors such as raw material costs, production economies of scale, and life-cycle costs should be evaluated during the material selection process [102].

Strength

The structural integrity of a rocket is critical to ensuring it can safely transport its payload through the extreme conditions of launch, ascent, and space travel. A successful design addresses various forces and stresses, ensuring the rocket's components remain secure and functional throughout its mission. Below is a detailed exploration of key structural considerations:

Load-Bearing Capacity: The rocket's structure must support the combined weight of all its components, including the engines, fuel tanks, payloads, and additional stages. This requires precise engineering to distribute these loads evenly across the vehicle's frame. Fuel, which accounts for a significant portion of a rocket's weight, adds complexity because its distribution changes as it burns during flight. The structure must maintain its integrity even as the centre of gravity shifts dynamically.

Additionally, during launch, the rocket endures a combination of thrust, gravitational forces, and aerodynamic drag. These dynamic loads result in significant stress on the structure. The rocket must be robust enough to prevent buckling or deformation under these conditions, especially during the critical moments of liftoff and initial ascent through Earth's dense atmosphere. To achieve this, engineers design the structure with load-bearing elements such as reinforced rings, trusses, and bulkheads that evenly distribute stress.

Thrust Transmission: Thrust generated by the rocket's engines must be transmitted through the entire structure to propel the vehicle upward. This requires a direct and efficient load path from the engines to the upper stages and payload. Any inefficiencies or weak points in this load path can lead to deformation, structural failure, or energy losses that compromise the mission.

To handle these forces, the engine mounts and the surrounding framework are heavily reinforced. Materials with high tensile strength, such as titanium or carbon composites, are often used to ensure that the structural elements can handle the thrust without cracking or bending. Engineers also consider stress concentrations—localized areas where forces might accumulate—and design the structure to distribute these stresses evenly, often using features like fillets, gussets, or tapered joints.

Aerodynamic Loads: During ascent, the rocket experiences significant aerodynamic forces, particularly in the lower atmosphere where air density is highest. Drag forces exert pressure on the rocket's surface, which can cause bending and compressive stresses in the structure. These aerodynamic loads are most intense during the period known as Max Q—the point where aerodynamic pressure is at its peak.

Max Q, or Maximum Dynamic Pressure, is a critical point during a rocket's ascent where the combination of atmospheric pressure and the rocket's velocity results in the greatest aerodynamic stress on the vehicle. It represents the moment when the rocket structure experiences the highest mechanical loads due to atmospheric drag, making it a crucial design and operational consideration in spaceflight.

Dynamic pressure (*q*) is a measure of the aerodynamic stress exerted on the rocket and is calculated using the formula:

$$q = \frac{1}{2}\rho v^2$$

Where:

- *p*: Air density
- *v*: Velocity of the rocket

Dynamic pressure increases with both the velocity of the rocket and the density of the air it travels through. As a rocket ascends, its velocity increases while the atmospheric density decreases. Max Q occurs at the altitude where these two factors combine to produce the peak dynamic pressure.

Max Q typically occurs at an altitude between 11 and 15 kilometres (6.8–9.3 miles) above sea level, approximately 30 to 60 seconds after liftoff. At this stage of the ascent, the rocket is traveling at high speed, and while the atmospheric density is decreasing, it remains substantial enough to generate significant drag forces. This combination of velocity and air density results in the peak dynamic pressure experienced by the rocket.

After passing through Max Q, the air density rapidly diminishes with increasing altitude. As a result, the dynamic pressure on the rocket decreases, even though the vehicle continues to accelerate. This reduction in aerodynamic stress allows the rocket to resume full engine thrust and focus on achieving the velocity required to reach orbit.

The design of rockets incorporates several critical features to handle the stresses of Max Q effectively. Streamlined shapes, including tapered nose cones and smooth, aerodynamic surfaces, are essential to reducing drag and optimizing performance during this phase. These designs ensure the rocket can efficiently move through the atmosphere while minimizing the effects of aerodynamic resistance.

To withstand the significant forces experienced during Max Q, components most susceptible to stress, such as the fuselage, fairings, and joints, are reinforced with advanced materials. Aluminium-lithium alloys, carbon composites, and titanium are commonly used to provide the necessary strength without adding excessive weight, ensuring the rocket maintains its structural integrity under peak dynamic pressure.

Vibrations caused by aerodynamic forces during Max Q pose additional challenges, particularly for sensitive instruments and payloads. Engineers address this issue by incorporating damping systems and insulation into the rocket's design. These features help absorb and mitigate vibrations, protecting critical components and ensuring the payload remains secure and functional throughout the ascent.

During Max Q, rockets face several operational challenges that require careful engineering to overcome. One of these is the intense acoustic loads generated by high-speed airflow, which creates vibrations that can stress both the rocket's structure and its payload. These vibrations, if not managed effectively, can compromise the integrity of the rocket and the sensitive instruments it carries.

Thermal effects are another significant challenge during Max Q. The friction between the rocket's surface and the atmosphere generates heat, which can lead to overheating of critical components. To address this, rockets are equipped with thermal protection systems, such as insulating materials or heat-resistant coatings, to safeguard against the extreme temperatures encountered during this phase of flight.

Maintaining stability and control is also crucial during Max Q, as aerodynamic forces reach their peak. Advanced guidance and control systems are employed to ensure the rocket remains on its intended trajectory. Engine gimbaling, a technique where the engines swivel slightly to adjust thrust direction, plays a key role in steering the rocket smoothly through this high-stress phase. These operational measures collectively ensure that the rocket can endure Max Q safely and continue its ascent toward orbit.

During SpaceX Falcon 9 launches, the announcers often highlight the process of throttling down the engines as the rocket approaches Max Q [103-105]. This deliberate reduction in thrust minimizes the aerodynamic stresses on the vehicle at the peak dynamic pressure [103-105]. Once the rocket passes through Max Q, the engines are gradually throttled back up to full power, allowing the rocket to continue its ascent efficiently [103-105].

The Saturn V, NASA's iconic moon rocket, also demonstrates the importance of managing Max Q [106-108]. This powerful rocket was engineered specifically to handle the intense aerodynamic stresses encountered during its ascent through the denser layers of Earth's atmosphere [106-108]. Its robust design ensured that it could endure these forces without compromising structural integrity, paving the way for the Apollo missions to reach the Moon successfully [106-108].

Both the Falcon 9 and Saturn V examples illustrate how rockets are designed and operated to navigate the challenges of Max Q effectively [103-108]. The careful management of this critical moment, through throttle control and structural engineering, is crucial for the successful launch and ascent of these rockets [103-108].

To withstand these loads, the rocket's body is designed to be both streamlined and structurally rigid. The outer shell, typically made from lightweight yet strong materials like aluminium-lithium alloys or carbon composites, plays a dual role in reducing drag and maintaining the structural integrity of the vehicle. Additional reinforcements, such as ribs and internal supports, are incorporated to prevent buckling or collapse under aerodynamic pressures.

Vibration and Acoustic Loads: Vibrations from the rocket engines and acoustic energy generated by high-speed airflow can create significant structural challenges. These vibrations and acoustic loads can lead to fatigue, particularly in sensitive components like avionics, payloads, or fuel lines. Vibrations from the engines, often called structural resonances, can amplify stresses if not properly managed, leading to potential failures.

To address this, damping mechanisms such as vibration isolators or tuned mass dampers are integrated into the rocket's design to absorb and dissipate these forces. The structure itself is also designed to minimize resonances, using materials and shapes that avoid frequencies matching the engine vibrations. Additionally, soundproofing measures, such as acoustic liners in the payload fairings, protect delicate payloads from the intense noise levels experienced during launch.

Material Strength

The materials used in a rocket's structure must provide the necessary strength to endure these extreme forces while remaining as lightweight as possible. Engineers often rely on advanced materials, including aluminium alloys, titanium, stainless steel, and carbon composites, to strike this balance.

- **Aluminium Alloys**: Lightweight and corrosion-resistant, these alloys are commonly used for structural components such as fuel tanks and outer shells.
- **Titanium**: Known for its high strength-to-weight ratio and resistance to high temperatures, titanium is used in critical areas requiring additional durability.
- **Stainless Steel**: Strong and resistant to cracking under stress, it is used in components like high-pressure tanks or reusable first stages (e.g., SpaceX Starship).
- **Carbon Composites**: Exceptionally lightweight and strong, these materials are ideal for reducing overall rocket weight while maintaining structural integrity, especially in payload fairings and support structures.

The choice of materials depends on the specific requirements of each stage of the rocket, with different materials selected based on their ability to withstand the unique stresses and environmental conditions encountered during flight.

Weight

The weight of a rocket is a fundamental consideration in its design, as it directly impacts its efficiency, payload capacity, and overall performance. Engineers must strike a delicate balance between minimizing the rocket's mass and ensuring its structural integrity. Various techniques and materials are employed to optimize weight without compromising the rocket's ability to endure the stresses of launch and space travel.

Minimizing the structural mass of a rocket is paramount to improving its efficiency and payload capacity. Every kilogram of unnecessary weight in the rocket's structure reduces the payload it can carry or increases the fuel required for propulsion. To address this, engineers use advanced lightweight materials such as carbon fibre composites and aluminium-lithium alloys. Carbon fibre composites offer exceptional strength while being significantly lighter than traditional metals, making them ideal for components like payload fairings and fuselage sections. Similarly, aluminium-lithium alloys combine the lightweight properties of aluminium with enhanced strength and durability, making them suitable for fuel tanks and structural supports.

In rocket design, the strength-to-weight ratio of materials is a critical parameter. Materials with high strength-to-weight ratios allow rockets to maintain structural integrity while minimizing mass. For example, titanium and carbon composites are often used in areas subjected to high stress, such as engine mounts and interstage adapters, due to their ability to withstand extreme forces without adding excessive weight. Optimizing this ratio enables engineers to construct rockets that are both robust and efficient, ensuring they can withstand the harsh conditions of launch and space travel while carrying maximum payloads.

A rocket's design incorporates multiple stages, each with its own structural and functional requirements. To optimize performance, each stage is designed to be as lightweight as possible while still capable of handling its specific loads and stresses. The first stage, responsible for lifting the entire rocket, is built to be strong yet efficient, as it is the heaviest and largest. Subsequent stages, which operate in higher altitudes and the vacuum of space, can be lighter because they handle lower aerodynamic loads. Stage separation allows the rocket to discard depleted stages, significantly reducing its weight during flight and enhancing the efficiency of the remaining stages. This modular approach ensures that only the essential mass is carried forward, maximizing fuel efficiency and payload delivery.

The weight of a rocket has a direct impact on its fuel efficiency. Lighter structures require less fuel to achieve the same level of propulsion, reducing overall costs and improving performance. Each kilogram of structural weight saved translates to a payload advantage, as more of the rocket's capacity can be allocated to its mission. This is especially critical in missions targeting higher altitudes or interplanetary destinations, where fuel demands are greater. By minimizing structural weight through advanced materials and efficient designs, rockets can achieve higher velocities and longer travel distances without requiring additional fuel.

To achieve the ideal balance between weight and structural integrity, engineers rely on sophisticated optimization techniques. Finite Element Analysis (FEA) is used to model the rocket's structure and simulate the forces it will encounter during launch and flight. This technique identifies areas of stress concentration and eliminates unnecessary material, ensuring the structure is both lightweight and resilient. Computational Fluid Dynamics (CFD) is employed to optimize the aerodynamic shape of the rocket, reducing drag and improving performance. These tools allow engineers to refine designs iteratively, achieving maximum efficiency while maintaining safety margins. By leveraging advanced computational techniques, designers can push the boundaries of lightweight construction while adhering to strict performance and safety requirements.

Weight is a critical factor in rocket design, influencing every aspect of performance, from payload capacity to fuel efficiency. By minimizing mass through the use of lightweight materials and optimizing strength-to-weight ratios, engineers can create rockets that are both efficient and robust. The modular nature of stage separation further reduces weight during flight, enhancing the overall efficiency of the vehicle. Advanced optimization techniques such as FEA and CFD ensure that every component of the rocket is designed with precision, eliminating unnecessary mass while maintaining safety and reliability. These considerations collectively enable rockets to deliver their payloads efficiently and meet the demanding requirements of space exploration and commercial missions.

Designing lightweight parts is a complex engineering process that requires a balance between strength, stiffness, manufacturability, and cost. The goal is to reduce mass while maintaining the required structural performance. This involves careful consideration of design requirements, material selection, advanced optimization techniques, and manufacturing constraints [109].

The first step in designing lightweight parts is understanding the primary requirement of the component: whether it is stiffness-driven or strength-driven. Stiffness-driven designs focus on minimizing deflection under load, optimizing for flexural modulus or vibration performance. These designs aim to prevent excessive compliance that could compromise functionality. On the other hand, strength-driven designs prioritize the ability to withstand loads without failure, tolerating greater displacement as long as the structural integrity is maintained. Identifying this distinction early in the design process is essential for selecting appropriate materials and design strategies [109].

Material selection for lightweight parts is a balance between cost, performance, and manufacturing methods. Aerospace and high-performance applications commonly use materials such as aluminium alloys, titanium, magnesium, stainless steel, and advanced nickel-based superalloys like Inconel and Hastelloy. Composites, including carbon fibre, fiberglass, and Kevlar, are increasingly favoured due to their exceptional strength-to-weight ratios. To choose the right material, engineers often use tools like Ashby charts, which visually compare material properties such as density, strength, and stiffness against cost or other parameters. These charts simplify trade-off evaluations, allowing designers to identify materials that best meet the part's performance requirements [109].

An Ashby Chart, like the one shown as Figure 26, is a powerful visual tool that enables engineers to compare material properties and identify suitable candidates for specific engineering applications. These charts plot two key material properties against each other, allowing for informed trade-offs between competing characteristics. The attached chart compares fracture toughness (K_{IC}) and elastic limit (σf), providing insights into a material's strength and its ability to resist crack propagation.

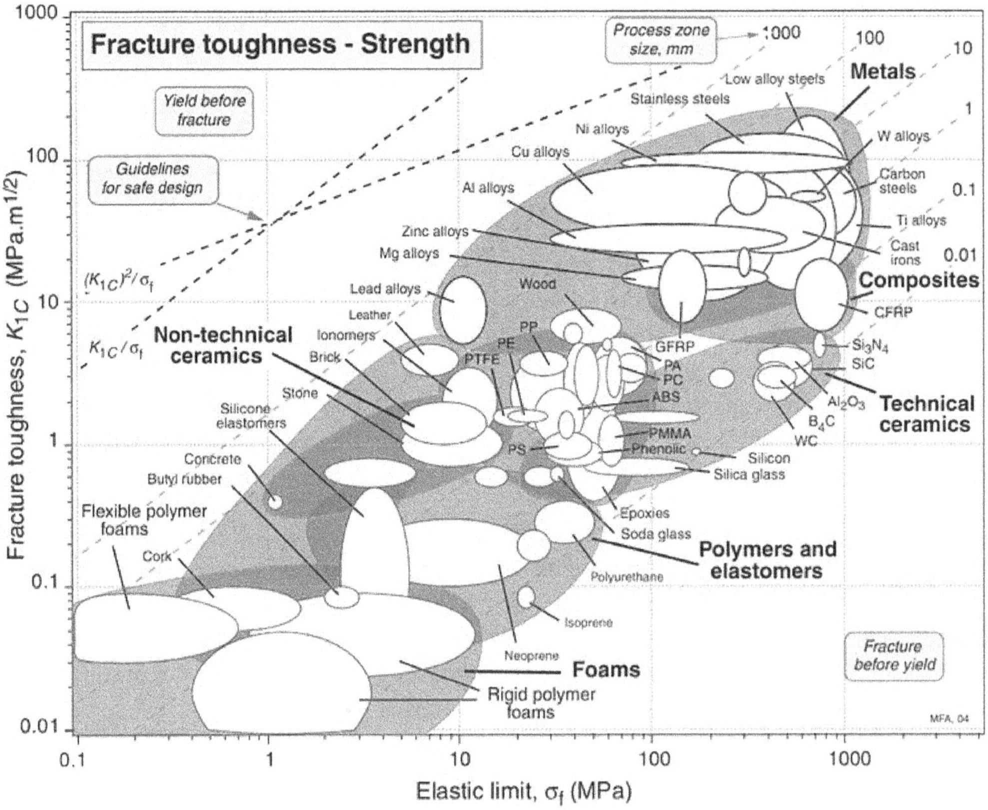

Figure 26: Fracture toughness vs. strength for various materials and material classes. M. F. Ashby, CC0, via Wikimedia Commons.

The axes of the chart are logarithmic, reflecting the exponential nature of the values. The x-axis represents the elastic limit, which measures the stress a material can endure without permanent deformation, expressed in megapascals (MPa). The y-axis measures fracture toughness, expressed in $\mathrm{MPa}\sqrt{\mathrm{m}}$, which indicates a material's ability to resist crack growth under stress. The logarithmic scaling makes it easier to compare materials with widely varying properties.

Materials on the chart are grouped into categories, such as metals, composites, technical ceramics, polymers, foams, and non-technical ceramics. Each category is represented by a cluster or bubble, which shows the range of properties for materials within that group. For example, metals, located in the upper-right corner, generally exhibit high elastic limits and fracture toughness, making them suitable for structural applications. Composites, located in the middle-right area, combine high strength with moderate toughness, making them ideal for aerospace and lightweight designs. Polymers and elastomers, located in the centre-left area, have lower strength and toughness but offer greater flexibility.

The chart also divides materials into regions based on their failure behaviour. Materials above the diagonal guideline are ductile, meaning they yield (deform plastically) before fracturing, while materials below the line are brittle, fracturing without significant deformation. This distinction helps engineers select materials based on application needs. Ductile materials are preferred for applications requiring energy absorption and toughness, while brittle materials may be better suited for environments requiring high stiffness and low deformation.

Each material type is represented by a bubble or ellipse, with the size indicating the range of properties. For example, stainless steels have a high elastic limit and fracture toughness, making them versatile for structural uses. Carbon fibre-reinforced plastics (CFRP) offer high stiffness with moderate toughness, ideal for aerospace applications. Ceramics, such as silicon nitride (Si_3N_4) and aluminium oxide (Al_2O_3), are extremely strong but brittle, with low fracture toughness.

The chart also includes diagonal guidelines, such as K_{IC}/σ_f, which indicate safe design thresholds. Materials above these lines can resist crack propagation under higher stresses, which is particularly valuable for safety-critical components. These guidelines help engineers design components that meet strict safety and reliability standards.

Ashby charts are used for material selection, trade-off analysis, and quick comparisons between material categories. For example, if an application demands both high strength and toughness, materials in the upper-right corner, such as steels or nickel alloys, are ideal. For lightweight designs, CFRP or aluminium alloys might be chosen for their high stiffness-to-weight ratios, even if they have slightly lower toughness than metals. The chart also facilitates comparisons, showing, for instance, that metals generally outperform polymers in both toughness and strength, though they are heavier and more expensive.

Ashby charts are invaluable for visualizing and comparing material properties. The chart provided focuses on fracture toughness and elastic limit, two critical factors in structural and safety-critical applications. By understanding the chart's axes, material groupings, and guidelines, engineers can effectively select materials that meet their specific requirements and design criteria. This ability to make data-driven decisions ensures optimized performance and safety in a variety of engineering applications.

Designing lightweight parts involves optimizing the distribution of material to achieve the highest performance with the least mass. This process begins with intuitive approaches informed by first principles, where designers visualize stress flows and sketch potential geometries. Hand calculations are often used to refine these designs, determining parameters such as wall thickness, beam profiles, and bolt sizing. While simple, this method remains effective for many aerospace components [109].

Finite Element Analysis (FEA) is a more detailed tool, allowing engineers to simulate stress, strain, and deformation under realistic conditions. FEA helps validate initial designs and refine them further by identifying stress concentrations and unnecessary material. For advanced designs, topology optimization is employed. This computational technique identifies the optimal material distribution within a given design space, often resulting in organic, complex shapes that maximize strength and stiffness. However, topology-optimized designs must be carefully reviewed to ensure they perform well under all expected load cases, not just the specified ones [109].

Certain geometries and structural elements are commonly used to achieve lightweight yet robust designs. Beams, webs, ribs, and isogrids are foundational patterns that distribute loads efficiently while minimizing weight. For example, isogrids, used in cylindrical structures like pressure vessels, offer high stiffness and strength while behaving like isotropic materials of equivalent thickness. Ribs and webs are frequently applied in machined or cast aerospace parts to enhance stiffness and vibration performance without adding significant mass [109].

Beams are foundational elements in the structural framework of rockets, designed to carry loads primarily through bending and shear. They provide the primary support for large spans and are used in areas such as the rocket fuselage, engine mounts, and payload adapters. Beams can have various cross-sectional geometries, such as I-beams, T-beams, or hollow rectangular sections, depending on the specific requirements for stiffness and weight reduction.

In the context of rockets, the second moment of area—a property that describes the beam's resistance to bending—is crucial. For instance, I-beams are highly efficient in bending loads due to their high second moment of inertia relative to their mass, making them ideal for stiffening rocket stages. However, in aerospace applications where torsional rigidity is equally important, beams are often combined with other design features like ribs or webs to handle multi-directional stresses.

Webs are thin, flat plates that connect beams or form part of a larger structural framework. Their primary function is to resist shear forces and distribute loads over a broader area without adding significant weight. In rockets, webs are often integrated into structural panels or used as connecting elements between key load-bearing components, such as between the skin of the rocket and internal bulkheads.

For example, in the fuselage, webs help distribute aerodynamic and structural loads evenly across the rocket's surface. This prevents localized stress concentrations that could lead to material failure. Additionally, in CNC-machined or cast aerospace parts, webs are used in combination with ribs to create a stiff yet lightweight structure that can resist vibrations and dynamic loads during flight.

Ribs are structural elements designed to provide additional stiffness to thin-walled structures or panels. They are often arranged in a lattice or grid-like pattern to prevent buckling and to improve resistance to vibrations caused by engine operation or aerodynamic forces. In rockets, ribs are typically found in areas where thin, lightweight materials are used to reduce mass, such as in the rocket's skin or fairings.

For instance, the payload fairing of a rocket often incorporates ribs to maintain its shape under aerodynamic pressure while minimizing its weight. Similarly, internal ribbing is used in fuel tanks and interstage adapters to stiffen the structures and ensure they can handle the compressive and tensile forces generated during launch and stage separation. The rib design must be carefully optimized to balance mass savings with structural strength, often using finite element analysis (FEA) to identify the optimal rib spacing and thickness.

Isogrids are highly efficient structural designs used in aerospace applications, particularly for cylindrical structures such as rocket fuel tanks, pressure vessels, and payload compartments. An isogrid consists of a lattice of intersecting ribs or webs arranged in an equilateral triangular pattern. This configuration provides exceptional stiffness and strength while minimizing weight.

Figure 27: Isogrid Structure.

The isogrid design is especially effective in resisting axial, bending, and torsional loads, making it a preferred choice for rockets. For example, NASA's Saturn V rocket used isogrid structures in its fuel tanks and fuselage to achieve high strength-to-weight ratios. The equilateral triangular pattern of an isogrid ensures that loads are distributed evenly across the structure, reducing the likelihood of stress concentrations that could lead to buckling or failure. Additionally, isogrids offer inherent redundancy, as the failure of a single rib does not compromise the integrity of the entire structure.

Modern advancements in manufacturing, such as CNC machining and additive manufacturing, have made it easier to produce isogrid structures with high precision. These methods allow engineers to tailor isogrid designs to specific mission requirements, optimizing the thickness and spacing of the grid to suit expected loads and environmental conditions.

In rocket design, beams, webs, ribs, and isogrids are often used in combination to create a cohesive structural framework. For instance, isogrids may form the primary structure of a fuel tank, with additional ribs and webs reinforcing high-stress areas or interfaces with other components. Beams may be integrated into the overall structure to carry concentrated loads, such as those generated by engine thrust or payload weight.

The optimization of these elements is guided by computational tools such as FEA and topology optimization. These tools allow engineers to simulate stress and load distributions across the rocket's structure, ensuring that material is placed only where it is needed. This approach reduces mass while maintaining the required strength and stiffness, which is critical in aerospace applications.

Composites have become a cornerstone of lightweight design due to their high strength-to-weight ratios and customizable properties. By varying the fibre orientation and resin types, designers can tailor composites to meet specific stiffness, strength, and deflection requirements. For instance, aircraft wings are often designed with composite materials to prevent flutter at high speeds, enabling a reduction in overall weight. Composites also offer corrosion resistance and durability, allowing for lower safety factors and longer lifespans. The use of sandwich panels, which combine composite skins with lightweight cores like honeycomb structures, is another effective way to achieve high stiffness while minimizing mass [109].

Manufacturing processes play a significant role in lightweight part design. CNC machining allows for precise material removal, enabling the creation of intricate geometries. However, machining thin-walled components can lead to challenges such as vibration and tool chatter, requiring careful fixturing and design for manufacturability (DFM). Additive manufacturing, particularly Direct Metal Laser Sintering (DMLS), facilitates the production of complex, lightweight designs, such as conformal ribbing or lattice structures, which are difficult to achieve with traditional methods. Techniques like friction stir welding are also employed in aerospace applications to join lightweight materials without compromising their strength [109].

Direct Metal Laser Sintering (DMLS) is an advanced additive manufacturing process used to create highly detailed and precise metal components directly from digital 3D models. This method employs a high-powered laser to selectively fuse fine metal powders layer by layer, building up a solid object in a controlled environment. Unlike traditional subtractive manufacturing methods, where material is removed from a larger block, DMLS enables the creation of complex geometries that would be difficult or impossible to achieve with conventional techniques.

The DMLS process begins with a CAD (Computer-Aided Design) model of the desired part, which is sliced into thin layers by specialized software. This layer data is then sent to the DMLS machine, where a recoater blade spreads a thin layer of fine metal powder—often as thin as 20 microns—across the build platform. A high-powered laser, controlled with precision by the machine, scans and melts the powder in the areas corresponding to the part's cross-section for that specific layer. Once the layer is complete, the platform lowers slightly, and another layer of powder is spread over the top. The process repeats until the entire part is built.

DMLS is renowned for its ability to produce parts with exceptional mechanical properties, as the process results in fully dense components comparable to those created through traditional methods like casting or forging. This makes it especially suitable for high-performance applications in industries such as aerospace, medical, and automotive, where material strength and durability are critical. The materials used in DMLS include a wide range of metals and alloys, such as titanium, aluminium, stainless steel, cobalt-chrome, and Inconel, offering designers a great deal of flexibility in choosing materials tailored to specific applications.

One of the key advantages of DMLS is its ability to produce intricate, lightweight designs that incorporate features such as lattice structures, internal channels, and organic shapes, which would be challenging or impossible to fabricate using conventional techniques. This capability allows engineers to optimize parts for specific performance criteria, such as reducing weight while maintaining strength or improving thermal management. Additionally, the additive nature of DMLS minimizes material waste, as unused powder can often be recycled for future builds, making it a more sustainable manufacturing option compared to traditional methods.

Despite its numerous advantages, DMLS also has limitations. The process is relatively slow compared to other manufacturing methods, especially for large parts, and the equipment and materials can be expensive. Additionally, the parts often require post-processing, such as heat treatment to relieve residual stresses, machining to improve surface finishes, or support structure removal to finalize the geometry. However, these challenges are outweighed by the unique capabilities of DMLS, particularly for applications requiring precision, customization, and material efficiency.

Mass budgeting is critical in large-scale projects like rockets or vehicles. Tracking the mass of each component helps identify areas where weight reduction efforts can yield the greatest impact. Engineers aim to reduce part count by combining multiple functions into single, monolithic structures, a principle evident in Tesla's integrally stiffened gigacastings or Formula One's stressed engine and gearbox designs. Reducing fastener counts and using innovative bonding techniques, such as adhesives combined with rivets, further simplifies assemblies and reduces weight [109].

Durability

Durability is a cornerstone of rocket design, ensuring that the structure can withstand the extreme conditions it faces during launch, flight, and possible recovery. A rocket must endure intense thermal, mechanical, and environmental stresses while maintaining structural integrity. For reusable rockets, durability is even more critical, as these systems are expected to perform across multiple missions with minimal maintenance. Below is a detailed exploration of the factors influencing durability in rocket design.

Rockets encounter extreme temperature variations throughout their mission profiles, and the structural materials must be engineered to withstand these conditions. During launch, certain components, such as engine nozzles and combustion chambers, are exposed to temperatures exceeding 3,000°C (5,432°F) due to the heat generated by the engines. Conversely, cryogenic fuel tanks must maintain structural integrity at temperatures as low as -253°C (-423°F) for liquid hydrogen and -183°C (-297°F) for liquid oxygen.

To manage these temperature extremes, thermal protection systems (TPS) are integrated into the design. These systems may include insulating foams to protect cryogenic tanks, ceramic coatings for heat shields, or ablative materials that burn away during re-entry to dissipate heat. For reusable rockets, thermal resistance is especially important as components must endure multiple cycles of intense heating and cooling without degradation.

Rockets are often launched from coastal sites where saltwater and atmospheric moisture can accelerate corrosion. Additionally, some rocket propellants and byproducts are highly corrosive. To counter these challenges, corrosion-resistant materials are used extensively. Stainless steel is a common choice due to its inherent resistance to rust and oxidation, while anodized aluminium is used for its protective oxide layer that prevents corrosion in harsh environments.

For reusable rockets, such as SpaceX's Falcon 9, corrosion resistance is even more critical. These rockets are exposed to environmental conditions during launch, recovery, and refurbishment cycles. Protective coatings and surface treatments, such as thermal sprays or anti-corrosion paints, further enhance the durability of structural components.

Rockets experience cyclic loading during various phases of their mission, including the thrust from engines, aerodynamic pressures during ascent, and mechanical stresses during stage separation and recovery. These repetitive forces can lead to material fatigue, which manifests as small cracks that propagate over time and can result in catastrophic failure if not addressed.

To ensure fatigue resistance, engineers design the rocket structure using materials with high fatigue thresholds, such as titanium and carbon composites. Components are also designed with safety margins and reinforced at stress points to prevent crack initiation. For reusable rockets, fatigue resistance is paramount, as these vehicles must endure repeated load cycles across multiple missions without significant structural deterioration.

Rockets face potential impacts from various sources, including debris during launch, micrometeoroids in space, and rough landings in the case of reusable components. The structural design must account for these hazards by incorporating reinforcements and damage-tolerant materials in critical areas.

For example, re-entry vehicles and reusable boosters are designed with robust landing legs and shock absorbers to withstand the impact of landing. Additionally, payload fairings may include reinforced sections to protect sensitive cargo. In space, micrometeoroid shielding is often implemented to prevent damage to critical systems and instruments, using materials like Kevlar or multilayer insulation.

The rise of reusable rockets, such as SpaceX's Falcon 9 and Starship, has significantly increased the emphasis on durability. These rockets are designed to withstand the extreme conditions of launch, flight, and recovery multiple times, reducing overall costs and improving efficiency.

Reusable components must incorporate high-durability materials and wear-resistant coatings to minimize the need for refurbishment. For instance, Falcon 9's first stage is equipped with robust thermal shielding and aerodynamic control surfaces to ensure it can re-enter the atmosphere and land safely. Modular designs also facilitate easy replacement of worn-out parts, extending the rocket's operational lifespan.

In addition to structural durability, reusability demands advanced maintenance practices and diagnostic tools to assess wear and tear after each mission. These measures help ensure that the rocket meets safety and performance standards over its intended lifecycle.

Environmental Factors

Rockets are subjected to a variety of environmental factors that challenge their structural integrity and operational efficiency. From the vacuum of space to intense radiation and dynamic mission requirements, the design must account for these conditions to ensure the rocket performs reliably. Below is a detailed discussion of the key environmental factors affecting rocket structures and how they are addressed.

Rockets operate in environments where atmospheric pressure varies dramatically, from the dense atmosphere at ground level to the near-total vacuum of space. The structure must be designed to withstand these pressure changes without deforming or failing.

During launch, the rocket ascends through the atmosphere at high speeds, encountering rapid pressure decreases. This creates stress on the outer shell, especially around joints, seals, and thin-walled structures such as fuel tanks. To address this, engineers select materials with high tensile and compressive strength, such as aluminium-lithium alloys and carbon composites, and design the structure to evenly distribute stress.

Once in space, the absence of atmospheric pressure introduces additional challenges. Internal components, such as fuel tanks and payload systems, must be engineered to maintain their structural integrity despite the lack of external pressure. For example, pressurized tanks are designed to resist collapse, using reinforcing ribs or optimized geometries. Similarly, payload fairings must ensure that sensitive instruments are protected from the effects of pressure fluctuations during both ascent and deployment.

In space, prolonged exposure to solar radiation and cosmic rays can degrade materials, especially in upper-stage components and payload systems. Solar radiation can cause thermal expansion and material fatigue, while cosmic rays can lead to microstructural damage in certain metals and composites. Over time, this exposure can compromise the performance and reliability of a rocket's structural elements.

To mitigate these effects, radiation-resistant materials and protective coatings are applied where necessary. For instance, materials such as titanium and carbon composites are less susceptible to radiation-induced degradation and are commonly used in critical components. Protective coatings, such as multilayer insulation (MLI) or thin films of gold or aluminium, are applied to reflect radiation and reduce heat absorption. In payload systems, radiation shielding may include materials like lead or Kevlar to protect sensitive electronics and instruments.

Radiation exposure is a particularly important consideration for missions involving extended durations in space, such as geostationary satellites or interplanetary probes. In these cases, engineers design the structural components to endure cumulative radiation effects over years or even decades.

Modern rockets are expected to serve a variety of missions, from launching satellites to carrying scientific instruments or human crews. This diversity in payloads and mission requirements demands a high degree of structural adaptability. Modular designs are essential to accommodate different configurations and environments without compromising performance.

A modular design allows for interchangeable components, such as payload fairings, upper stages, or propulsion systems, to be customized for specific missions. For example, a rocket designed for low Earth orbit (LEO) missions may use a simpler payload fairing, while one intended for interplanetary travel may require additional thermal protection or radiation shielding. This flexibility reduces manufacturing costs and increases the rocket's utility across a broader range of applications.

In addition to payload adaptability, rockets must also be structurally compatible with various launch environments. For instance, rockets launched from coastal sites may require enhanced corrosion resistance, while those designed for deep-space missions must include robust thermal and radiation protection. Engineers achieve this adaptability by incorporating modular interfaces and standardized components, ensuring the rocket can be configured quickly and efficiently for different mission profiles.

Integration and Compatibility

Integration and compatibility are essential considerations in rocket design, ensuring that all components work seamlessly together to achieve mission objectives. From securely housing payloads to managing stage connections, engine mounting, and aerodynamic efficiency, the structural design must facilitate smooth interactions between subsystems while withstanding the intense forces of launch and flight. Below is a detailed exploration of these critical aspects.

The payload is the primary purpose of the rocket's mission, whether it is a satellite, a crewed capsule, or scientific instruments. The structural design must securely house the payload, protecting it from mechanical stresses during launch and ensuring its safe deployment in orbit or beyond. Payloads are often subjected to intense vibrations, acoustic forces, and acceleration during ascent, as well as extreme temperature changes and radiation in space.

To address these challenges, the payload bay is designed with precision to provide a secure and stable environment. Advanced materials like carbon composites and aluminium alloys are used to create lightweight but robust structures. Padding or vibration-dampening systems are often incorporated to reduce mechanical stresses on sensitive instruments. The payload fairing, which encapsulates the payload, adds an additional layer of protection by shielding it from aerodynamic forces and environmental exposure during launch. For payload deployment, systems like hinges, springs, or pyrotechnic release mechanisms ensure smooth and accurate separation.

The interfaces between rocket stages play a critical role in the integration of the launch vehicle. These connections must transmit the forces generated by the engines to the upper stages while ensuring that the rocket remains stable and intact during flight. Stage connections must also facilitate clean and reliable separation to ensure that depleted stages do not interfere with the continued ascent of the rocket.

To achieve this, engineers design robust but lightweight connection mechanisms that can handle the mechanical loads of launch and flight. Commonly used materials include high-strength aluminium and titanium alloys. Separation mechanisms, such as explosive bolts, pneumatic actuators, or spring-loaded systems, are incorporated into these connections to enable smooth detachment of stages at the appropriate times. These systems are rigorously tested to ensure reliability, as a failed stage separation could compromise the mission or even result in vehicle loss.

The engines are the heart of the rocket, providing the thrust needed to overcome gravity and propel the vehicle into space. Engine mounts must be designed to securely attach the engines to the rocket's structure while withstanding the intense forces generated during ignition and sustained thrust. This includes handling vibrations, high-frequency oscillations, and thermal expansion without transmitting excessive stress to adjacent structural components.

Engine mounting systems are typically constructed from materials with high tensile and thermal strength, such as titanium or stainless steel. The mounts are often designed with flexible connections or dampening elements to absorb and dissipate vibrations, reducing the risk of structural fatigue or damage. The integration of engine mounts must also account for thrust vectoring, where the engines swivel slightly to steer the rocket. This requires precise alignment and reinforced attachment points to handle the dynamic forces of flight.

Aerodynamic efficiency is critical for minimizing drag and improving the rocket's performance during ascent through the atmosphere. The structural design must ensure a streamlined shape that reduces air resistance, allowing the rocket to achieve higher speeds with less fuel consumption. The nose cone and payload fairing are specifically designed to optimize airflow around the rocket, reducing turbulence and pressure drag.

Smooth surfaces and precise construction techniques are essential to achieving aerodynamic efficiency. Materials like aluminium-lithium alloys and carbon composites are frequently used for their lightweight properties and ability to maintain smooth finishes. Fairings are often designed to split and jettison in two halves once the rocket reaches a sufficiently high altitude, minimizing weight while exposing the payload for deployment.

In addition to fairings, other aerodynamic features, such as fins or control surfaces, may be included in the design. These components provide stability and control during the early phases of flight when the rocket is traveling through denser atmospheric layers. Modern rockets often use engine gimbaling for steering, reducing the need for large external fins and further improving aerodynamic efficiency.

Advances in Manufacturing, Including 3D Printing and Automation

The field of rocket construction has undergone significant transformations thanks to advances in manufacturing technologies like 3D printing (additive manufacturing) and automation. These innovations are redefining how rockets are designed, built, and tested, enabling faster production, cost efficiency, and the creation of highly complex and optimized components. This paradigm shift is making space exploration more accessible and sustainable while also improving the performance and reliability of rockets.

3D printing is one of the most transformative advancements in rocket manufacturing. Additive manufacturing techniques, such as Direct Metal Laser Sintering (DMLS), Electron Beam Melting (EBM), and Fused Deposition Modelling (FDM), allow engineers to create complex geometries directly from digital models [110, 111]. For rockets, this means that components like engine nozzles, combustion chambers, and structural supports can be manufactured as single pieces rather than assemblies of multiple parts [110, 111]. This reduces weight by eliminating the need for fasteners and welds, minimizes weak points in the structure, and enhances overall reliability [110, 111]. For example, SpaceX has utilized 3D printing to produce parts for its Raptor engines, achieving significant reductions in production time and weight compared to traditional methods [110, 111].

Another advantage of 3D printing in rocket construction is its ability to optimize designs for specific missions [112, 113]. Engineers can incorporate lattice structures, conformal cooling channels, and organic shapes that improve performance while minimizing material usage [112, 113]. Such optimization is particularly crucial in the aerospace industry, where every gram saved translates to increased payload capacity or reduced fuel consumption [112, 113]. Furthermore, additive manufacturing allows for rapid prototyping, enabling engineers to iterate and test designs quickly, shortening development cycles and reducing costs [112, 113].

Automation has also become a cornerstone of modern rocket construction, revolutionizing both production processes and system integration [114, 115]. Automated manufacturing tools, such as robotic arms, CNC machines, and laser welding systems, ensure precision and repeatability, which are critical in the high-stakes environment of rocket building [114, 115]. Automation minimizes human error, reduces labour costs, and accelerates production timelines [114, 115]. For instance, automated filament winding machines are used to produce composite rocket motor casings and fuel tanks, ensuring consistent quality and eliminating the variability inherent in manual processes [114, 115].

In addition to manufacturing, automation plays a key role in the assembly and integration of rocket systems [114, 115]. Advanced robotics and automated systems streamline the process of joining critical components, such as attaching engines to rocket stages or integrating payloads into the payload fairing [114, 115]. This not only improves efficiency but also enhances safety, as robotic systems can perform tasks in hazardous environments, such as handling propellants or working within confined spaces [114, 115].

The combination of 3D printing and automation has enabled the development of entirely new approaches to rocket design [116, 117]. For example, companies like Rocket Lab have leveraged these technologies to create rockets like the Electron, which features carbon composite structures and 3D-printed Rutherford engines [116, 117]. These advances allow smaller companies to compete in the growing commercial space industry, offering cost-effective solutions for launching small satellites and other payloads [116, 117].

Even established aerospace giants like NASA, Boeing, and Lockheed Martin are integrating these advancements into their programs [118]. NASA's Artemis program, for instance, utilizes 3D-printed components in the Space Launch System (SLS) and robotic automation to assemble large segments of the rocket [118]. Similarly, SpaceX has introduced automation in its manufacturing lines for the Starship rocket, enabling the rapid production of stainless steel structures with minimal manual intervention [118].

While these technologies offer numerous benefits, they also come with challenges. 3D printing requires specialized materials and post-processing to ensure components meet the stringent standards of aerospace applications [119]. Similarly, implementing automation requires significant upfront investment in equipment and software [120]. However, the long-term benefits—faster production, lower costs, and improved reliability—outweigh these challenges, making them indispensable for modern rocket construction [120].

PART 2

Designing Rockets

Chapter 4

Rocket System Components

Key Subsystems: Propulsion, Guidance, Control, and Payload

Rockets are highly complex vehicles made up of several subsystems, each designed to perform a critical function during launch, ascent, and mission execution. Among these, propulsion, guidance, control, and payload subsystems are the most essential, as they collectively ensure the rocket's ability to achieve its mission objectives.

Propulsion Subsystem

The propulsion subsystem is the heart of a rocket, providing the thrust necessary to overcome Earth's gravity and propel the rocket into space. This subsystem consists of engines, propellant tanks, and fuel delivery systems, all of which work together to generate thrust through the expulsion of high-speed exhaust gases.

There are two primary types of rocket propulsion systems: liquid and solid. Liquid propulsion systems, such as those used in the Saturn V or SpaceX's Falcon 9, mix fuel (e.g., liquid hydrogen or RP-1) and oxidizer (e.g., liquid oxygen) in a combustion chamber to produce a controlled and adjustable thrust. Liquid propulsion is versatile, allowing engines to throttle, restart, and shut down as needed.

The process of mixing fuel (such as liquid hydrogen or RP-1) with an oxidizer (such as liquid oxygen) is at the core of how a rocket engine generates thrust. This process occurs within the rocket's combustion chamber, where the two substances are combined under precise conditions to produce a controlled, high-energy chemical reaction that expels exhaust gases at extremely high speeds. The basic principle is rooted in Newton's Third Law of Motion: for every action, there is an equal and opposite reaction. The expulsion of exhaust gases generates the thrust needed to propel the rocket.

Mixing a fuel and an oxidizer, such as liquid hydrogen and liquid oxygen, creates a bipropellant that produces thrust for rockets [121]. Liquid hydrogen and liquid oxygen is a combination that has one of the highest specific impulses of any commonly used rocket fuel, meaning it produces a significant amount of thrust per unit of fuel burned [121]. However, liquid hydrogen has a low density, which necessitates large, heavy tanks to store enough fuel for launch from Earth's surface [121].

RP-1 and liquid oxygen is a more convenient combination than liquid hydrogen and liquid oxygen because RP-1 can be stored at room temperature and has a higher density [121]. RP-1 is a highly refined kerosene that is similar to Jet A and JP-8 but is manufactured to stricter standards [121].

Injectors are used to mix and atomize the propellants in the combustion chamber to create stable and efficient combustion [122]. The design of the injector depends on the type of propellants being used and the engine's operational needs [122].

Some propellants, like unsymmetrical dimethylhydrazine and dinitrogen tetroxide, ignite spontaneously when mixed [121]. These propellants are used in upper stages, launch escape systems, and amateur rocketry [121].

High-performance hydrogen-oxygen engines often use titanium alloys for components due to their strength and heat resistance [85, 123]. Nickel-base alloys are typically used for the oxygen pumps because titanium is not compatible with oxygen [85].

To begin, the fuel and oxidizer are stored in separate tanks within the rocket. A fuel is a substance that burns to release energy, while an oxidizer provides the oxygen necessary for combustion. In space, where there is no atmospheric oxygen, rockets must carry their own oxidizer to sustain the combustion process. The combination of a fuel and an oxidizer is referred to as a propellant. Liquid hydrogen, a cryogenic fuel, and RP-1, a refined form of kerosene, are commonly used fuels. Liquid oxygen (LOX), also cryogenic, is the most common oxidizer.

The fuel and oxidizer are pumped into the combustion chamber at high pressures using turbopumps. These pumps ensure that both substances flow into the chamber at the correct rate and in the precise proportions required for efficient combustion. The ratio of fuel to oxidizer, known as the mixture ratio, is critical. For example, in a rocket using liquid hydrogen and liquid oxygen, the optimal mixture ratio is approximately 6:1 (six parts oxygen to one part hydrogen by weight). This ensures that the chemical reaction releases the maximum amount of energy without leaving unburned fuel or oxidizer.

Inside the combustion chamber, the fuel and oxidizer are combined and ignited, typically by an electric spark or a hypergolic reaction (where the substances ignite upon contact, as in some rocket designs). The combustion process is highly exothermic, releasing a tremendous amount of heat. This heat rapidly converts the fuel and oxidizer into high-pressure, high-temperature gases. These gases expand and are expelled at high velocity through the rocket's nozzle, which is designed to accelerate the gases to supersonic speeds and direct them downward, creating thrust.

Different fuels and oxidizers have unique properties that affect the rocket's performance. Liquid hydrogen and liquid oxygen, for instance, produce one of the most efficient chemical reactions, with a high specific impulse (a measure of efficiency). However, they require cryogenic storage and handling, as both substances must be kept at extremely low temperatures to remain in liquid form. RP-1 and liquid oxygen, on the other hand, are easier to handle and store but produce slightly less efficient combustion. The choice of propellant depends on the specific mission requirements, such as the rocket's size, payload, and destination.

The mixing and combustion process must occur with precise timing and control to maintain stability. Any imbalance in the mixture ratio, fuel flow, or combustion conditions can lead to inefficiencies, incomplete combustion, or catastrophic failures like engine explosions. Modern rocket engines use sophisticated sensors and control systems to monitor and adjust these parameters in real-time, ensuring optimal performance.

On the other hand, solid propulsion systems, commonly used in boosters like those of the Space Shuttle, consist of a pre-mixed propellant that burns continuously once ignited. Solid rockets are simpler and more robust but lack the ability to throttle or restart.

Modern advancements in propulsion include hybrid engines, which combine the benefits of both liquid and solid systems, and electric propulsion, which is ideal for in-space manoeuvres. The propulsion subsystem is meticulously engineered to optimize thrust, efficiency, and reliability while minimizing weight, making it a critical component in rocket design.

Hybrid engines merge the advantages of liquid and solid propulsion systems, creating a versatile and efficient solution for specific rocket applications. In a hybrid rocket engine, the fuel is typically stored in a solid form, while the oxidizer is stored in liquid or gaseous form. These two components are mixed and ignited in the combustion chamber to produce thrust. A commonly used combination is hydroxyl-terminated polybutadiene (HTPB) as the solid fuel and liquid oxygen (LOX) as the oxidizer.

HTPB is a commonly used solid fuel in hybrid rocket engines [124, 125]. It has attractive characteristics such as high thermal stability and good mechanical performance [125]. When combined with liquid oxygen (LOX) as the oxidizer, it forms an effective propellant combination for rocket applications [124].

LOX is a cryogenic liquid oxidizer that is widely used in liquid-propellant rocket engines due to its desirable properties such as high density, high vaporization temperature, and superior cooling capability compared to other oxidizers like hydrogen [124, 126]. The combination of HTPB as the solid fuel and LOX as the oxidizer is a well-established and commonly used propellant system in hybrid rocket engines [124, 126].

The combustion of HTPB and LOX in a hybrid rocket engine involves the injection of the liquid oxidizer (LOX) into the combustion chamber, where it reacts with the solid fuel (HTPB) to produce thrust [124, 126, 127]. The regression rate, or the rate at which the solid fuel is consumed, is an important parameter in the performance of hybrid rocket engines, and it can be influenced by factors such as the design of the injector and the distribution of the oxidizer flow [125, 127, 128].

Hybrid engines operate by injecting the oxidizer into the combustion chamber, where it reacts with the solid fuel along its exposed surface. This design allows for better control of the combustion process compared to solid rockets, as the flow of the liquid oxidizer can be throttled, stopped, or restarted during flight. This feature provides greater flexibility in mission planning and makes hybrid engines safer to operate than fully solid rocket motors, which cannot be throttled or shut down once ignited.

One key advantage of hybrid engines is their simplicity and cost-effectiveness compared to liquid engines, as they require fewer moving parts, such as turbopumps. Additionally, they are more stable than solid rockets, with a reduced risk of explosion during storage or handling. However, hybrid engines typically have lower performance (specific impulse) than pure liquid engines and are less commonly used for heavy-lift or high-performance applications. They are often found in smaller launch vehicles, sounding rockets, or as propulsion systems for experimental and suborbital missions.

Electric propulsion represents a cutting-edge approach to in-space manoeuvres, offering unmatched efficiency for long-duration missions. Unlike chemical propulsion, which relies on high-energy combustion, electric propulsion systems use electrical energy to accelerate ions or plasma to generate thrust. These systems are not suited for launch but excel in the vacuum of space, where their high efficiency and low fuel consumption are critical.

One of the most common types of electric propulsion is the ion thruster, which uses electric fields to accelerate positively charged ions to high speeds. The system starts by ionizing a propellant, such as xenon gas, using an electron bombardment process. The resulting ions are then accelerated through an electric field, and the expelled ions produce thrust. Although the thrust generated by ion thrusters is very low, they are incredibly efficient, achieving specific impulses several times greater than traditional chemical rockets.

An ion thruster is an advanced form of propulsion technology that generates thrust by accelerating charged particles (ions) using electric fields. Unlike traditional chemical rockets that rely on high-temperature combustion to create thrust, ion thrusters use electricity to achieve incredibly efficient propulsion, making them ideal for long-duration space missions where fuel economy is critical. Here's how an ion thruster works, step by step:

The first step in the operation of an ion thruster is to ionize the propellant, which is typically a neutral gas such as xenon. Xenon is preferred because of its inert properties, high atomic weight, and ease of ionization. Ionization occurs when xenon atoms are bombarded by high-energy electrons, which are generated by a hollow cathode emitting an electron stream. When these high-energy electrons collide with the xenon atoms, they strip away one or more electrons, leaving behind positively charged xenon ions.

The result is a plasma—a hot, electrically charged gas composed of positive xenon ions and free electrons. This ionized plasma forms the basis for the thrust generation process.

Once the xenon gas has been ionized, the positively charged xenon ions are accelerated through a strong electric field. This is achieved using two grids, known as the accelerator grid and the screen grid, which are located at the exit of the ion thruster.

The screen grid is positively charged, while the accelerator grid is negatively charged. This creates an intense electric field between the two grids. The positively charged xenon ions are attracted toward the negatively charged accelerator grid, gaining significant velocity as they pass through it. By the time they exit the thruster, the ions are traveling at speeds of up to 30 km/s (67,000 mph), which generates thrust as they are expelled into space.

As the positively charged xenon ions are ejected from the thruster, they create a net positive charge within the spacecraft. This charge imbalance could eventually interfere with the thruster's operation if left uncorrected. To prevent this, a device called a neutralizer is placed at the exit of the thruster.

The neutralizer emits free electrons into the ion exhaust, combining with the positively charged xenon ions to create a neutral plume of gas. This neutralization ensures that the spacecraft remains electrically balanced and that the exhaust does not interfere with the spacecraft's onboard electronics.

The high-speed expulsion of ions produces thrust based on Newton's Third Law of Motion, which states that every action has an equal and opposite reaction. As the ions are accelerated and ejected out of the thruster at extremely high speeds, the spacecraft experiences a corresponding force in the opposite direction. Although the thrust generated by an ion thruster is relatively low compared to chemical rockets—typically on the order of millinewtons (mN)—the efficiency of the system allows it to operate continuously over long periods, gradually building up significant velocity.

One of the key advantages of ion thrusters is their high specific impulse, which is a measure of propulsion efficiency. Specific impulse refers to the amount of thrust produced per unit of propellant consumed. Ion thrusters have specific impulses ranging from 1,500 to 10,000 seconds, far exceeding that of chemical rockets, which typically range from 200 to 400 seconds. This efficiency makes ion thrusters ideal for missions requiring significant changes in velocity, such as interplanetary travel, station-keeping, or orbital transfers.

Ion thrusters are particularly valuable for space missions because they consume far less propellant than chemical rockets. This allows spacecraft to carry smaller fuel tanks, leaving more room for scientific instruments or other payloads. Additionally, their ability to operate continuously over months or even years makes them suitable for deep-space exploration missions, such as NASA's Dawn mission, which explored the asteroid belt, and ESA's BepiColombo, which is en route to Mercury.

Figure 28: Mercury Ion Thruster. The U.S. National Archives, Public Domain, via NARA & DVIDS Public Domain Archive.

Despite their efficiency, ion thrusters have certain limitations. The most notable is their low thrust output, which makes them unsuitable for launching rockets or achieving rapid accelerations. For example, an ion thruster can take days or weeks to achieve the same velocity change that a chemical rocket can achieve in minutes. Additionally, ion thrusters require significant electrical power to operate, typically supplied by solar panels or nuclear power sources, which can limit their effectiveness in regions of space far from the Sun.

Another widely used electric propulsion system is the Hall-effect thruster, which generates thrust by trapping electrons in a magnetic field and using them to ionize and accelerate a propellant. Hall thrusters are commonly used for satellite station-keeping, orbital adjustments, and deep-space exploration missions.

A Hall-effect thruster is an advanced electric propulsion system that uses electric and magnetic fields to ionize and accelerate a propellant, typically xenon gas, to generate thrust. This type of thruster is widely used for in-space operations such as orbital adjustments, station-keeping, and interplanetary missions due to its high efficiency and reliability. Unlike traditional chemical rockets that rely on combustion, Hall-effect thrusters harness the principles of the Hall effect to create a plasma that produces thrust.

The operation of a Hall-effect thruster begins with the injection of xenon gas into a cylindrical discharge channel. A high voltage applied between an anode inside the channel and a cathode outside creates an electric field along the axis of the thruster. When the xenon atoms pass through this field, high-energy electrons emitted by the cathode collide with them, ionizing the atoms into positively charged xenon ions and free electrons, thereby forming a plasma.

The magnetic field generated by magnetic coils within the thruster traps the electrons in a circular motion around the axis of the discharge channel. This circular motion, induced by the Hall effect, increases the likelihood of collisions between electrons and xenon atoms, enhancing the ionization process. While the electrons are confined by the magnetic field, the heavier xenon ions remain largely unaffected, allowing them to move freely within the electric field.

The electric field accelerates the xenon ions toward the exit of the discharge channel at extremely high speeds, often reaching 15-20 kilometres per second. This rapid expulsion of ions generates thrust according to Newton's Third Law of Motion, where the action of ion ejection produces an equal and opposite reaction that propels the spacecraft forward. At the same time, the cathode emits free electrons into the ion exhaust plume, neutralizing the positively charged ions. This neutralization ensures that the spacecraft remains electrically balanced and prevents interference with onboard systems.

Figure 29: Ten KW Kilowatt T-220 Hall Effect Thruster Life Test. Defense Visual Information Distribution Service, Public Domain, via Picryl.

Hall-effect thrusters are valued for their high specific impulse, which measures propulsion efficiency. They consume significantly less propellant compared to chemical rockets, allowing spacecraft to carry smaller fuel loads and dedicate more mass to payloads. Their ability to operate continuously for long durations makes them ideal for missions requiring precise manoeuvring and gradual velocity changes, such as maintaining satellite orbits or conducting interplanetary transfers.

However, Hall-effect thrusters have limitations. Their thrust is relatively low, often in the range of millinewtons, making them unsuitable for launching rockets or rapid accelerations. They also require substantial electrical power, typically provided by solar panels, which can limit their effectiveness in regions of space far from the Sun. Additionally, the high-energy plasma can erode components within the discharge channel over time, potentially shortening the thruster's operational lifespan.

Hall-effect thrusters are widely used in commercial and scientific applications. They are commonly employed in satellites for station-keeping and orbital adjustments, as well as in space exploration missions such as ESA's BepiColombo mission to Mercury and NASA's Psyche mission to the asteroid Psyche. Their combination of efficiency, simplicity, and adaptability makes them an essential technology in modern space exploration, particularly for long-duration missions that prioritize fuel efficiency and precise manoeuvring. As advancements in materials and power systems continue, Hall-effect thrusters are expected to play an even greater role in the future of space propulsion.

Electric propulsion systems are highly efficient because they consume much less propellant than chemical systems, making them ideal for missions where minimizing mass is critical, such as interplanetary travel. They are currently used in spacecraft like NASA's Dawn mission to the asteroid belt and in commercial satellite platforms to maintain precise orbits over extended periods.

Figure 30: Against a backdrop of clouds on the horizon, the Delta II rocket carrying NASA's Dawn spacecraft rises from the smoke and fire on the launch pad to begin its 1.7-billion-mile journey through the inner solar system to study a pair of asteroids. NASA, Public Domain, via GetArchive.

However, electric propulsion systems have limitations. The low thrust levels mean they cannot be used for rapid accelerations or escaping Earth's gravity, making them unsuitable for launch vehicles. Additionally, they require significant electrical power, often provided by solar panels, which can limit their use in regions of space far from the Sun.

Guidance Subsystem

The guidance subsystem ensures that the rocket follows its intended trajectory from launch to orbit or its final destination. It includes sensors, onboard computers, and navigation equipment that work together to calculate the rocket's position, velocity, and orientation in real-time.

One of the primary functions of the guidance system is to continuously determine the desired path and issue commands to the control subsystem to keep the rocket on track. For this, the guidance subsystem relies on inputs from various sensors, such as gyroscopes, accelerometers, star trackers, and GPS receivers. For missions beyond Earth's orbit, celestial navigation methods, which use the positions of stars or planets, are also employed.

The guidance system of a rocket plays a pivotal role in ensuring stability during launch and controlling its trajectory throughout flight. Its two main tasks—providing stability and facilitating control during manoeuvres—are vital for achieving mission objectives, whether the rocket is intercepting a target or reaching orbit. To achieve these goals, the system employs various control methods to adjust the rocket's motion and orientation. These methods are based on the principles of forces acting on the rocket, and they utilize torque to create rotational movements about the rocket's centre of gravity (or centre of mass) [129].

One of the earliest and still widely used methods for controlling rockets in flight involves movable fins. These fins are located at the rear of the rocket and are used to adjust the aerodynamic forces acting on the vehicle. As the rocket moves through the atmosphere, these aerodynamic forces act through the centre of pressure, which is typically offset from the centre of gravity. This offset generates a torque that rotates the rocket. For instance, if the trailing edge of a fin is deflected to the right, the resulting aerodynamic force will cause the rocket's nose to turn to the right. Movable fins are especially effective for rockets traveling through the lower atmosphere, where aerodynamic forces are significant, and are still commonly found in air-to-air missiles and other smaller rockets [129].

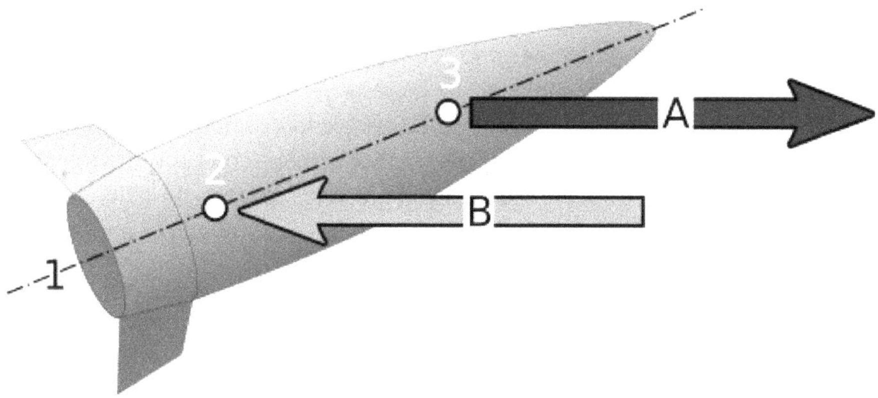

Figure 31: Effect of the aerodynamic moment on a wing-stabilized projectile. 1: Axis of projectile. 2: Center of Pressure 3: Center of gravity. A: Direction of flight. B: Aerodynamic moment. MagentaGreen, CC BY-SA 4.0, via Wikimedia Commons.

In modern rockets, gimbaled thrust systems are the most common method of control. In these systems, the nozzle of the main rocket engine can be swivelled or rotated to change the direction of thrust. This creates a torque relative to the rocket's centre of gravity, allowing precise control over its orientation. For example, if the gimbaled nozzle is pivoted to the right, the resulting thrust force causes the nose of the rocket to move to the right. Gimbaled thrust systems are highly effective because they provide reliable and continuous control, even outside the atmosphere where aerodynamic forces are negligible. This method is used in many modern launch vehicles, including SpaceX's Falcon 9 and NASA's Saturn V, as it allows for precise trajectory adjustments during all phases of flight [129].

Older rockets, such as the Atlas missile, used vernier rockets for control. These small auxiliary engines were located at the base of the main rocket and provided fine adjustments to the rocket's orientation by generating torque. For instance, firing a vernier rocket on the right side of the main rocket would cause the nose to pivot to the right. While effective, this method added extra weight due to the additional fuel and plumbing required, which made it less desirable as rockets evolved. As a result, vernier rockets have largely been replaced by gimbaled thrust systems, which are lighter and more efficient [129].

Early rockets like the V2 and Redstone employed thrust vanes, a rudimentary but effective control method. These vanes were small surfaces placed directly within the exhaust stream of the main rocket engine. By deflecting the exhaust flow, they created torque that could adjust the rocket's orientation. For instance, deflecting a thrust vane to the right would cause the exhaust flow to push the nose of the rocket in the same direction. While relatively simple, thrust vanes were limited in their efficiency and added drag to the exhaust stream, making them less favourable as propulsion and control technologies advanced [129].

Each of these control methods has been developed and refined to address specific challenges in rocket flight. Early systems, like thrust vanes and vernier rockets, provided foundational approaches to rocket control but were eventually replaced by more advanced and efficient systems, such as gimbaled thrust and movable fins. By combining aerodynamic forces, torque generation, and precise adjustments to the rocket's centre of mass, these methods ensure that rockets remain stable and responsive during their journey through the atmosphere and beyond. As guidance technologies continue to evolve, they contribute significantly to the increasing reliability and sophistication of modern rocketry [129].

Advanced guidance systems employ algorithms such as Kalman filters to process sensor data and reduce errors.

The Kalman filter is an algorithm that provides an efficient way to estimate the state of a dynamic system in real-time, even when the system is affected by noise or uncertainty [130-132]. It is widely used in various applications such as control systems, navigation, robotics, and signal processing to estimate variables like position, velocity, and orientation [130-132].

The Kalman filter works on the principle of predicting and correcting state estimates iteratively to minimize error [130, 132, 133]. It is particularly effective in systems where measurements are noisy and the process being measured evolves dynamically over time [130-132]. For example, in rocket guidance or spacecraft navigation, the Kalman filter combines data from sensors like gyroscopes, accelerometers, and GPS with a mathematical model of the vehicle's motion to provide accurate real-time estimates of position and velocity [130, 132, 134].

There are various types of Kalman filters, such as standard Kalman filters, extended Kalman filters, unscented Kalman filters, and ensemble Kalman filters [131, 132]. These different variants have been developed to address the limitations of the standard Kalman filter, such as the assumption of Gaussian noise and linearity [135, 136].

The Kalman filter is an optimal linear estimator that minimizes the mean-square error (MSE) and operates with the fundamental assumption that both the process and measurement noises can be represented by a Gaussian distribution of random variables [133, 135]. When the Kalman filter is applied to a non-linear system, this Gaussian assumption becomes a poor approximation, leading to the development of the extended Kalman filter and the unscented Kalman filter [135, 136].

A Kalman filter operates in two main steps: prediction and update (correction). These steps are repeated iteratively as new measurements are received, allowing the filter to refine its estimates over time.

1. Prediction Step: In the prediction step, the Kalman filter uses a mathematical model of the system to estimate the state of the system at the next time step and predict the associated uncertainty. The system's state might include variables such as position, velocity, or acceleration. The prediction is based on the known dynamics of the system, represented by a set of linear equations, and assumes some process noise due to uncertainties in the model or external disturbances.

The prediction equations involve two parts:

- **State Prediction**: Using the system model, the current state is projected forward in time to estimate the next state.
- **Error Covariance Prediction**: The uncertainty in the predicted state is also estimated, accounting for the process noise and the accuracy of the model.

2. Update (Correction) Step: Once a new measurement is available (e.g., from a sensor), the Kalman filter compares the predicted state to the measured value and updates its estimate to reduce the error. This step incorporates the measurement into the state estimate, weighting it according to its reliability. Measurements with higher uncertainty are given less weight, while more reliable measurements are given more weight.

The update process involves:

- **Calculating the Kalman Gain**: This is a weighting factor that determines how much the new measurement influences the updated estimate. It depends on the uncertainty in the predicted state and the uncertainty in the measurement.
- **Updating the State Estimate**: The predicted state is corrected by adding the difference (or residual) between the measured value and the predicted value, scaled by the Kalman gain.
- **Updating the Error Covariance**: The uncertainty in the updated state estimate is recalculated to reflect the incorporation of the new measurement.

Figure 32 illustrates the basic concept of Kalman filtering, which combines predictions from a mathematical model with noisy measurements to produce an optimal estimate of a system's state.

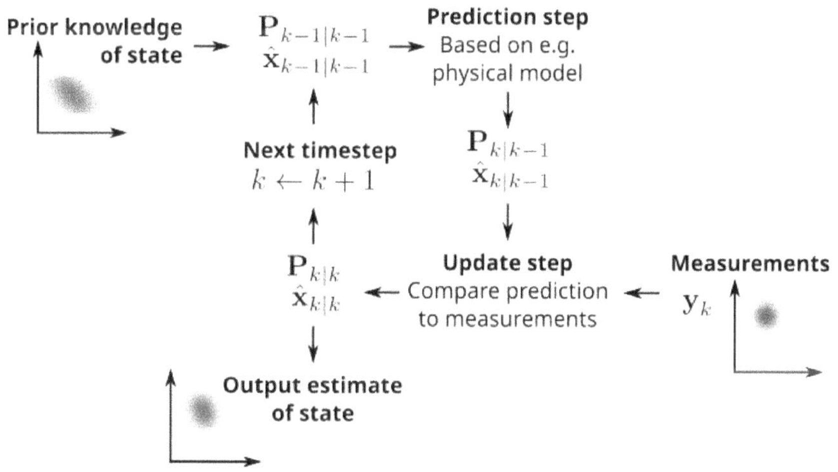

Figure 32: The diagram explains the basic steps of Kalman filtering: prediction and update. It also illustrates how the filter keeps track of not only the mean value of the state, but also the estimated variance. Petteri Aimonen, CC0, via Wikimedia Commons.

The Kalman filtering process consists of three main phases: prediction, measurement, and correction. In the prediction phase, the filter estimates the system's next state based on the current state and a mathematical model of the system's dynamics. This predicted state is represented by blue ovals in the diagram, which also display the associated uncertainty as error ellipses. The size of these ellipses reflects the level of uncertainty, which is influenced by process noise and the limitations of the predictive model.

During the measurement phase, the system acquires a new data point, which inherently contains noise and uncertainty. These measurements are depicted in the diagram as individual data points with their uncertainties also represented as blue ellipses. This illustrates the variability and imprecision typically associated with real-world measurements, arising from factors such as sensor errors or environmental disturbances.

The correction phase is where the Kalman filter refines its estimate by combining the predicted state with the new measurement. This is achieved through a weighted averaging process, where the weights are determined by the relative uncertainties of the prediction and the measurement. The result is a corrected state estimate that minimizes overall error. This corrected estimate is more accurate than either the prediction or the measurement alone, as it leverages the strengths of both while compensating for their individual weaknesses.

Figure 32 highlights key concepts central to the Kalman filter's functionality. The prediction uncertainty, shown as ellipses around the predicted states, indicates how much confidence the filter has in its model-based prediction. Larger ellipses represent greater uncertainty, reflecting the challenges of relying solely on predictions in dynamic and noisy systems. Measurement points and their associated uncertainties, also shown as ellipses, illustrate the variability inherent in observational data, emphasizing the need to reconcile these inputs with predictions.

The fusion of prediction and measurement is represented by smaller corrected ellipses, which demonstrate the filter's ability to refine the state estimate by reducing uncertainty. The smaller size of these ellipses compared to the prediction or measurement uncertainties individually highlights the effectiveness of the Kalman filter in producing more accurate and reliable state estimates. This iterative process of prediction, measurement, and correction is what makes the Kalman filter a powerful tool for real-time state estimation in dynamic systems.

The Kalman filter is a mathematical algorithm used to estimate the state of a dynamic system based on noisy measurements and a predictive model. The key components of the Kalman filter include several equations that operate in a recursive manner, iteratively predicting and updating the system's state while minimizing error. These equations are critical in the Kalman filter's ability to fuse predictions and measurements effectively. The following breakdown explains these key mathematical components in detail.

The state prediction equation provides the estimated state of the system at the next time step. It is given by:

$$\hat{x}^{k}_{|k-1} = F_k \hat{x}^{k-1}_{|k-1} + B_k u_k$$

Here, $\hat{x}^{k}_{|k-1}$ represents the predicted state of the system at time step *k*, given the state estimate from the previous time step $\hat{x}^{k-1}_{|k-1}$. The state transition matrix F_k describes how the state evolves from time *k*−1 to time *k*, based on the system dynamics or a mathematical model. The term $B_k u_k$ accounts for any control inputs u_k, such as external forces or actions applied to the system that affect its state. Essentially, this equation models how the system's state is expected to change, based on the previous state and any control actions.

The error covariance prediction equation describes the uncertainty in the predicted state. It is expressed as:

$$P^{k}_{|k-1} = F_k P^{k-1}_{|k-1} F_k^T + Q_k$$

In this equation, $P^{k}_{|k-1}$ is the predicted error covariance matrix, which represents the uncertainty in the predicted state $\hat{x}^{k}_{|k-1}$. The term $P^{k}_{|k-1}$ is the error covariance from the previous time step. The matrix $F_k P^{k-1}_{|k-1} F_k^T$ accounts for the propagation of uncertainty due to the system dynamics. The additional term Q_k represents process noise, or the uncertainty introduced by model inaccuracies or unmodeled factors that affect the system's behaviour. This equation helps track how the uncertainty in the system state evolves over time.

The Kalman gain equation calculates the optimal weight that should be given to the new measurement when updating the predicted state. It is given by:

$$K_k = P^{k}_{|k-1} H_k^T (H_k P^{k}_{|k-1} H_k^T + R_k)^{-1}$$

The Kalman gain K_k is a critical factor in determining how much the predicted state should be corrected by the new measurement. It is calculated based on the predicted error covariance $P^{k}_{|k-1'}$ the measurement model matrix H_k, and the measurement noise covariance R_k. The matrix H_k maps the predicted state to the measurement space, essentially transforming the state prediction into the predicted measurement. The Kalman gain ensures that the updated state estimate balances the relative uncertainties between the prediction and the measurement, giving more weight to the source with lower uncertainty.

The state update equation corrects the predicted state using the actual measurement. It is given by:

$$\hat{x}^k_{|k} = \hat{x}^k_{|k-1} + K_k(z_k - H_k\hat{x}^k_{|k-1})$$

Here, $\hat{x}^k_{|k}$ is the updated state estimate, where z_k is the actual measured value at time step k, and $H_k\hat{x}^k_{|k-1}$ is the predicted measurement based on the model. The term $(z_k - H_k\hat{x}^k_{|k-1})$ represents the measurement residual, or the difference between the actual measurement and the predicted measurement. The Kalman gain K_k scales the residual to determine how much the state estimate should be corrected, with more weight given to the measurement when the uncertainty in the prediction is high and vice versa.

Finally, the error covariance update equation recalculates the uncertainty in the updated state estimate. It is given by:

$$P^k_{|k} = (I - K_kH_k)P^k_{|k-1}$$

In this equation, $P^k_{|k}$ is the updated error covariance matrix, which represents the uncertainty in the final state estimate $\hat{x}^k_{|k}$. The term $(I - K_kH_k)$ is a matrix that adjusts the covariance based on the Kalman gain. This update reduces the uncertainty in the estimate, as the correction from the measurement refines the state estimate. This equation ensures that the filter's uncertainty is properly adjusted after each update.

Practical Example: Kalman Filter in Rocket Altitude Estimation

Imagine a rocket is ascending, and we want to estimate its altitude and velocity in real-time using noisy data from two sensors: an altimeter, which directly measures altitude but with noise, and an accelerometer, which measures acceleration that can be integrated over time to estimate changes in velocity and altitude. The Kalman filter combines these measurements with a predictive model to optimally estimate the rocket's state.

Problem Setup

State Variables: The rocket's state is represented as:

$$\mathbf{x}_k = \begin{bmatrix} h_k \\ v_k \end{bmatrix}$$

where h_k is the altitude, and v_k is the velocity at time step k.

System Dynamics: The relationship between the state at consecutive time steps is given by:

$$\mathbf{x}_k = \mathbf{F}_k\mathbf{x}_{k-1} + \mathbf{B}_ku_k + \mathbf{w}_k$$

where:

$\mathbf{F}_k = \begin{bmatrix} 1 & \Delta t \\ 0 & 1 \end{bmatrix}$ is the state transition matrix, which models how altitude and velocity evolve over time.

$\mathbf{B}_k = \begin{bmatrix} 0.5\Delta t^2 \\ \Delta t \end{bmatrix}$ accounts for the effect of control input (acceleration).

u_k is the control input (measured acceleration).

W_k represents process noise, accounting for uncertainties in the model.

Measurement Model: The altimeter measures the altitude directly, modelled as:

$$z_k = \mathbf{H}_k \mathbf{x}_k + v_k$$

where:

$\mathbf{H}_k = \begin{bmatrix} 1 & 0 \end{bmatrix}$ maps the state vector to the altitude.

V_k is the measurement noise, assumed to follow a Gaussian distribution with zero mean and covariance R_k.

Error Covariances:

Process noise covariance (Q_k): Models uncertainties in the dynamics.

Measurement noise covariance (R_k): Represents sensor noise (e.g., altimeter inaccuracies).

Error covariance (P_k): Tracks uncertainty in the state estimate.

Initial Conditions and Known Parameters

Initial state estimate: $\hat{\mathbf{x}}_{0|0} = \begin{bmatrix} 100 \\ 20 \end{bmatrix}$ (altitude in meters, velocity in m/s)

Initial error covariance: $\mathbf{P}_{0|0} = \begin{bmatrix} 10 & 0 \\ 0 & 5 \end{bmatrix}$

Process noise covariance: $\mathbf{Q}_k = \begin{bmatrix} 0.1 & 0 \\ 0 & 0.1 \end{bmatrix}$

Measurement noise covariance: $R_k = 2\,\mathrm{m}^2$

Measured acceleration: $u_1 = 2\,\mathrm{m/s}^2$

Altimeter reading at *t*=1: $z_1 = 120\,\text{m}$

Time step: $\Delta t = 1\,\text{s}$

Step-by-Step Calculation

1. Prediction Step

The Kalman filter first predicts the state and its uncertainty based on the dynamics model.

State Prediction:

$$\hat{\mathbf{x}}_{1|0} = \mathbf{F}_k \hat{\mathbf{x}}_{0|0} + \mathbf{B}_k u_k$$

Substitute values:

$$\mathbf{F}_k = \begin{bmatrix} 1 & 1 \\ 0 & 1 \end{bmatrix}, \quad \mathbf{B}_k = \begin{bmatrix} 0.5 \\ 1 \end{bmatrix}, \quad u_k = 2$$

$$\hat{\mathbf{x}}_{1|0} = \begin{bmatrix} 1 & 1 \\ 0 & 1 \end{bmatrix} \begin{bmatrix} 100 \\ 20 \end{bmatrix} + \begin{bmatrix} 0.5 \\ 1 \end{bmatrix} \cdot 2$$

$$\hat{\mathbf{x}}_{1|0} = \begin{bmatrix} 120 \\ 22 \end{bmatrix}$$

Error Covariance Prediction:

$$\mathbf{P}_{1|0} = \mathbf{F}_k \mathbf{P}_{0|0} \mathbf{F}_k^\top + \mathbf{Q}_k$$

Substitute:

$$\mathbf{P}_{1|0} = \begin{bmatrix} 1 & 1 \\ 0 & 1 \end{bmatrix} \begin{bmatrix} 10 & 0 \\ 0 & 5 \end{bmatrix} \begin{bmatrix} 1 & 0 \\ 1 & 1 \end{bmatrix} + \begin{bmatrix} 0.1 & 0 \\ 0 & 0.1 \end{bmatrix}$$

$$\mathbf{P}_{1|0} = \begin{bmatrix} 15.1 & 5 \\ 5 & 5.1 \end{bmatrix}$$

2. Update Step

The Kalman filter updates the state estimate and error covariance using the new measurement.

Kalman Gain:

$$\mathbf{K}_k = \mathbf{P}_{1|0} \mathbf{H}_k^\top (\mathbf{H}_k \mathbf{P}_{1|0} \mathbf{H}_k^\top + R_k)^{-1}$$

Substitute:

$$\mathbf{H}_k = \begin{bmatrix} 1 & 0 \end{bmatrix}, \quad R_k = 2$$

$$\mathbf{K}_k = \begin{bmatrix} 15.1 & 5 \\ 5 & 5.1 \end{bmatrix} \begin{bmatrix} 1 \\ 0 \end{bmatrix} \left(\begin{bmatrix} 1 & 0 \end{bmatrix} \begin{bmatrix} 15.1 & 5 \\ 5 & 5.1 \end{bmatrix} \begin{bmatrix} 1 \\ 0 \end{bmatrix} + 2 \right)^{-1}$$

$$\mathbf{K}_k = \begin{bmatrix} 15.1 \\ 5 \end{bmatrix} (15.1 + 2)^{-1}$$

$$\mathbf{K}_k = \begin{bmatrix} 0.883 \\ 0.294 \end{bmatrix}$$

State Update:

$$\hat{\mathbf{x}}_{1|1} = \hat{\mathbf{x}}_{1|0} + \mathbf{K}_k (z_k - \mathbf{H}_k \hat{\mathbf{x}}_{1|0})$$

Substitute:

$$z_k = 120, \quad \mathbf{H}_k = \begin{bmatrix} 1 & 0 \end{bmatrix}, \quad \hat{\mathbf{x}}_{1|0} = \begin{bmatrix} 120 \\ 22 \end{bmatrix}$$

$$\hat{\mathbf{x}}_{1|1} = \begin{bmatrix} 120 \\ 22 \end{bmatrix} + \begin{bmatrix} 0.883 \\ 0.294 \end{bmatrix} (120 - 120)$$

$$\hat{\mathbf{x}}_{1|1} = \begin{bmatrix} 120 \\ 22 \end{bmatrix}$$

Error Covariance Update:

$$\mathbf{P}_{1|1} = (\mathbf{I} - \mathbf{K}_k \mathbf{H}_k) \mathbf{P}_{1|0}$$

Substitute:

$$\mathbf{P}_{1|1} = \left(\begin{bmatrix} 1 & 0 \\ 0 & 1 \end{bmatrix} - \begin{bmatrix} 0.883 \\ 0.294 \end{bmatrix} \begin{bmatrix} 1 & 0 \end{bmatrix} \right) \begin{bmatrix} 15.1 & 5 \\ 5 & 5.1 \end{bmatrix}$$

$$\mathbf{P}_{1|1} = \begin{bmatrix} 1.766 & 0.588 \\ 0.588 & 3.529 \end{bmatrix}$$

Final Results

At t=1:

The updated state estimate is:

$$\hat{\mathbf{x}}_{1|1} = \begin{bmatrix} 120 \\ 22 \end{bmatrix}$$

This represents an altitude of 120 m and a velocity of 22 m/s.

The updated error covariance matrix is:

$$\mathbf{P}_{1|1} = \begin{bmatrix} 1.766 & 0.588 \\ 0.588 & 3.529 \end{bmatrix}$$

This indicates reduced uncertainty in the altitude and velocity estimates compared to the initial prediction.

The Kalman filter effectively combines noisy sensor data to refine the rocket's state estimate, improving accuracy and reliability in real-time.

The importance of the guidance subsystem cannot be overstated, as even minor deviations in trajectory during launch can result in mission failure. Autonomous guidance systems, like those used in SpaceX's Falcon 9, have made significant advancements in accuracy and adaptability, enabling reusable rockets to land precisely on Earth or barges after completing their missions.

Control Subsystem

The control subsystem is responsible for maintaining stability and ensuring the rocket can respond to guidance commands to follow its trajectory accurately. It includes actuators, thrust vectoring mechanisms, and aerodynamic surfaces that allow the rocket to adjust its orientation and direction during flight.

Control is achieved by manipulating the rocket's attitude (its pitch, yaw, and roll) and thrust vector. In liquid propulsion systems, thrust vector control (TVC) is commonly used, where the engines swivel slightly to change the direction of the thrust. For solid rockets or boosters, other methods such as movable nozzles, jet vanes, or small auxiliary thrusters are used to achieve similar results.

During the ascent through Earth's atmosphere, the control subsystem also manages aerodynamic forces. Rockets may use fins, flaps, or grid fins for stability and manoeuvrability during this phase. In addition to hardware, control systems rely heavily on software to process real-time data and execute precise adjustments. Modern rockets employ advanced autopilot systems that reduce human intervention and improve flight precision.

For reusable rockets, control systems are even more sophisticated, as they must manage descent and landing operations. SpaceX's Falcon 9, for instance, uses grid fins and retro-thrust engines to guide and stabilize its first stage during its return to Earth.

Payload Subsystem

The payload subsystem represents the purpose of the rocket's mission. It includes the cargo being carried into space, such as satellites, scientific instruments, crewed capsules, or supplies for the International Space Station (ISS). In some cases, the payload can also be an exploratory spacecraft, such as a Mars rover or an interplanetary probe.

Figure 33: Rocket Payload Under Construction. National Parks Gallery, Public Domain, via Picryl.

The payload is typically housed within a payload fairing, a protective shell that shields it from aerodynamic forces and environmental conditions during launch. The fairing is jettisoned once the rocket exits the denser layers of Earth's atmosphere to reduce weight and expose the payload for deployment.

Figure 34: The payload fairing containing NOAA's GOES-T satellite is hoisted into place atop the Atlas V rocket on Feb. 17, 2022. NOAA Satellites, Public Domain, via Flickr.

Payload systems are designed to ensure the safe delivery and functionality of the cargo. For instance, scientific instruments often require thermal protection, vibration isolation, and radiation shielding to remain operational during the harsh conditions of space travel. Crewed payloads, such as those carried by SpaceX's Dragon or NASA's Orion, must include life-support systems, radiation protection, and re-entry shielding.

In commercial satellite launches, maximizing payload capacity is a key priority, as it directly impacts the mission's profitability. Advances in lightweight materials and modular designs have enabled rockets to carry heavier or more numerous payloads without significantly increasing launch costs.

Integration of Subsystems

The success of a rocket mission depends on the seamless integration of propulsion, guidance, control, and payload subsystems. These systems must work in harmony to ensure the rocket achieves the correct trajectory, maintains stability, and delivers its payload accurately. For example, guidance and control systems must compensate for any thrust imbalances in the propulsion subsystem, while the payload subsystem must be securely integrated to avoid damage from vibrations or accelerations.

Modern rockets leverage automation and advanced software to integrate these systems more efficiently. Pre-launch simulations and tests are conducted to verify that the subsystems can handle real-world conditions. This integration is especially critical for reusable rockets, where subsystems must perform flawlessly during multiple launch and recovery cycles.

Staging and Modular Design for Efficiency

Staging Design

Rocket staging and modular design are fundamental engineering strategies that maximize the efficiency and performance of launch vehicles. These approaches address the key challenges of rocket propulsion, weight reduction, and adaptability to diverse missions. By separating the rocket into multiple stages or modular components, engineers can optimize the vehicle's structure, reduce unnecessary mass, and improve fuel efficiency.

Figure 35: Fragment of unfinished rocket. SpaceX, CC0, via Pexels.

Rocket staging is a key concept that significantly enhances the efficiency of rocket-based space transportation [137, 138]. It involves dividing the rocket into two or more stages, each with its own propulsion system and fuel [137, 138]. As each stage completes its function, it is jettisoned, leaving the remaining stages to continue the ascent [137, 138].

The primary advantages of rocket staging are:

1. Reducing Dead Weight: In a single-stage rocket, the entire structure, including empty fuel tanks and spent engines, must be carried throughout the mission. In a multi-stage rocket, the empty tanks and engines of each stage are discarded once their fuel is exhausted, eliminating unnecessary weight and allowing the remaining stages to accelerate more efficiently [137, 138].

2. Optimized Propulsion for Each Phase: Each stage is designed to operate within specific environmental conditions. The first stage is optimized for dense atmospheric conditions near the surface, producing significant thrust to overcome gravity and atmospheric drag, while subsequent stages are designed to operate in thinner atmospheres or in the vacuum of space, using smaller engines tailored for higher efficiency in low-pressure conditions [137, 138].

3. Fuel Efficiency and Payload Capacity: By shedding weight with each stage, multi-stage rockets require less total fuel compared to single-stage vehicles for the same mission. This weight reduction allows for greater payload capacity, as demonstrated by the iconic Saturn V rocket, which used three stages to efficiently carry the Apollo spacecraft to the Moon [137, 138].

Examples of staged rocket systems include the Falcon 9 from SpaceX, which utilizes a two-stage design, and the Saturn V, which had three stages [137, 138].

The concept of rocket staging has been applied to various rocket designs, including solid-propellant rockets [139], hybrid rockets [140], and air-launched rockets [141]. Researchers have also explored optimization techniques, such as multi-objective genetic algorithms [101, 142, 143], to design and optimize multi-stage rocket systems.

Figure 36: Four Falcon 9 first stages under construction at SpaceX headquarters in Hawthorne, CA. SpaceX, CC0, via Wikimedia Commons.

Impact of Rocket Staging on the Construction Process

Rocket staging is a critical design principle in rocket construction that enables efficient propulsion and optimal performance during launch. By dividing the rocket into two or more stages, each with its own propulsion system, fuel tanks, and specific structural designs, engineers address the challenges of reducing weight and optimizing propulsion for varying atmospheric conditions. This method fundamentally influences the rocket's construction process in terms of design complexity, structural optimization, and overall mission success.

Staging introduces additional complexity into the rocket construction process, as each stage must function as an independent propulsion system while seamlessly integrating with the other stages. Engineers must design each stage with its own engines, fuel tanks, and structural supports. These components are interconnected through stage interfaces, which require precision engineering to ensure smooth separation at predetermined points during ascent. Separation mechanisms, such as explosive bolts, pneumatic actuators, or retro rockets, are integrated into the design to guarantee that each stage jettisons cleanly without disrupting the rocket's trajectory.

The structural design of each stage is unique and tailored to its role in the mission. For instance, the first stage is built to withstand the immense thrust and aerodynamic forces during liftoff, requiring robust construction and high-thrust engines. In contrast, upper stages are lighter and equipped with engines optimized for efficiency in thin atmospheres or the vacuum of space. This necessitates a modular approach to construction, where stages are manufactured as separate components and later assembled to form the complete launch vehicle.

One of the primary advantages of staging is the reduction of dead weight, which significantly influences the construction process. In single-stage rockets, the entire structure, including empty fuel tanks and spent engines, must be carried throughout the mission. This imposes a substantial weight penalty, reducing payload capacity and fuel efficiency. Staging solves this problem by allowing engineers to design each stage to be discarded after its fuel is depleted.

From a construction standpoint, this means each stage must be engineered to perform its specific function while remaining as lightweight as possible. Materials such as aluminium alloys, carbon composites, and titanium are commonly used to minimize weight while maintaining structural strength. Engineers use techniques such as finite element analysis (FEA) to ensure that each stage can endure its respective loads and stresses without unnecessary mass. Weight savings are prioritized in upper stages, where even small reductions translate to significant fuel efficiency gains.

Staging allows each propulsion system to be designed for the specific atmospheric conditions encountered at different altitudes. The first stage, tasked with overcoming Earth's gravity and atmospheric drag, is equipped with large, powerful engines capable of producing immense thrust. These engines are constructed to endure high pressures and temperatures, requiring robust materials like nickel alloys and advanced cooling systems.

In contrast, upper stages operate in thinner atmospheres or space, where efficiency matters more than raw thrust. Engines for these stages are typically smaller, lighter, and optimized for vacuum conditions with high specific impulse (efficiency). Their nozzles are extended to enhance performance in low-pressure environments, and their structures are designed to carry reduced loads. The construction process must account for this variation, ensuring smooth transitions between propulsion systems while maintaining overall stability and alignment during flight.

The construction of a staged rocket is inherently tied to fuel efficiency and payload capacity. By shedding weight as each stage is jettisoned, multi-stage rockets require less total fuel to achieve their mission goals compared to single-stage rockets. This allows engineers to dedicate a larger proportion of the total vehicle mass to the payload.

The payload integration process is carefully considered during the rocket's design and assembly. Upper stages are structurally lighter and optimized to carry satellites, crew capsules, or exploration equipment. Engineers design payload fairings to protect these sensitive components from aerodynamic forces and vibrations during launch. The payload section, situated atop the final stage, is integrated last, ensuring compatibility with the vehicle's structural and aerodynamic design.

Staging impacts the manufacturing process by requiring each stage to be built as a separate module. Each stage undergoes individual testing, including propulsion system validation, structural integrity checks, and stage separation trials. This modular approach simplifies the construction workflow, as components for different stages can be manufactured simultaneously and integrated later. Modern rockets like the Falcon 9 and Saturn V highlight the benefits of modular construction, where each stage is assembled, tested, and optimized independently before final integration.

During assembly, precise alignment of the stages is crucial to ensure proper stage separation and stability during flight. Staging mechanisms, such as pyrotechnic bolts or pneumatic pushers, are installed to guarantee clean detachment without damaging the remaining structure. Once assembled, the entire rocket undergoes full-scale integration tests to verify the performance of all systems, including thrust alignment, separation timing, and fuel flow between stages.

Modular Design

Modular design is a well-established approach in the aerospace industry, particularly for the development of rockets and launch vehicles [144-146]. This design philosophy involves constructing rockets as a collection of standardized, interchangeable components or modules [144, 145]. This modular approach offers several key advantages that have made it a widely adopted practice in the field of rocketry.

One of the primary benefits of modular design is the increased flexibility it provides across different missions [144-146]. By using standardized modules, rocket configurations can be easily adapted to accommodate various payloads and mission requirements. For instance, different upper stages or payload adapters can be swapped in and out to enable the same core rocket system to be used for delivering satellites to low Earth orbit or deep space probes to distant destinations [144, 145].

Another significant advantage of modular design is the potential for cost reduction through reusability [144-146]. Many modular rocket components, such as the first stage, are designed to be recoverable, refurbished, and reused for multiple launches. This dramatically lowers the overall cost of launching payloads compared to traditional expendable rocket systems [144, 145]. For example, SpaceX's Falcon 9 rocket employs a modular first stage that can be recovered, refurbished, and reused, enabling significant cost savings [144].

Modular design also facilitates standardization and mass production, which can further reduce costs and development time [144-146]. By producing standardized modules in bulk, manufacturers can streamline their manufacturing processes, taking advantage of economies of scale and advanced manufacturing techniques like 3D printing [144, 145]. This is exemplified by Rocket Lab's Electron rocket, which utilizes 3D-printed modular components to enable rapid and cost-effective production [144].

Additionally, modular designs offer benefits in terms of maintenance and upgrades [144, 145]. The ability to isolate and replace specific components without overhauling the entire rocket system is particularly advantageous for reusable systems, as worn-out modules can be swapped out after recovery [144, 145]. Furthermore, modular components can be individually upgraded as technology advances, allowing for incremental improvements to the overall rocket system [144, 145].

Impact of Modular Design on the Rocket Construction Process

Modular design has revolutionized the rocket construction process by introducing efficiency, adaptability, and cost-effectiveness. By breaking down rockets into standardized, interchangeable components or modules, engineers streamline manufacturing, simplify assembly, and improve flexibility to meet diverse mission requirements. This approach addresses many of the challenges associated with traditional rocket construction, particularly for modern reusable launch vehicles.

Modular design allows rockets to be reconfigured easily to accommodate different missions, payloads, and destinations. This flexibility is achieved by swapping or modifying specific components such as upper stages, payload adapters, or boosters. For example, a mission to low Earth orbit (LEO) may require a simpler and lighter configuration compared to a mission aimed at placing a spacecraft in geostationary orbit or deep space. By reusing a core rocket structure and attaching mission-specific modules, manufacturers can tailor rockets to meet varying mission demands without overhauling the entire design.

This adaptability impacts the construction process by enabling parallel development of modules. While the core rocket or booster stage is standardized, engineers can simultaneously develop and test specialized components, such as payload fairings or additional upper stages. For instance, SpaceX's Falcon 9 first stage remains largely unchanged between missions, but the payload adapters and upper stages can be customized based on customer needs. This modular flexibility saves time and resources, as only certain components need to be adjusted for each mission.

One of the most significant benefits of modular design is the ability to recover and reuse critical rocket components, dramatically reducing launch costs. Traditionally, rockets were built as single-use vehicles, with all components discarded after a mission. Modular systems, however, are designed for reusability, meaning that key modules—such as the first stage—can be recovered, refurbished, and reused multiple times.

For example, SpaceX's Falcon 9 first stage is modular, equipped with landing legs and grid fins that enable controlled descent and recovery. After each launch, the first stage is retrieved, inspected, and refurbished for reuse, eliminating the need to rebuild this expensive component for every mission. This approach not only reduces the cost per launch but also impacts the construction process by shifting focus toward durability and ease of maintenance for reusable modules. Engineers design modular components to withstand multiple flight cycles with minimal refurbishment, incorporating features like wear-resistant materials and simplified recovery systems.

Modular design relies on the standardization of components, which has a profound impact on manufacturing processes. Instead of building a unique rocket for each mission, manufacturers produce standardized modules—such as fuel tanks, engines, or structural components—in bulk. This allows for economies of scale, reducing both development time and production costs.

Figure 37: Modular construction. SpaceX, CC0, via Pexels.

Standardized modules streamline assembly, as engineers can integrate pre-manufactured components quickly and efficiently. For example, Rocket Lab's Electron rocket employs modular components, many of which are produced using advanced techniques like 3D printing. The use of 3D-printed components ensures rapid, repeatable production while maintaining precision and consistency across modules. Such innovations make it possible to manufacture rockets at a faster pace and lower cost, supporting the growing demand for satellite launches and commercial missions.

Standardization also simplifies quality control and testing. Since each module is identical, testing procedures can be applied uniformly, ensuring reliability and reducing the risk of failure. This approach enhances scalability, allowing manufacturers to produce rockets in larger quantities without sacrificing performance.

Modular systems greatly simplify maintenance and upgrades, as individual components can be isolated, inspected, and replaced without disassembling the entire rocket. This is particularly valuable for reusable rockets, where components are recovered after flight and require servicing. Engineers can quickly identify worn-out or damaged modules, swap them with replacements, and prepare the rocket for its next launch. This modular approach reduces downtime and allows rockets to be turned around for subsequent missions in a shorter timeframe.

Additionally, modular components can be upgraded individually as technology advances. For instance, improved engines, lighter fuel tanks, or upgraded avionics can be integrated into an existing rocket design without redesigning the entire vehicle. This incremental upgrade capability enhances the long-term performance and competitiveness of modular rockets. A notable example is SpaceX's Falcon 9, which has undergone multiple upgrades (such as the Block 5 version) while retaining its modular architecture. Such iterative improvements are more cost-effective and feasible in a modular framework.

The modular approach fundamentally reshapes how rockets are designed, built, and operated. It introduces a clear division of labour, with specialized teams focusing on the development of specific modules. Modules can be manufactured in parallel, reducing the overall construction timeline. For example, the first stage, upper stage, and payload fairing can be built and tested independently before being integrated into the final launch vehicle.

Moreover, modularity simplifies assembly at the launch site. Pre-tested modules are transported to the facility and assembled in a streamlined process, minimizing the complexity and risk associated with on-site construction. This modular workflow is particularly advantageous for commercial launch providers aiming to deliver frequent, reliable launches.

Environmental Considerations: Heat, Vibration, and Vacuum Effects

Rocket design and construction must account for extreme environmental conditions throughout a mission, from the moment of ignition on Earth to the harshness of outer space. Factors such as intense heat, severe vibrations, and vacuum effects place immense demands on the structural integrity, materials, and overall systems of the rocket. Failure to address these conditions can compromise the mission, lead to structural failure, or damage critical payloads.

Heat Effects

Rocket design faces significant challenges in managing extreme temperatures generated during various stages of operation [147, 148]. The primary sources of heat are combustion in the engine, aerodynamic heating during atmospheric flight, and re-entry into the Earth's atmosphere for reusable rockets [147].

During launch, the rocket engines produce immense heat, with temperatures in the combustion chamber exceeding 3000-6000°C depending on the fuel and oxidizer [147]. Materials used in the engine components, such as Inconel, titanium alloys, or regeneratively cooled copper chambers, must withstand these extreme conditions without melting or deforming [147]. Engineers implement thermal management systems, including regenerative cooling, where cryogenic fuels are circulated around the engine nozzle to absorb heat before combustion [147].

As the rocket ascends through the atmosphere, high-speed motion causes air molecules to compress and generate frictional heat on the vehicle's exterior [147]. This heating effect is most pronounced during Max Q, the point of maximum dynamic pressure [147]. Materials like ablative coatings or heat-resistant composites, such as carbon-carbon, are used to protect the surface from structural weakening caused by thermal stress [147].

For reusable rockets or spacecraft returning to Earth, re-entry generates extreme temperatures due to compression and friction with the atmosphere [147, 148]. Reusable vehicles like SpaceX's Falcon 9 and crewed capsules like NASA's Orion require advanced thermal protection systems (TPS) [147, 148]. Ablative materials, ceramic tiles, or reinforced carbon-carbon shields are common solutions that dissipate heat while protecting underlying structures [147, 148].

Addressing heat effects requires rigorous testing and material selection, ensuring that thermal stresses do not cause deformation, cracking, or failure of critical components [147, 148]. Researchers have studied the effects of surface discontinuities, chemical reactions, and microstructural changes in materials under thermal stress [147, 148]. Numerical simulations and experimental investigations, such as arc-jet testing, are crucial for evaluating the performance of thermal protection systems [148-150].

Vibration Effects

Vibrations during rocket launch and flight pose a critical challenge in rocket design. Rockets experience intense vibrations primarily due to engine thrust, combustion instabilities, and aerodynamic forces [151, 152].

Engine-Induced Vibrations: The powerful thrust generated by rocket engines creates significant vibrations that propagate through the rocket's structure. Combustion instabilities—fluctuations in pressure within the engine—can exacerbate these vibrations, posing risks to sensitive systems. To mitigate this, engineers optimize engine designs, using dampening systems or baffles to stabilize combustion [152, 153].

Aerodynamic Forces: During the ascent phase, aerodynamic interactions with the vehicle's surface cause oscillations and flutter, particularly around thin or protruding structures such as fins and fairings. Aerodynamic shaping and reinforced structural supports are used to prevent flutter-induced damage [154, 155].

Resonance and Structural Fatigue: If vibrations align with the natural frequency of the rocket or its components, resonance can occur, amplifying the vibrations and causing catastrophic failures. Engineers conduct modal analysis to identify these natural frequencies and design the rocket to avoid resonance. Reinforcements such as stiffeners, ribs, and vibration dampers are incorporated into critical sections to improve stability [156, 157].

Vibration testing is an integral part of the rocket construction process, often performed on vibration tables to simulate launch conditions. Special care is taken to protect sensitive payloads, like satellites and scientific instruments, using shock-absorbing mounts and isolators to minimize the impact of mechanical stresses [158, 159].

Vacuum Effects

As rockets transition from Earth's atmosphere to space, they encounter the unique challenges of the vacuum environment. These challenges require specialized design and construction considerations to ensure the successful operation of the rocket systems. The key challenges and design considerations are as follows:

Material Outgassing: In a vacuum, certain materials can release trapped gases, a phenomenon known as outgassing [137]. This can lead to contamination of sensitive payloads, such as optical lenses or electronic components. Engineers select low-outgassing materials, such as epoxies, polymers, and adhesives, that are specially formulated for space use and comply with NASA or ESA outgassing standards [137].

Thermal Regulation: In the vacuum of space, heat transfer occurs primarily through radiation, as there is no air for conduction or convection [160]. This means that areas exposed to sunlight can experience extreme heating, while shaded regions can plunge to cryogenic temperatures. Engineers use thermal coatings, multi-layer insulation (MLI), and radiators to regulate temperatures and prevent overheating or freezing of critical systems [160].

Structural Considerations: Without atmospheric pressure, external loads on the rocket's structure decrease, but internal pressure differences remain significant [161]. Propellant tanks and pressurized systems must be carefully designed to maintain structural integrity in a vacuum. Thin-walled tanks made from lightweight, high-strength materials like aluminium-lithium alloys are commonly used to balance weight and strength [161].

Engine Efficiency: Rocket engines are more efficient in a vacuum because there is no atmospheric resistance to exhaust gases [162]. For this reason, upper-stage engines are equipped with larger nozzles to expand exhaust gases more efficiently in the absence of air pressure. This improves specific impulse (fuel efficiency) and thrust output [162].

Component Reliability: In a vacuum, lubricants can evaporate or freeze, and electronics are prone to malfunction due to radiation exposure [163]. Engineers address these challenges using vacuum-compatible lubricants, radiation-hardened electronics, and hermetically sealed systems to ensure reliability during extended space missions [163].

Chapter 5

Types of Rockets

Rocket designers make a strategic choice between single-stage and multi-stage rockets based on the mission objectives, technical requirements, and cost-effectiveness of the launch vehicle. Each design offers unique advantages and disadvantages, which are carefully evaluated to meet the specific needs of the mission. Here are the key considerations that influence this decision:

Mission Objectives and Performance Requirements: The primary factor in selecting between single-stage and multi-stage rockets is the mission's performance requirements, including payload mass, target orbit, and delta-v (change in velocity). Reaching low Earth orbit (LEO) or interplanetary destinations typically requires extremely high velocities that are challenging for single-stage rockets to achieve due to the constraints of the rocket equation. Multi-stage rockets excel in such scenarios because they discard unnecessary mass (spent stages) during the ascent, enabling each subsequent stage to operate more efficiently.

For missions with lower velocity requirements or smaller payloads, such as suborbital flights or lightweight satellites, single-stage rockets may be sufficient. These simpler designs reduce operational complexity and manufacturing costs, making them an attractive option for specific use cases like sounding rockets.

Mass and Efficiency Considerations: A critical limitation of single-stage rockets is their **wet-to-dry mass ratio**, governed by the Tsiolkovsky rocket equation. To achieve orbital velocity, a single-stage rocket requires an extremely high proportion of propellant relative to its dry mass (structural components, engines, and payload). This leaves little room for payload capacity, especially for heavier payloads or higher orbits.

Multi-stage rockets address this limitation by splitting the delta-v requirement into smaller segments. Each stage is optimized for its specific phase of flight, and spent stages are jettisoned to reduce dead weight. This results in a higher overall payload capacity and efficiency for missions requiring significant energy, such as reaching geostationary orbits or conducting interplanetary transfers.

Complexity vs. Simplicity: Single-stage rockets are simpler to design, manufacture, and operate because they lack the staging mechanisms required in multi-stage designs. This simplicity reduces the likelihood of failure during staging events, making single-stage rockets inherently more reliable for certain missions. They also have fewer moving parts, which lowers maintenance and manufacturing costs.

Multi-stage rockets, while more efficient, introduce added complexity. Staging events, such as separation and ignition of upper stages, are critical points of potential failure. Designers must incorporate robust systems to ensure safe and reliable transitions between stages. This complexity increases both the development time and operational costs of multi-stage rockets, but the benefits in payload capacity and performance often outweigh these drawbacks for high-energy missions.

Cost Considerations: Cost is a significant factor in the selection process. Single-stage rockets have lower development and manufacturing costs because they require fewer components and less advanced technology. They are ideal for budget-sensitive missions with limited performance demands, such as educational or research-based suborbital launches.

Multi-stage rockets, while more expensive to develop and manufacture, offer better cost efficiency for high-performance missions. By optimizing each stage for a specific phase of flight, these rockets deliver larger payloads or reach higher orbits, reducing the cost per kilogram of payload delivered. Moreover, modern advancements in reusable stages (e.g., SpaceX's Falcon 9 first stage) have made multi-stage rockets more cost-effective over multiple launches.

Reusability and Sustainability: The push for sustainable space exploration has influenced the choice between single- and multi-stage rockets. Reusable single-stage rockets, such as the hypothetical single-stage-to-orbit (SSTO) vehicles, could revolutionize space travel by eliminating the need for disposable stages. However, no practical SSTO design has been demonstrated due to the extreme performance requirements.

In contrast, reusable multi-stage rockets, like the Falcon 9 or Starship, have already proven their economic and operational viability. By recovering and reusing lower stages, these rockets significantly reduce costs while retaining the performance benefits of staging. This balance of reusability and efficiency often makes multi-stage rockets the preferred choice for modern space missions.

Classification of Rockets by Type

Rockets can be classified based on several criteria, including their propellant type, stage configuration, mission objectives, and launch environment. These classifications provide a framework for understanding the wide variety of rockets used in different applications, from military and scientific endeavours to commercial space exploration. Below is a detailed explanation of the primary categories by which rockets are classified [106, 164-166]:

1. By Propellant Type

The type of propellant used significantly impacts a rocket's performance, cost, and design. Rockets are commonly divided into the following categories based on their propulsion systems:

- **Solid-Propellant Rockets:** These rockets use a solid mixture of fuel and oxidizer that is molded into a specific shape. Solid rockets are simple, reliable, and capable of delivering high thrust, making them suitable for military applications, space launch boosters, and small payload missions. Examples include the Space Shuttle's Solid Rocket Boosters (SRBs) and military missiles like the Minuteman III.

- **Liquid-Propellant Rockets:** These rockets use liquid fuel and oxidizer stored in separate tanks, mixed and ignited in a combustion chamber. They offer high efficiency and controllability, making them ideal for space exploration and orbital missions. They can be further divided into:
    - **Cryogenic Rockets** (e.g., using liquid hydrogen and liquid oxygen, as seen in the Space Shuttle's main engines).
    - **Hypergolic Rockets** (e.g., using fuel-oxidizer combinations that ignite on contact, such as in the Apollo Lunar Module).
- **Hybrid-Propellant Rockets:** These rockets combine a solid fuel with a liquid or gaseous oxidizer. Hybrid designs offer a balance between the simplicity of solid rockets and the controllability of liquid rockets. They are used in experimental and commercial applications, such as Virgin Galactic's SpaceShipTwo.
- **Electric Rockets:** Electric propulsion systems, such as ion thrusters and Hall-effect thrusters, use electricity to accelerate charged particles, providing low but efficient thrust. These are primarily used for in-space propulsion, such as on communication satellites and deep-space probes.

2. By Stage Configuration

The number of stages in a rocket influences its design and mission profile. Stages are sections of the rocket that detach after their fuel is spent.

- **Single-Stage Rockets:** These rockets consist of a single stage that carries all the propellant and payload. While simpler and more reliable, they are limited in performance and are typically used for suborbital missions or small payloads.
- **Multi-Stage Rockets:** Multi-stage rockets use two or more stages, each with its engines and propellant. The spent stages are jettisoned during flight to reduce weight, allowing the remaining stages to accelerate more efficiently. Common configurations include:
    - **Two-stage-to-orbit (TSTO):** Used for payload delivery to low Earth orbit (LEO).
    - **Three-stage-to-orbit:** Necessary for higher orbits or interplanetary missions.
    - **Four-stage rockets:** Rare but used for specialized missions requiring extreme velocity, such as escaping Earth's gravity entirely.

3. By Mission Objectives

Rockets are classified by the purpose they serve, which influences their design and payload configuration.

- **Military Rockets:** These rockets are designed for defence and offense, including ballistic missiles and anti-missile systems. Examples include intercontinental ballistic missiles (ICBMs) like the Trident II.
- **Scientific Rockets:** These rockets are used for research and exploration. Subcategories include:
    - **Sounding Rockets:** Small rockets for atmospheric and suborbital studies.
    - **Space Exploration Rockets:** Vehicles like the Saturn V and Space Launch System (SLS), designed for interplanetary missions and deep-space exploration.
- **Commercial Rockets:** These rockets are designed for economic purposes, such as launching satellites, conducting space tourism, or delivering cargo to the International Space Station. Examples include SpaceX's Falcon 9 and Rocket Lab's Electron.
- **Experimental Rockets:** These are used to test new technologies, propulsion systems, or aerodynamic designs. Examples include Blue Origin's New Shepard and SpaceX's Starship prototypes.

4. By Launch Environment

The environment in which a rocket operates determines its design and classification.

- **Ground-Launched Rockets:** These rockets are launched from stationary platforms on the ground, often from dedicated spaceports. Examples include the Falcon 9 and Atlas V.
- **Air-Launched Rockets:** Air-launched rockets are carried to high altitudes by an aircraft before ignition. This approach reduces drag and saves energy. Examples include Northrop Grumman's Pegasus and Virgin Orbit's LauncherOne.
- **Sea-Launched Rockets:** Sea launches are conducted from ocean platforms or submarines. They provide flexibility in launch locations and are used for both military and commercial purposes. Examples include the Russian Shtil and Sea Launch's Zenit-3SL.
- **Space-Launched Rockets:** Rockets launched from space are used for interplanetary travel or satellite deployment. An example is a spacecraft using its onboard rocket stage to transfer to higher orbits or escape trajectories.

5. By Reusability

Modern rocketry emphasizes reusability to reduce costs and improve sustainability.

- **Expendable Rockets:** Traditional rockets, like the Delta IV and Ariane 5, are designed for single use. Once their mission is complete, their stages are discarded and are not recovered.
- **Reusable Rockets:** Reusable rockets, like SpaceX's Falcon 9 and Starship, are designed to return stages for refurbishment and reuse, drastically lowering the cost per launch.

6. By Size and Payload Capacity

Rockets can also be classified based on their physical size and the weight of the payload they can deliver.

- **Small Rockets:** Designed for lightweight payloads, typically under 2,000 kg. Examples include Rocket Lab's Electron and Astra's Rocket 3.
- **Medium-Lift Rockets:** Capable of carrying payloads between 2,000 and 20,000 kg. Examples include the Soyuz rocket and SpaceX's Falcon 9.
- **Heavy-Lift Rockets:** These rockets can carry payloads above 20,000 kg to LEO. Examples include the Delta IV Heavy and Ariane 5.
- **Super Heavy-Lift Rockets:** Rockets designed for very large payloads, often exceeding 50,000 kg, such as NASA's SLS and SpaceX's Starship.

Rockets are classified based on their propellant type, stage configuration, mission objectives, launch environment, reusability, and size. Each type serves a distinct role in advancing space exploration, defence, or commercial activities. By carefully analysing mission requirements and technical constraints, rocket designers select the appropriate type to optimize performance and cost-efficiency. This classification framework highlights the diversity and specialization within modern rocketry, enabling tailored solutions for various applications.

Single-Stage and Multi-Stage Rockets

The main difference between single-stage and multi-stage rockets is that multi-stage rockets can reach the high speeds required for space flight, while single-stage rockets face significant challenges in achieving this.

Multi-stage rockets have multiple stages that fire in succession, with each stage jettisoned after it has expended its fuel. This design allows for a reduction in weight as the rocket ascends, enabling it to achieve the necessary velocity for orbital insertion [167, 168]. Multi-stage rockets are more complex to design and launch than single-stage rockets, requiring precise timing between stages [167, 168].

On the other hand, single-stage-to-orbit (SSTO) vehicles are designed to reach orbit from the surface of a body without dropping any hardware. However, no Earth-launched SSTO vehicles have successfully flown yet [169, 170]. The physical difficulty of designing a fully reusable SSTO launch vehicle is exacerbated by the significant coupling between various design disciplines, where nearly every subsystem design decision has far-reaching consequences [167, 171].

Many papers have discussed the challenges of designing a viable SSTO launch vehicle [167, 170-174]. A key challenge is that achieving orbit with a practical single-stage vehicle using a pure chemical rocket is a marginal possibility [175]. Researchers have explored various approaches to address this, such as using deeply precooled air-breathing engines combined with a rocket mode of operation for the later stages of ascent [175], or utilizing high-altitude launch sites to provide an advantage for SSTO vehicles [176].

In contrast, multi-stage rockets are a more established and viable approach for reaching the high speeds required for space flight [167, 168, 172, 173]. The use of multiple stages allows the rocket to shed weight as it progresses, enabling it to achieve the necessary velocity for orbital insertion [167, 168].

A multistage rocket, also known as a step rocket, is a launch vehicle that uses two or more stages, each equipped with its own engines and propellant. These stages are either stacked on top of each other in a tandem or serial configuration or arranged side-by-side in a parallel configuration. Effectively, a multistage rocket is a combination of multiple rockets, each contributing to the overall mission by firing sequentially or simultaneously, depending on the design. While two-stage rockets are common, rockets with as many as five stages have been successfully launched, showcasing the versatility and complexity of multistage designs.

The primary advantage of a multistage rocket lies in its ability to shed mass as it ascends. When a stage exhausts its propellant, it is jettisoned, significantly reducing the weight of the remaining rocket. This reduction in mass allows the remaining stages to achieve greater acceleration and efficiency with their available thrust. Furthermore, each stage can be optimized for the specific atmospheric conditions it encounters. For instance, the first stage, operating in the dense lower atmosphere, is designed for maximum thrust to overcome drag and gravity. In contrast, upper stages are tailored for high-altitude or vacuum conditions, often featuring larger nozzles and engines optimized for greater efficiency in low-pressure environments. This staged approach allows the rocket to achieve the final velocity and altitude required for its mission with maximum efficiency.

In tandem or serial staging, the stages are arranged vertically, with the largest and most powerful stage at the bottom. Subsequent upper stages, which are smaller, sit atop the lower stages. The first stage ignites and lifts the entire rocket off the ground. Once its fuel is depleted, it detaches and falls away, leaving the second stage to ignite and continue the ascent. This process can be repeated with additional stages until the payload reaches its desired orbit or trajectory. In some advanced designs, the upper stage may ignite before separation to assist in the detachment process. This method, known as hot staging, uses the thrust from the upper stage to positively separate the spent lower stage, reducing the reliance on mechanical separation mechanisms.

Parallel staging, on the other hand, involves the use of additional boosters, often referred to as "stage 0," which are attached alongside the main first stage. These boosters, which can be either solid or liquid-fuelled, fire simultaneously with the first stage engines to provide additional thrust during liftoff. Once their fuel is exhausted, the boosters are jettisoned, typically using explosive bolts or separation charges, allowing the main stage to continue its operation. Parallel staging is particularly useful for heavy-lift vehicles, as it allows the rocket to carry larger payloads by supplementing the main stage's thrust during the most demanding phase of the ascent.

Staging is a critical process in multistage rockets, ensuring smooth transitions between stages and maximizing the rocket's efficiency. The separation of spent stages is often achieved using explosive bolts, separation charges, or mechanical interstage structures designed to disengage cleanly and avoid damage to the remaining stages. The sequencing of stage activation and detachment is carefully timed to maintain the rocket's trajectory and stability throughout the flight.

Multistage rockets are the only designs that have successfully achieved orbital speeds, as the energy required to reach orbit exceeds the capabilities of single-stage designs. While single-stage-to-orbit (SSTO) concepts remain an area of interest due to their simplicity and potential reusability, they have not yet been demonstrated due to the significant challenges in balancing fuel efficiency, structural integrity, and payload capacity. Multistage rockets, therefore, remain the backbone of modern space exploration and satellite deployment, offering a proven and reliable means of overcoming Earth's gravity and delivering payloads to space.

Multi-stage rockets are engineering marvels designed to overcome the physical limitations imposed by the classical rocket equation. This equation governs the relationship between a rocket's velocity change (delta-v), its mass, and the effective exhaust velocity of its engines. The rocket equation reveals that a single-stage rocket would require an impractically high fuelled-to-dry mass ratio to achieve the velocities necessary for orbital or interplanetary missions. Multi-stage rockets solve this problem by dividing the required delta-v into manageable fractions across multiple stages.

The classical rocket equation, $\Delta v = v_e \ln\left(\frac{m_0}{m_f}\right)$, illustrates how a rocket's velocity change depends on the exhaust velocity (v_e), the initial mass (m_0), and the final mass (m_f). Here, m_0 includes the rocket's structural mass, fuel, and payload, while m_f is the mass remaining after all fuel is burned. Achieving high delta-v requires a high initial-to-final mass ratio, which is limited by material constraints and the rocket's structural integrity.

By introducing multiple stages, each with its own propulsion system and fuel, multi-stage rockets reduce the mass progressively during flight. When a stage expends its fuel, its structural mass (now dead weight) is jettisoned, leaving the remaining rocket lighter and more efficient. This allows the subsequent stages to achieve greater acceleration with their own fuel reserves, solving the mass ratio limitation imposed by the rocket equation. For example, as each lower stage sheds its empty tanks and engines, the payload and upper stages continue to accelerate without the burden of spent components.

Multi-staging offers several key advantages. First, it allows each stage to be optimized for specific atmospheric and operational conditions. The lower stages, operating in Earth's dense atmosphere, use engines designed for high thrust and low specific impulse, ensuring sufficient power to overcome gravity and drag. Upper stages, operating in near-vacuum conditions, are equipped with engines optimized for high specific impulse, delivering superior fuel efficiency. This tailored approach maximizes the performance of each stage, making the entire rocket more efficient.

Additionally, multi-staging permits structural optimization. Lower stages bear the weight of all subsequent stages, requiring robust structural designs to withstand immense loads and stresses during liftoff. Conversely, upper stages are lighter and structurally simpler, as they only need to support their own mass and the payload. This reduces the total weight of the rocket, further enhancing efficiency. Multi-staging also minimizes fuel consumption by reducing the mass of each stage progressively, allowing the rocket to achieve higher velocities with less total propellant.

Despite its benefits, multi-staging introduces complexity to the rocket construction process. Each stage adds mechanical and operational intricacies, including the need for reliable stage separation mechanisms, ignition systems for upper stages, and interstage connections. Separation events are critical points of failure, as issues such as incomplete detachment, ignition failure, or stage collision can jeopardize the mission. Additionally, the engines and fuel of lower stages must lift not only the payload but also the structural mass of unused upper stages, reducing overall efficiency until those stages are jettisoned.

Furthermore, the need to optimize each stage for specific conditions requires sophisticated engineering and testing. Lower stages must endure aerodynamic forces and vibrations during ascent, while upper stages must operate reliably in the vacuum of space. Despite these challenges, the efficiency gains provided by multi-staging are so significant that nearly all rockets used for orbital missions employ this design.

Rocket efficiency is often measured by specific impulse (I_{sp}), defined as the thrust produced per unit rate of propellant consumption. Specific impulse reflects how efficiently a rocket engine converts fuel into thrust. Engines used in the lower stages of a multi-stage rocket typically prioritize high thrust over efficiency, enabling rapid acceleration through the dense lower atmosphere. In contrast, upper-stage engines are designed for higher specific impulse, as the reduced atmospheric drag at higher altitudes allows for greater emphasis on fuel economy.

Another important metric is the thrust-to-weight ratio (TWR), which indicates how much thrust a rocket produces relative to its weight. The TWR of a launch vehicle typically ranges from 1.3 to 2.0, ensuring enough thrust to lift the rocket off the ground while maintaining stability. These metrics, along with burn time, burnout height, and burnout velocity, are critical for designing each stage of a rocket and ensuring it performs optimally under its specific operating conditions.

Designing an efficient multi-stage rocket involves balancing several dimensionless performance ratios. These include the initial-to-final mass ratio (η), structural ratio (ϵ), and payload ratio (λ). The mass ratio quantifies how much of the initial mass is consumed as fuel, while the structural ratio measures the fraction of a stage's mass that consists of its empty structure. The payload ratio, on the other hand, indicates how much of the rocket's mass is devoted to the payload. Together, these ratios help engineers compare different rocket designs and optimize configurations for specific missions.

By analysing these ratios, engineers can identify trade-offs between structural integrity, fuel efficiency, and payload capacity. For example, increasing the structural strength of a stage may reduce its payload capacity, while maximizing the payload ratio may require sacrificing some structural durability. Advanced simulation tools and iterative design processes are used to refine these trade-offs and achieve the optimal balance for each mission.

Multi-stage rockets remain the only proven method for achieving orbital speeds, as single-stage-to-orbit designs have not yet been realized due to their stringent mass ratio requirements. While multi-staging is highly effective for reaching orbit and beyond, it is not without limitations. The added complexity of staging mechanisms increases development and manufacturing costs, while each separation event poses a potential risk to mission success. Nonetheless, the advantages of multi-staging, including reduced mass, optimized propulsion, and improved efficiency, outweigh these challenges, making it the standard approach for modern space exploration and satellite deployment.

Component Selection and Sizing for Multistage Rockets

The design and sizing of rocket components are fundamental steps in creating a functional and efficient multistage rocket. This process involves calculating the propellant mass, determining the structural weight, and balancing performance with practical considerations such as fuel tank volume and manufacturing constraints. The Saturn family of rockets, which famously carried the Apollo spacecraft to the Moon, serves as a prime example of how precise calculations and engineering trade-offs are applied in rocket design.

Propellant Mass Calculation: The first step in sizing a rocket involves calculating the amount of propellant needed for the mission. This is achieved using the rocket equation:

$$m_p = \frac{I_{\text{tot}}}{g I_{\text{sp}}}$$

Here, I_{tot} is the total impulse required (in Newton-seconds), g is Earth's gravitational constant, and I_{sp} is the specific impulse of the rocket engine. This equation provides the mass of propellant necessary to achieve the desired thrust and velocity change (delta-v) for the rocket's stage.

Once the propellant mass is determined, its volume can be calculated by dividing the mass by the fuel's density, which is typically well-documented for various propellants. The storage requirements for the fuel tanks are then derived from this volume. This step is crucial for determining the physical dimensions of the rocket and integrating the fuel tanks into the overall design.

Structural Mass Estimation: In addition to propellant mass, the rocket's structural mass must be accounted for, including the mass of engines, electronics, instruments, power systems, and other components. While precise values for these components are typically known during later design phases, early-stage designs use simplified approximations, often relying on mass fractions. For example, modern solid rocket motors generally consist of 91–94% propellant by total mass, with the remaining percentage representing the structural components. Designers also account for residual propellant, which refers to the small percentage of fuel left in the tanks after consumption. This is typically estimated at around 5% of the total propellant mass and is factored into performance calculations.

For preliminary estimates, the mass of small components such as stage separation initiators or safe-and-arm devices can often be considered negligible compared to the overall mass. However, these components are included in detailed design phases to refine the total system mass.

Liquid Bipropellant Systems: The design process becomes more intricate when dealing with liquid bipropellant systems. These rockets require two separate tanks—one for fuel and one for oxidizer. The relationship between the two is governed by the mixture ratio, defined as:

$$\frac{O}{F} = \frac{m_{\text{ox}}}{m_{\text{fuel}}}$$

Where m_{ox} is the oxidizer mass, and m_{fuel} is the fuel mass. The mixture ratio impacts not only the size of the tanks but also the rocket's specific impulse, which measures its fuel efficiency. Achieving the ideal mixture ratio involves balancing performance with practical design considerations.

For instance, the mixture ratio might be adjusted to achieve equal-sized fuel and oxidizer tanks, simplifying manufacturing and integration. While this may slightly compromise the specific impulse, the cost and time savings in manufacturing can outweigh the performance drawbacks. Alternatively, if the rocket's design constraints prioritize minimizing tank volume—such as when using low-density fuels like liquid hydrogen—the mixture ratio can be adjusted to favour an oxidizer-rich composition. This reduces tank size requirements, albeit at the cost of reduced engine efficiency.

Trade-Offs in Tank Design: The choice of fuel and oxidizer plays a critical role in defining the overall design of the tanks and the rocket. For example, using liquid hydrogen as fuel requires significantly larger tanks due to its low density. To counterbalance this, the mixture ratio can be oxidizer-rich, reducing the required fuel volume. Conversely, dense propellants like RP-1 (a refined kerosene) result in smaller tanks but require additional structural reinforcement to handle higher stresses during launch.

Engineers must also consider manufacturing and integration constraints. Equal-sized tanks are easier to construct and integrate, reducing complexity and cost. However, this design simplicity might lead to a slight reduction in performance if the ideal mixture ratio for specific impulse is sacrificed. The final choice of tank design and mixture ratio reflects a compromise between maximizing engine efficiency and meeting physical and operational constraints.

Balancing Efficiency and Practicality: Rocket sizing ultimately involves balancing competing factors, including performance, manufacturing ease, and mission constraints. Designers must optimize the relationship between structural mass, propellant mass, and payload capacity to ensure the rocket can achieve its desired performance goals within physical and economic limits. Modern design processes leverage advanced modelling and simulation tools to evaluate these trade-offs and refine initial estimates.

The Saturn rockets demonstrated these principles on a grand scale. Their designers carefully calculated propellant and structural requirements for each stage, incorporating trade-offs to maximize performance and reliability. The use of separate tanks for liquid oxygen and liquid hydrogen, optimized mixture ratios, and modular designs ensured that each stage met its specific requirements while contributing to the overall mission success.

Component selection and sizing are critical aspects of rocket design, requiring precise calculations and thoughtful trade-offs to achieve a balance between efficiency, performance, and practicality. From estimating propellant mass to determining tank sizes and adjusting mixture ratios, every decision impacts the rocket's structural layout and mission capabilities. By combining mathematical rigor with engineering intuition, designers create rockets that not only meet performance targets but also align with manufacturing and operational constraints, exemplifying the sophistication of modern aerospace engineering.

Optimal Staging and Restricted Staging in Rocket Design

The design and configuration of rocket stages directly impact the performance, efficiency, and cost-effectiveness of a launch vehicle. Two approaches to staging—optimal staging and restricted staging—provide frameworks for designing multistage rockets, each tailored to different engineering goals and constraints. Optimal staging prioritizes maximum efficiency and payload delivery, while restricted staging offers simplicity for conceptual designs or preliminary analysis.

Maximizing Payload Efficiency: The primary objective of optimal staging is to maximize the payload ratio, ensuring the largest possible payload reaches the desired burnout velocity while minimizing the non-payload mass. The non-payload mass includes the structural components, engines, and propellant of each stage. This approach assumes a correlation between the total rocket mass at liftoff and the cost of the launch, making efficiency critical for reducing overall mission expenses.

Optimal staging follows several guiding principles to achieve the best performance. Early stages, operating in the dense atmosphere near the Earth's surface, should use engines with lower specific impulse (I_{sp}) but high thrust, enabling the rocket to overcome gravity and atmospheric drag efficiently. Conversely, later stages, which operate in thinner or no atmosphere, should prioritize higher I_{sp}, sacrificing thrust for greater fuel efficiency. This strategy ensures that each stage contributes to the overall delta-v requirement in a manner suited to its specific operating conditions. Lower-I_{sp} stages should contribute proportionally less delta-v, reducing the need for excessive propellant in the earlier stages.

An important consideration in optimal staging is the decreasing size of successive stages. Each stage must not only propel the payload but also the mass of all subsequent stages, creating a cascading reduction in size as the rocket ascends. Additionally, stages are not necessarily designed to provide equal delta-v contributions. Instead, delta-v partitioning is determined through methods such as analytical algorithms or trial-and-error approaches, balancing efficiency with the practical constraints of the mission.

The overall payload ratio for a multistage rocket is the product of the individual payload ratios of each stage:

$$\lambda = \prod_{i=1}^{n} \lambda_i$$

Here, n is the number of stages, and λ_i represents the payload ratio of each stage. While similar stages with consistent ratios simplify calculations, achieving maximum efficiency often requires varied designs and delta-v contributions across stages. Engineers typically start by designing the final stage, calculating its initial mass (which serves as the payload for the preceding stage), and iteratively working backward to size all earlier stages.

Simplified Assumptions for Conceptual Design: Restricted staging simplifies the complexity of multistage rocket design by assuming consistent properties across all stages. Each stage is presumed to have the same specific impulse, structural ratio, and payload ratio, with only the total mass of each stage decreasing as the rocket ascends. While this assumption may not yield the most efficient design, it significantly simplifies the calculations for burnout velocity, burnout time, burnout altitude, and stage mass, making it an ideal approach for conceptual or preliminary designs.

One of the key insights from restricted staging is the relationship between the number of stages and the burnout velocity. As the number of stages increases while keeping all other parameters constant, the rocket achieves progressively higher burnout velocities. This occurs because each stage sheds its structural mass as it burns out, leaving less mass for the subsequent stages to accelerate. However, the principle of diminishing returns applies: as the number of stages grows, the incremental increase in burnout velocity decreases. The velocity gain asymptotically approaches a limit, meaning that beyond a certain point, adding more stages provides negligible improvement in performance.

Practical Constraints on Staging: While restricted staging illustrates the benefits of increasing stage numbers, real-world rockets rarely use more than three stages. This limitation arises from the practical challenges associated with adding stages. Each additional stage introduces greater structural complexity, weight, and failure risk, such as stage separation failures or ignition malfunctions. Moreover, the increased engineering and manufacturing costs outweigh the modest gains in burnout velocity beyond a certain point.

Optimal staging, on the other hand, requires careful tailoring of each stage to maximize efficiency and payload delivery. While more complex than restricted staging, it ensures the best use of resources for missions requiring high performance, such as launching payloads to geostationary orbit or conducting interplanetary missions.

Optimal staging and restricted staging represent two approaches to designing multistage rockets, each suited to different phases of the design process. Optimal staging maximizes payload efficiency by tailoring each stage's size, delta-v contribution, and specific impulse to its operating conditions. In contrast, restricted staging offers simplicity and ease of calculation, making it useful for conceptual designs or initial studies. Both approaches highlight the trade-offs between performance, complexity, and cost, underscoring the need for careful planning and engineering to achieve mission success.

Hot-Staging in Rocket Design and Operations

Hot-staging is a specialized type of rocket staging where the engines of the next stage ignite before the separation of the preceding stage is complete. Unlike conventional staging, where the next stage's engines fire only after detachment, hot-staging involves a brief overlap during which the spent stage is still partially attached. This method has distinct advantages in terms of simplicity, efficiency, and payload capacity, but also introduces unique engineering challenges.

In a hot-staging operation, the first stage (or preceding stage) throttles down its engines to reduce thrust as it nears the end of its burn. Before the complete separation of this stage, the engines of the next stage ignite. The partial overlap in operation ensures a seamless transition of thrust and eliminates any gaps in propulsion. The remaining thrust from the earlier stage assists in the separation process, pushing the spent stage away from the rocket. The acceleration provided by the nearly exhausted stage also helps settle the propellants at the bottom of the tanks in the next stage, removing the need for ullage motors that are typically required to position the propellants before ignition.

One of the primary benefits of hot-staging is the reduction in staging complexity. By using the thrust of the upcoming stage to aid separation, the need for additional separation mechanisms or ullage motors is eliminated. This simplification not only saves weight but also improves reliability by reducing potential points of failure during the staging process.

Hot-staging also provides a slight boost in payload capacity. The overlapping thrust allows the rocket to maintain a more continuous acceleration profile, avoiding the slight deceleration that may occur between stages in traditional staging methods. This continuous propulsion results in a marginal increase in overall efficiency, allowing for slightly heavier payloads or extended range.

Hot-staging has been employed in several rockets, particularly those from the Soviet-era Russian space program. Rockets such as the Soyuz and Proton-M make extensive use of this staging technique, demonstrating its reliability and effectiveness. The ill-fated N1 Moon rocket, while never achieving a successful mission, was also designed to utilize hot-staging, although none of its test flights progressed far enough to implement the technique.

Figure 38: A Soyuz booster rocket launches the Soyuz MS-11 spacecraft from the Baikonur Cosmodrome in Kazakhstan on Monday, Dec. 3, 2018. Defense Visual Information Distribution Service, Public Domain, via National Archives and Defense Visual Information Distribution Service.

The Titan family of rockets, beginning with the Titan II, also incorporated hot-staging into their designs. This adaptation allowed these rockets to achieve high performance and reliability for both civilian and military applications. More recently, SpaceX retrofitted their Starship rocket to include hot-staging after its first test flight. This made Starship not only the largest rocket ever to use hot-staging but also the first reusable launch vehicle to incorporate the technique, showcasing its potential to improve the performance of modern, reusable rocket systems.

Despite its advantages, hot-staging introduces specific engineering challenges. The ignition of the next stage's engines while the previous stage is still attached requires careful thermal and structural management. The exhaust from the ignited engines can damage the lower stage or the interstage structure. To mitigate this, rockets employing hot-staging often include protective measures, such as heat shielding or vented interstage structures, to direct exhaust gases away from vulnerable components.

Additionally, the timing of stage ignition and separation must be precisely controlled. A premature ignition can lead to structural damage, while a delay in separation can result in inefficiencies or loss of thrust. These timing requirements demand sophisticated control systems and rigorous testing to ensure safe and effective hot-staging operations.

Upper Stages: The High-Altitude and Space-Bound Workhorses

The upper stage of a rocket plays a critical role in the final phases of a mission, tasked with completing orbital injection or delivering payloads into higher energy orbits, such as geostationary transfer orbit (GTO) or even escape velocity for interplanetary missions. Unlike lower stages, upper stages are designed to operate in conditions with minimal or no atmospheric pressure, significantly influencing their design, propulsion systems, and functionality.

One of the defining features of upper stages is their adaptation for high-altitude and vacuum conditions. In the absence of atmospheric pressure, these stages can employ engine nozzles with optimal vacuum expansion ratios, maximizing the efficiency of exhaust gas expansion and thus improving the specific impulse of the engines. This contrasts with lower stages, which must contend with atmospheric drag and pressure, requiring smaller nozzles optimized for sea-level operation.

The combustion chambers of upper stages often operate at lower pressures compared to those in lower stages. This is because the absence of atmospheric pressure at high altitudes reduces the need for extremely high chamber pressures to maintain efficiency. These design choices allow engineers to prioritize lightweight structures and simpler engine systems, which are advantageous for maximizing payload capacity and reducing costs.

Upper stages utilize a variety of propulsion systems tailored to their specific mission requirements and the characteristics of the propellants they carry. Many upper stages rely on pressure-fed engines, which eliminate the need for complex turbopumps. In pressure-fed systems, such as those using hypergolic propellants like the Delta-K or the Ariane 5 ES second stage, pressurized gas forces the propellant into the combustion chamber. This simplicity enhances reliability, making these systems particularly attractive for missions requiring high precision, such as orbital adjustments or satellite positioning.

Other upper stages use advanced liquid-fuelled engines to achieve high efficiency. For example, the Centaur upper stage employs liquid hydrogen expander cycle engines, where the engine's heat is used to convert cryogenic hydrogen into a gas, which powers the turbopumps. Similarly, the Delta Cryogenic Second Stage (DCSS) uses a similar expander cycle engine. In contrast, some systems, like the Ariane 5 ECA's HM7B engine or the Saturn V's S-IVB stage with its J-2 engine, operate using gas generator cycles. These engines burn a small amount of fuel to drive the turbopumps and are optimized for delivering high thrust at high efficiency in vacuum conditions.

Figure 39: Workers monitor the second stage of a Delta IV as a crane lifts it from its transporter in the Horizontal Integration Facility at Launch Complex 37 on Cape Canaveral Air Force Station in Florida. NASA, Public Domain, via Picryl.

The primary function of an upper stage is to perform orbital injection or to accelerate payloads into desired trajectories beyond Earth's orbit. For example, upper stages are often tasked with delivering satellites into GTO or transferring scientific payloads into heliocentric or interplanetary trajectories. This requires precise control over thrust, burn duration, and trajectory adjustments, which upper stages achieve through advanced propulsion and guidance systems.

Certain upper stages, known as space tugs, specialize in transferring payloads from one orbit to another. For instance, the Fregat upper stage, commonly used with Soyuz rockets, is designed to move payloads from low Earth orbit (LEO) to GTO or beyond. Space tugs like Fregat are equipped with additional manoeuvring capabilities, enabling them to perform complex orbital transfers or deploy multiple payloads into distinct orbits.

While upper stages benefit from operating in a vacuum, they face unique challenges. The lack of atmospheric pressure necessitates precise thermal management, as radiative cooling becomes the primary mechanism for dissipating heat. Additionally, cryogenic propellants like liquid hydrogen and liquid oxygen must remain at extremely low temperatures throughout the mission, requiring robust insulation and cooling systems to prevent boil-off.

Another challenge is the need for precise guidance and control. Upper stages often carry sophisticated navigation and avionics systems to ensure accurate orbital insertion and payload delivery. These systems must account for the gravitational influences of celestial bodies, the rocket's velocity, and any deviations from the planned trajectory.

Rocket Staging Systems in Use

This following table provides a clear comparison of the various staging systems, their characteristics, and notable examples.

Staging System	Description	Examples
Two-stage-to-orbit (TSTO)	A spacecraft system using two distinct stages for propulsion, achieving orbital velocity. Intermediate between a three-stage-to-orbit launcher and a hypothetical single-stage-to-orbit (SSTO) launcher.	- NASA-ESA Mars Ascent Vehicle (MAV) (planned launch from Mars in 2028)
Three-stage-to-orbit	A commonly used rocket system with three distinct stages providing propulsion consecutively to achieve orbital velocity. Intermediate between a four-stage-to-orbit launcher and a two-stage-to-orbit launcher.	- Saturn V - Vanguard - Ariane 4 (optional boosters) - Ariane 2 - GSLV (three stages and boosters) - PSLV (four stages) - Zenit-3SL - Proton (optional fourth stage) - Long March 5 (optional boosters and third stage)
Examples of two stages with boosters	These designs include strap-on boosters ("stage-0") with two core stages. Boosters and first stage fire simultaneously, with boosters jettisoned a few minutes into flight to reduce weight.	- US Space Shuttle (SRB + External Tank + OMS) - Angara A5 - Falcon Heavy - Ariane 5 - Atlas V 551 - H-IIA, H-IIB - Space Launch System - Soyuz - Long March 2E, 2F, 3B - Delta II/III - Titan IV
Four-stage-to-orbit	A system using four distinct stages for propulsion consecutively to achieve orbital velocity. Most commonly associated with solid-propellant launch systems.	- Ariane 1 - PSLV - Minotaur IV - Minotaur V (five stages)

Staging System	Description	Examples
		- ASLV (five stages) - Proton (optional fourth stage)
Examples of three stages with boosters	Includes strap-on boosters ("stage-0") with three core stages. Boosters and first stage fire simultaneously, jettisoned a few minutes into flight to reduce weight.	- Long March 5 (optional boosters and optional third stage)

Reusable Vs. Expendable Rockets

The design philosophy of a rocket—whether reusable or expendable—has significant implications for cost, performance, engineering complexity, and the sustainability of space exploration. Reusable and expendable rockets both have their own advantages and disadvantages, and the choice between them depends on the specific mission requirements, budget constraints, and technological capabilities [177-179].

Expendable rockets are designed for single-use operations, where the entire rocket or its individual stages are discarded after fulfilling their purpose [177, 179]. This traditional approach has been the backbone of space exploration for decades, powering missions like the Apollo Moon landings (Saturn V), Mars exploration (Delta IV), and satellite deployments (Ariane 5) [177, 179]. The key advantage of expendable rockets is their simplicity in design, as engineers can focus entirely on maximizing performance and payload capacity without accounting for recovery or refurbishment systems [177, 179]. This allows them to carry heavier payloads relative to their total weight, making them well-suited for high-energy missions requiring significant delta-v, such as interplanetary missions or payloads destined for geostationary orbits [177, 179]. However, expendable rockets come with the notable downside of high operational costs, as a new rocket must be built for every launch, resulting in significant material, manufacturing, and labour costs [177, 179]. Additionally, the environmental impact of discarded rocket stages raises sustainability concerns [180].

Several countries and organizations continue to rely on ELVs for their space programs:

1. **Arianespace**: The Vega C and Ariane 6 are leading examples of expendable European rockets, optimized for small to medium payloads.
2. **China**: China's Long March series, including the heavy-lift Long March 5 and quick-response Long March 11, remain at the forefront of its space program.
3. **United States**: NASA's Space Launch System (SLS) and ULA's Atlas V represent high-profile ELV programs for crewed and scientific missions.
4. **India**: ISRO's GSLV and PSLV rockets have been successful in deploying satellites and conducting exploratory missions.

5. **Russia**: The Soyuz and Proton rockets have been reliable workhorses, with the Proton offering heavy-lift capabilities.
6. **Other Nations**: Countries like Iran and Israel also rely on expendable systems like the Safir and Shavit rockets for satellite launches.

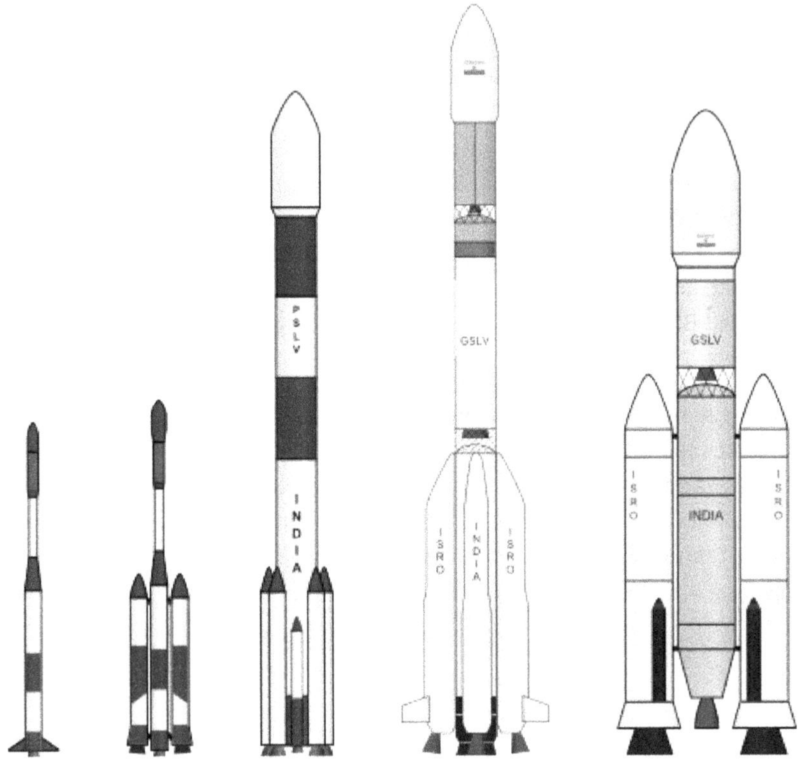

Figure 40: Comparison of Indian carrier rockets. Left to right: SLV, ASLV, PSLV, GSLV, LVM 3. GW_Simulations, Attribution, via Wikimedia Commons.

On the other hand, reusable rockets are designed to return key components to Earth for refurbishment and reuse in subsequent missions [177-179, 181]. This innovative approach has been popularized by companies like SpaceX, whose Falcon 9 and Starship rockets have revolutionized spaceflight economics [177, 179, 181, 182]. The primary advantage of reusable rockets is cost-efficiency, as the reuse of stages can significantly reduce launch costs, enabling more frequent and affordable launches [177, 179, 181]. This cost reduction has opened up new opportunities for commercial space ventures, including satellite megaconstellations (e.g., Starlink) and space tourism [177, 182]. Additionally, reusable systems promote sustainability, as fewer materials are wasted, and discarded stages no longer contribute to space debris or environmental pollution [180]. Reusable rockets also enable faster launch cadences, as turnaround times between launches are reduced with optimized refurbishment processes [177, 179, 181]. However, reusability comes with engineering challenges, as designing rockets to withstand the stresses of both launch and recovery requires heavier structures and additional systems for guidance, thermal protection, and landing, which can reduce payload capacity compared to expendable rockets, especially for high-energy missions [177, 179, 183, 184].

The choice between reusable and expendable rockets often depends on the specific mission profile [177, 179, 181]. Expendable rockets excel in missions requiring maximum payload capacity or those involving destinations where recovery is impractical, such as deep-space missions [177, 179]. For example, NASA's Space Launch System (SLS) is an expendable rocket designed to support crewed lunar missions and beyond, where every kilogram of payload is crucial [177]. Reusable rockets, on the other hand, are ideal for commercial satellite deployments, cargo resupply missions, and missions to low Earth orbit (LEO), where recovery and refurbishment are feasible [177, 179, 181, 182]. For instance, SpaceX's Falcon 9 and Falcon Heavy have demonstrated that reusability is highly effective for frequent, lower-cost missions [177, 179, 182]. Additionally, reusable systems like SpaceX's Starship and Blue Origin's New Shepard are paving the way for future space tourism and interplanetary exploration [177, 179].

Reusable rockets have reshaped the economics of spaceflight, enabling private companies and smaller nations to access space affordably [177, 179, 181]. By lowering the cost per kilogram of payload, reusability has spurred growth in industries like satellite internet, remote sensing, and Earth observation [177, 179]. This shift has also encouraged innovation, with companies investing in advanced recovery technologies like propulsive landing, parachute systems, and aerodynamic gliding [177, 179].

Figure 41: SpaceX Starship booster during it's landing approach to the tower on IFT-5. Steve Jurvetson, CC BY 2.0, via Wikimedia Commons.

Expendable rockets, while more costly on a per-launch basis, remain essential for missions requiring maximum performance, as their simpler design allows them to focus entirely on achieving higher payload capacities or navigating complex trajectories, making them indispensable for specific scientific and exploratory missions [177, 179, 185, 186].

Reusable System

A Reusable Launch Vehicle (RLV) is designed with the capability to recover and reuse key components, such as rocket stages, engines, and boosters, while delivering payloads from Earth to outer space. This approach differs from expendable launch vehicles, which are discarded after a single use. The primary benefit of RLVs lies in their potential to significantly reduce the cost of launches, as reusable components eliminate the need for manufacturing new parts for every mission. However, these benefits are tempered by the added complexities of recovery, refurbishment, and the increased mass associated with reusable systems.

Reusable rockets often include additional avionics, heat shields, and propellant, making them heavier than expendable counterparts. The need to withstand re-entry and navigate the atmosphere necessitates features such as grid fins, retrorockets, and sometimes parachutes. These components ensure that reusable stages or spacecraft can decelerate and be controlled for safe recovery. Some designs, such as spaceplanes, leverage aerodynamic principles to glide back to Earth, requiring specialized recovery infrastructure like runways. Vertical landing systems, such as those used by SpaceX, depend on retrorockets and landing legs, often requiring precise autonomous navigation. These systems may also demand extensive ground infrastructure, including facilities like autonomous drone ships or mass drivers, which accelerate vehicles at launch.

The concept of reusable launch vehicles dates back to science fiction of the early 20th century, but it became a reality in the 1970s with the development of the Space Shuttle. While the Space Shuttle was partially reusable, its operational and refurbishment costs ultimately fell short of expectations, limiting its cost-effectiveness. During the late 20th century, interest in reusable systems declined, with many concepts remaining at the prototype stage. The resurgence of private space companies in the 2000s and 2010s, such as SpaceX and Blue Origin, revolutionized the development and deployment of reusable systems. Vehicles like SpaceShipOne, Falcon 9, and New Shepard demonstrated that reusability could significantly lower costs while maintaining reliability. Today, upcoming systems such as Starship, New Glenn, and Ariane Next further cement reusability as a standard in the space industry.

Fully reusable launch systems are designed to recover and reuse all major components, including all rocket stages. As of 2024, SpaceX's Starship represents the most advanced example of this concept, having successfully completed multiple test flights that included recovering both the Super Heavy booster and the Starship upper stage. Other companies, such as Blue Origin with Project Jarvis and Stoke Space, are also pursuing fully reusable systems. Fully reusable designs aim to minimize operational costs while maximizing the lifespan of rocket components, although this requires advanced engineering to handle the stress of multiple launches and landings.

Most reusable systems in operation today are partially reusable. These vehicles typically recover and reuse the first stage or specific components, while upper stages remain expendable. For example, SpaceX's Falcon 9 reuses its first-stage boosters, and the Space Shuttle reused its solid rocket boosters and main engines but discarded the external fuel tank. ULA's Vulcan Centaur plans to reuse its first-stage engines via splashdown recovery, demonstrating another partially reusable approach.

The recovery process for reusable rockets introduces design challenges and operational costs. Systems must include additional mass for landing mechanisms, such as heat shields, grid fins, or parachutes, which can reduce payload capacity. After recovery, components often require inspection and refurbishment to ensure they are flight-ready, a process that can vary in cost and complexity depending on the system. For instance, SpaceX has demonstrated rapid turnaround times for Falcon 9 boosters, achieving up to 22 flights with a single engine. However, older systems like the Space Shuttle required extensive refurbishment, limiting cost savings.

Reusable systems employ various recovery techniques to ensure safe landings. Parachutes and airbags, as used in the Space Shuttle's solid rocket boosters, allow for controlled deceleration and impact. Horizontal landing systems, such as those used by spaceplanes like the Shuttle Orbiter, enable gliding landings on runways. Vertical landing systems, pioneered by SpaceX with the Falcon 9, use retrorockets for precise, powered landings, requiring additional fuel reserves but offering high reusability potential.

Emerging technologies, such as inflatable heat shields, aim to expand the reusability of components traditionally discarded, such as engines and tanks. For example, retrofitting expendable systems like the Space Launch System (SLS) with inflatable heat shields could enable engine recovery, reducing costs. Aerodynamic innovations, such as lifting body designs and delta wings, further enhance the efficiency of reusable systems by reducing landing mass penalties.

The rise of reusable launch vehicles has fundamentally transformed the space industry. By reducing launch costs, they have enabled a higher cadence of missions, expanded access to space, and opened new commercial opportunities, such as satellite constellations and space tourism. Infrastructure upgrades, such as those at Cape Canaveral, reflect the industry's commitment to supporting reusable systems. As technology advances, fully reusable systems are expected to dominate the market, driving further innovation and lowering the barriers to space exploration.

Vertical Launch Vs. Horizontal Launch Systems

Vertical launch systems are the most traditional and widely used method for rocket launches. In this configuration, the rocket is positioned upright on a launch pad and takes off vertically, propelled by powerful engines that overcome gravity and atmospheric drag. This method is favoured for launching payloads to orbit because it allows the rocket to ascend quickly through the densest part of the atmosphere, minimizing drag-related energy losses.

A vertical launch system requires a robust infrastructure, including a launch pad, fuelling systems, and integration towers to assemble and service the rocket. This setup supports large, heavy-lift rockets designed to carry significant payloads into space. Vertical launches are efficient for reaching orbit because the rocket expends most of its energy directly countering gravity and then transitioning to a horizontal trajectory for orbital insertion.

One advantage of vertical launch systems is their scalability. They can accommodate a wide range of rocket sizes, from small satellite launchers to massive heavy-lift rockets like the Saturn V or SpaceX's Starship. Moreover, vertical launches allow rockets to carry their full fuel load and payload without the need for additional structural support to counter lateral forces during take-off. However, vertical launch systems require precise engineering and control systems to ensure stability and alignment during ascent.

Horizontal launch systems, on the other hand, involve a vehicle that takes off horizontally, much like an airplane. This approach is typically used with air-launched rockets or spaceplanes. In an air-launch scenario, a carrier aircraft carries the rocket to a high altitude before releasing it, at which point the rocket ignites and ascends to orbit. Spaceplanes, such as the Space Shuttle or Virgin Galactic's SpaceShipTwo, combine aviation and rocketry, allowing for horizontal take-off or landing.

Horizontal launches offer several advantages, especially for smaller payloads or reusable systems. By starting at higher altitudes, air-launched rockets avoid the densest part of the atmosphere, reducing drag and saving fuel. Horizontal launches also allow for greater flexibility in launch locations since they do not require fixed infrastructure like a vertical launch pad. This can be particularly advantageous for military or commercial operations that require rapid deployment or varied trajectories.

Spaceplanes and horizontally launched systems often feature reusable designs, which reduce overall costs. These systems can return to Earth and land on conventional runways, simplifying recovery and refurbishment. However, horizontal launches are generally limited in payload capacity due to the structural and aerodynamic constraints of the carrier aircraft or spaceplane.

The choice between vertical and horizontal launch systems depends on the mission requirements and the scale of the payload. Vertical launches are best suited for heavy-lift missions, such as launching large satellites, space station modules, or crewed spacecraft. Horizontal launches are ideal for smaller payloads, experimental missions, or space tourism, where reusability and cost-efficiency are priorities.

Vertical systems require extensive infrastructure, including launch pads and ground support systems, while horizontal systems rely on runways and carrier vehicles. This makes vertical systems less flexible in terms of launch locations but better suited for large-scale missions. Horizontal systems, by contrast, can operate from a variety of airfields, offering more operational flexibility.

Some systems combine elements of both vertical and horizontal launches. For example, systems like the Pegasus rocket are air-launched horizontally from an aircraft but ascend vertically after ignition. SpaceX's Starship is another example; it launches vertically but incorporates horizontal manoeuvring capabilities during re-entry and landing.

As space exploration evolves, both vertical and horizontal launch systems are likely to coexist, serving complementary roles. Vertical launches will remain the cornerstone of heavy-lift operations, while horizontal launches will expand accessibility and reduce costs for smaller payloads and reusable missions. Advances in propulsion, materials, and design may further blur the lines between these two approaches, enabling innovative hybrid systems that capitalize on the strengths of both methods.

Examples of Vertical, Horizontal, and Hybrid Launch Systems

Vertical Launch Systems

Vertical launches are the most common approach for space missions and are utilized by many governmental and commercial space agencies for their simplicity and effectiveness in reaching orbit.

1. NASA (United States):

- **Saturn V**: Used for Apollo moon missions, the Saturn V was a heavy-lift vertical launch system.
- **Space Launch System (SLS)**: NASA's flagship vertical launch vehicle for the Artemis program, designed for crewed lunar missions and beyond.

2. SpaceX:

- **Falcon 9**: A partially reusable vertical launch system, widely used for satellite launches, ISS resupply missions, and crewed missions.
- **Starship**: SpaceX's fully reusable rocket under development also employs a vertical launch configuration for its initial ascent.

3. Roscosmos (Russia):

- **Soyuz Rocket**: A workhorse vertical launch system used for crewed and cargo missions to the ISS and satellite launches.

4. ISRO (India):

- **GSLV Mk III (LVM3)**: India's vertical heavy-lift launch system used for deploying satellites into geostationary orbits and crewed missions.

5. Arianespace (Europe):

- **Ariane 5 and 6**: European heavy-lift launch systems used for deploying satellites and scientific missions into high Earth orbits.

Figure 42: The James Webb Space Telescope lifted off on an Ariane 5 rocket from Europe's Spaceport in French Guiana. ESA - S. Corvaja, CC BY 4.0, via Wikimedia Commons.

Horizontal Launch Systems

Horizontal launch systems are less common and are often employed for reusable spaceplanes or air-launched rockets.

1. Virgin Galactic:

- **SpaceShipTwo**: A spaceplane launched horizontally from the carrier aircraft **WhiteKnightTwo**, designed for suborbital space tourism.

Figure 43: WhiteKnightTwo (VMS Eve) and SpaceShipTwo (VSS Enterprise). Robert Sullivan, Public Domain, via Picryl.

2. Stratolaunch:

- **Roc and Talon-A**: Stratolaunch's carrier aircraft **Roc** air-launches payloads such as the hypersonic test vehicle **Talon-A**.

3. Northrop Grumman:

- **Pegasus Rocket**: An air-launched rocket released from an aircraft at high altitude for launching small satellites.

Figure 44: Pegasus XL CYGNSS Mate to L-1011. NASA, CC0, via Picryl.

4. Scaled Composites (Virgin Galactic's Parent Company):

- **SpaceShipOne**: The first privately-funded spacecraft to reach space, utilizing a horizontal launch system.

5. Reaction Engines (UK):

- **Skylon (Concept)**: A single-stage-to-orbit (SSTO) spaceplane with a horizontal takeoff and landing design, currently under development.

Hybrid Launch Systems

Hybrid systems combine elements of both vertical and horizontal launches, often leveraging the benefits of air-launch techniques with vertical propulsion.

1. SpaceX Starship (Partially Hybrid):

- **Launch and Landing**: While Starship launches vertically, its re-entry and landing employ horizontal gliding and aerodynamics to assist in recovery, showcasing hybrid capabilities.

2. DARPA and Northrop Grumman:

- **XS-1 Experimental Spaceplane (Now Cancelled)**: Designed to take off horizontally with jet assistance, transitioning to rocket propulsion for orbital insertion.

3. Orbital Sciences Corporation (Acquired by Northrop Grumman):

- **Pegasus Rocket**: The Pegasus system involves horizontal launch from a carrier aircraft, followed by vertical rocket propulsion to orbit.

4. Blue Origin's New Shepard:

- **Capsule and Booster**: Although primarily vertical, the reusable capsule employs parachutes and aerodynamic control for horizontal manoeuvrability during landing.

5. Sierra Space:

- **Dream Chaser**: Launched vertically on rockets like ULA's Atlas V, it lands horizontally on runways, combining vertical and horizontal methodologies.

Summary by Type

Launch Type	Organizations/Systems
Vertical	NASA (Saturn V, SLS), SpaceX (Falcon 9, Starship), ISRO (GSLV Mk III), Roscosmos (Soyuz)
Horizontal	Virgin Galactic (SpaceShipTwo), Northrop Grumman (Pegasus), Stratolaunch (Roc, Talon-A)
Hybrid	SpaceX (Starship), Northrop Grumman (XS-1), Sierra Space (Dream Chaser)

Each launch system type is optimized for specific mission requirements, with vertical systems dominating heavy-lift and crewed missions, horizontal systems excelling in reusability and flexibility, and hybrid systems blending the best of both approaches.

Infrastructure Requirements for Vertical, Horizontal, and Hybrid Launch Systems

Vertical Launch Systems

Vertical launch systems require significant ground infrastructure designed to support the rocket from assembly to liftoff. These systems are the most common and well-established in the industry, and their infrastructure reflects decades of refinement.

Launch Pads and Towers: Vertical rockets need specialized launch pads equipped with strong supports to hold the vehicle upright during fuelling, pre-flight checks, and launch. Towers adjacent to the rocket provide access for engineers and house umbilical connections for fuel, power, and communication lines. These towers are often equipped with swing arms or retractable platforms that move away shortly before liftoff.

Figure 45: This aerial view shows the Delta II launch pads at Complex 17 on Cape Canaveral Air Force Station in Florida, rimmed by the blue Atlantic Ocean in the background. NASA, CC0, via Picryl.

Fuelling and Propellant Storage: Large storage tanks are needed for cryogenic fuels (such as liquid oxygen and hydrogen), kerosene (RP-1), or hypergolic propellants. Advanced plumbing systems connect these tanks to the rocket for fuelling, with additional infrastructure to maintain cryogenic temperatures.

Flame Trenches and Sound Suppression Systems: To manage the intense heat and thrust during launch, flame trenches are built beneath the rocket to deflect exhaust gases away from the pad. Water deluge systems suppress the sound energy and thermal stress to protect the rocket and infrastructure during liftoff.

Transport and Integration Facilities: Rockets are typically assembled and integrated horizontally in a Vehicle Assembly Building (VAB) or similar facility. Once ready, they are transported vertically or horizontally to the launch pad using mobile launch platforms or specialized crawlers, such as NASA's Crawler-Transporter.

Telemetry and Control Systems: Ground-based telemetry systems track the rocket's performance during launch, and mission control centres monitor all launch parameters. Antennas and communication arrays are critical for maintaining a connection with the rocket.

Horizontal Launch Systems

Horizontal launch systems have unique infrastructure requirements due to their reliance on runways or carrier vehicles. These systems are often more flexible and require less specialized ground equipment.

Runways: Horizontal launch systems require long, reinforced runways similar to those used by large aircraft. These runways need to support the weight of the carrier aircraft or spaceplane during take-off and landing. Runway design must accommodate high speeds and the stresses from heavy payloads and unique landing dynamics.

Hangars and Integration Facilities: Instead of vertical assembly buildings, horizontal launch systems use hangars to house and integrate components. These facilities are used to assemble and attach the rocket or spaceplane to its carrier aircraft. Hangars must also provide space for maintenance and refurbishment of reusable components.

Carrier Aircraft and Support Infrastructure: Air-launched systems like Northrop Grumman's Pegasus require infrastructure to support the carrier aircraft, including fuelling stations, maintenance facilities, and specialized cranes for loading rockets. For example, Virgin Galactic's WhiteKnightTwo mothership needs hangars equipped with custom docking and servicing systems.

Ground Support Equipment: Horizontal systems may require unique equipment for transporting rockets and spaceplanes between facilities. These can include trolleys or lifting platforms designed to secure and transport the vehicle safely.

Reduced Fuelling Requirements: Fuelling infrastructure for horizontal systems is often simpler than vertical systems. For example, the carrier aircraft is fuelled conventionally, while the onboard rocket stages may require smaller-scale fuelling systems.

Hybrid Launch Systems

Hybrid launch systems blend vertical and horizontal approaches, demanding a combination of infrastructure elements from both methodologies. These systems are versatile but require more complex setups.

Dual-Purpose Launch Pads and Runways: Hybrid systems often require both a vertical launch pad for initial liftoff and a runway for recovery or secondary stages. For instance, systems like SpaceX's

Starship need a launch tower for vertical ascent but also require reinforced landing platforms for booster recovery.

Recovery and Refurbishment Facilities: Hybrid systems often involve reusable components, such as boosters that land vertically or spaceplanes that return horizontally. These components necessitate recovery facilities, which may include drone ships for offshore landings or runways for horizontal re-entry vehicles.

Advanced Propellant Management: Hybrid systems must accommodate both vertical fuelling for rocket stages and aircraft-style fuelling for carrier systems or spaceplanes. This may require separate storage and handling facilities for different propellant types.

Reusability and Maintenance: Hybrid systems like SpaceX's Starship rely on refurbishment and maintenance facilities for rapid turnaround between launches. These facilities include clean rooms, advanced diagnostic equipment, and precision manufacturing tools.

Landing Pads and Control Systems: Landing pads for hybrid systems must be equipped with shock absorption, guidance beacons, and telemetry systems to assist with controlled landings of reusable stages. SpaceX's autonomous drone ships exemplify infrastructure optimized for recovering vertically landing boosters.

Chapter 6

Propulsion Technologies

Solid Vs. Liquid Propellants

Rocket propulsion relies on chemical propellants to generate thrust, with solid and liquid propellants being the two primary types. Each has unique characteristics, advantages, and drawbacks that influence their use in various applications.

Key Comparisons

Feature	Solid Propellants	Liquid Propellants
Thrust Control	Non-throttleable; cannot be shut off	Throttleable; can be reignited and shut down
Complexity	Simple design with no moving parts	Mechanically complex with pumps, valves, and plumbing
Efficiency (Specific Impulse)	Lower efficiency, typically around 200-300 seconds	Higher efficiency, typically around 300-450 seconds
Storage and Handling	Easy to store and handle; long shelf life	Requires careful handling; cryogenics need insulation
Ignition	Single ignition event; cannot restart	Can be restarted multiple times
Applications	Boosters, missiles, low-cost rockets	Upper stages, crewed missions, deep-space exploration
Safety	Risk of accidental ignition; stable when stored properly	Complex systems prone to leaks or mechanical failure

Solid Propellants

Solid propellants consist of a homogeneous or composite mixture of fuel and oxidizer [187, 188]. Homogeneous propellants have a uniform blend, while composite propellants use separate particles of fuel (e.g., powdered aluminium) and oxidizer (e.g., ammonium perchlorate) suspended in a polymer binder [187, 188]. The resulting solid mass is shaped into a grain with specific patterns that determine the burn rate and thrust profile [187, 188].

Solid rocket propellants encompass a wide range of formulations and applications, from traditional black powder to cutting-edge electric solid propellants. Each family of propellants is tailored for specific performance characteristics, operational needs, and safety considerations.

Black powder (gunpowder) is one of the oldest propellant types and is composed of charcoal (fuel), potassium nitrate (oxidizer), and sulphur (a catalyst and secondary fuel). While its use is now largely confined to low-power model rocketry, black powder remains significant due to its historical importance and ease of production. It is typically pressed into a dense slug to control burn rates, but its low specific impulse (~80 seconds) and susceptibility to grain fractures limit its use to motors producing less than 40 newtons of thrust.

Zinc–sulphur (ZS) propellants, commonly known as "micrograin" propellants, combine powdered zinc and sulphur. These propellants are primarily used by amateur rocketeers for their novelty and dramatic visual effects, as they produce large, fiery exhaust plumes. However, their poor efficiency and extremely rapid burn rates (~2 m/s) make them impractical for serious applications, as most combustion occurs outside the motor.

Candy propellants are a simple and popular choice for amateur and experimental rocketry. These are typically composed of a sugar fuel, such as sucrose or sorbitol, combined with potassium nitrate as the oxidizer. Melted and cast into molds, candy propellants produce modest performance with a specific impulse of around 130 seconds. Their ease of production makes them a popular choice for hobbyists, though they are unsuitable for commercial or high-performance missions.

Double-base (DB) propellants are formed by combining two monopropellant components, typically nitro-glycerine and nitrocellulose, into a gel that is solidified with additives. These propellants are notable for producing minimal smoke and achieving a medium-to-high specific impulse (~235 seconds). Adding metal powders like aluminium can increase performance to ~250 seconds, but this can introduce opaque smoke in the exhaust. DB propellants are widely used in applications requiring clean-burning characteristics, such as tactical missiles.

Composite propellants represent one of the most versatile and widely used categories. They consist of powdered oxidizers like ammonium nitrate (AN) or ammonium perchlorate (AP), mixed with powdered metal fuels (usually aluminium) and immobilized in a rubbery binder such as hydroxyl-terminated polybutadiene (HTPB). Ammonium nitrate-based composites (ANCP) offer moderate performance (~210 seconds Isp), while ammonium perchlorate-based composites (APCP) provide high performance, with vacuum-specific impulses reaching up to 304 seconds when optimized with high-area-ratio nozzles. These propellants are used in a range of applications, from military rockets to space launch vehicles like the Space Shuttle Solid Rocket Boosters. New chlorine-free oxidizers, such as ammonium dinitramide (ADN), are being explored as environmentally friendly alternatives.

High-energy composite (HEC) propellants enhance standard composite formulations by adding high-energy explosives like RDX or HMX. These additives provide modest improvements in specific impulse but increase the risks associated with handling and storage. As a result, HEC propellants are typically reserved for specialized military applications.

Composite-modified double-base (CMDB) propellants combine elements of DB and composite formulations. By adding ammonium perchlorate and powdered aluminium to a nitrocellulose/nitro-glycerine base, CMDB propellants achieve improved specific impulse and combustion stability. Advanced formulations, such as NEPE-75 used in Trident II D-5 missiles, replace most of the ammonium perchlorate with polyethylene glycol-bound HMX to further enhance performance.

Minimum-signature (smokeless) propellants are a cutting-edge area of research. These formulations, often using CL-20 nitroamine, offer significantly higher energy density and specific impulse than traditional formulations. They produce a nearly invisible exhaust with minimal smoke, making them ideal for tactical applications where stealth is critical. CL-20 is non-polluting, shock-insensitive, and classified as a safer alternative to HMX, though its high cost has limited widespread adoption.

Electric solid propellants (ESPs) represent a revolutionary advancement in solid rocket technology. These plastisol-based propellants can be precisely ignited and throttled using electrical currents, enabling unprecedented control over thrust. Unlike conventional propellants, ESPs are insensitive to flames and electrical sparks, making them safer and more versatile for a variety of applications.

Each family of solid rocket propellants offers unique advantages and trade-offs, making them suitable for different missions, performance requirements, and safety considerations. From historical black powder to cutting-edge ESPs, the diversity of solid propellant technologies continues to drive innovation in rocketry.

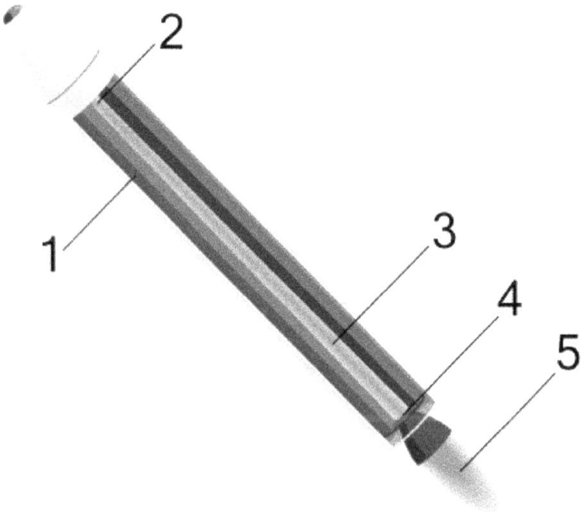

Figure 46: A simplified diagram of a solid-fuel rocket. 1 - A solid fuel-oxidizer mixture (propellant) is packed into the rocket, with a cylindrical hole in the middle; 2 - An igniter combusts the surface of the propellant; 3 -The cylindrical hole in the propellant acts as a combustion chamber; 4- The hot exhaust is choked at the throat, which, among other things, dictates the amount of thrust produced; 5- Exhaust exits the rocket. Pbroks13, CC BY-SA 4.0, via Wikimedia Commons.

As shown in Figure 46, a simple solid rocket motor is a fundamental propulsion system characterized by its straightforward design and reliable operation. Its primary components include a casing, nozzle, grain (propellant charge), and an igniter. These elements work together to produce thrust by burning the solid propellant and expelling exhaust gases at high speed.

The solid grain within the rocket serves as the fuel and oxidizer combined into a single solid mass. This grain burns in a predictable manner, guided by principles such as Taylor–Culick flow, which describes the behaviour of gas flow within the motor. The nozzle plays a critical role in managing this flow, with dimensions carefully calculated to maintain a specified chamber pressure while optimizing the conversion of gas energy into thrust. This balance ensures that the motor operates efficiently and reliably throughout its burn.

Taylor–Culick flow is a crucial concept in the field of internal fluid dynamics, specifically within the operation of solid rocket motors. It explains the behaviour of gas flow generated from the combustion of solid propellant grains, describing how these gases move through the motor chamber and toward the nozzle. This flow arises because the combustion process releases gases from the burning surface of the propellant, which then must be channelled through the chamber efficiently to produce thrust.

The flow exhibits both axial and radial components. The axial component drives the gas toward the nozzle, facilitating the generation of thrust, which is the primary objective of the rocket motor. Simultaneously, the radial component ensures that gases are replenished from the burning surface of the propellant grain, maintaining a consistent flow throughout the chamber. These components work in tandem to ensure the smooth and predictable operation of the motor.

Boundary layers play a vital role in Taylor–Culick flow. These layers develop along the chamber walls and the burning surface of the propellant. They influence key factors such as heat transfer, pressure distribution, and the potential for erosive burning. Proper management of these boundary layers is critical for maintaining the structural integrity of the motor and ensuring efficient combustion.

Taylor–Culick flow operates under the fundamental principles of mass, momentum, and energy conservation. The mass conservation principle ensures that the volume of gas produced by the combustion process aligns with the flow rate exiting through the nozzle. Momentum conservation governs the interplay between pressure gradients and gas velocity, ensuring efficient thrust production. Energy conservation accounts for the thermal and kinetic energy of the gases, optimizing the motor's overall performance.

In well-designed solid rocket motors, the flow reaches a steady-state behaviour during operation. At this stage, the chamber pressure remains relatively constant, ensuring predictable motor performance. This steady-state is vital for the reliability and efficiency of the rocket, as fluctuations in pressure could lead to instability or inefficiency.

The velocity profile of Taylor–Culick flow is parabolic in nature. The highest velocity is found at the centre of the chamber, where viscous effects are minimal. Near the walls, the velocity decreases due to the influence of viscosity within the boundary layers. This parabolic profile ensures that gases are uniformly directed toward the nozzle, reducing turbulence and enhancing thrust efficiency. This predictable and stable flow behaviour is essential for the optimal operation of solid rocket motors, enabling them to achieve the desired performance with minimal energy loss.

The flow description in the context of Taylor–Culick flow involves understanding the internal gas dynamics within a cylindrical chamber, such as those found in solid rocket motors. The mathematical framework governing this flow is based on an axisymmetric and inviscid (neglecting viscosity) fluid model. The flow is described using a stream function, ψ, which simplifies the representation of the velocity field in cylindrical coordinates.

The governing equation, known as the Hicks equation, reduces to:

$$\frac{\partial^2 \psi}{\partial r^2} - \frac{1}{r}\frac{\partial \psi}{\partial r} + \frac{\partial^2 \psi}{\partial z^2} = -r^2 f(\psi)$$

where:

- ψ is the stream function.
- r is the radial distance from the axis of the cylinder.
- z is the axial distance from the closed end of the cylinder.
- $f(\psi) = \pi^2 \psi$ is a specific function chosen to satisfy the solution conditions.

This equation models the behaviour of the gas flow as it moves radially outward from the chamber walls (due to the combustion of the propellant) and axially toward the chamber exit (the nozzle). The solution to this equation, satisfying appropriate boundary conditions, is given by:

$$\psi = aUz \sin\left(\frac{\pi r^2}{2a^2}\right)$$

where:

- a is the radius of the cylinder.
- U is the injection velocity of the gas at the chamber wall.

This formula predicts the flow field inside the chamber, with the sinusoidal term representing the radial velocity component due to the gas injection, and the linear z-dependence indicating axial transport toward the nozzle. Experimental results have verified the accuracy of this solution, confirming its utility in describing Taylor–Culick flow for conditions where the axial distance z is much greater than the radius a (i.e., $z \gg a$).

Key Features of the Flow:

1. **Boundary Layer Considerations**: While the model is derived for inviscid flow, it satisfies the no-slip boundary condition at the chamber walls. This is because any viscous boundary layer

at the walls is effectively "blown off" by the continuous gas injection. Therefore, the flow is referred to as "quasi-viscous."

2. **Applicability Limits**: The Taylor–Culick profile is valid for regions where $z \gg a$. Near the closed end of the cylinder ($z \sim a$), boundary layer separation occurs, and the solution is less accurate.

3. **Experimental Verification**: Despite the simplified assumptions in the mathematical derivation, such as neglecting viscosity and assuming axisymmetric conditions, experimental observations confirm the reliability of this model for describing internal rocket motor gas dynamics.

The solution provides a practical tool for predicting the behaviour of exhaust gases within solid rocket motors. It helps engineers design rocket chambers and nozzles to optimize thrust by accurately accounting for how gases flow through the system.

Once ignited, a simple solid rocket motor cannot be turned off or throttled, as it contains all the ingredients required for combustion within a sealed chamber [187, 188]. This characteristic makes it highly reliable but less flexible compared to liquid rocket engines. However, advancements in solid rocket technology have led to the development of more sophisticated designs. Some modern solid rocket motors can be throttled, extinguished, and even re-ignited through the manipulation of nozzle geometry or the use of vent ports. Additionally, pulsed rocket motors have been designed to burn in discrete segments, allowing controlled ignition of each segment on demand, which enhances their utility in tactical and space applications.

Modern solid rocket motor designs also incorporate features that enhance performance, control, and safety. For instance, steerable nozzles allow for precise guidance and trajectory adjustments during flight. Avionics systems provide real-time data and enable advanced control mechanisms. Recovery hardware, such as parachutes, can be integrated to facilitate the recovery and reuse of certain rocket components. Self-destruct mechanisms ensure safe termination of the rocket's flight if necessary, particularly in cases of malfunction or deviation from the intended path.

Further enhancements include auxiliary power units (APUs) for operational support, controllable tactical motors for precise targeting, and divert and attitude control motors for fine adjustments in position and orientation. Thermal management materials are used to protect the motor casing and internal components from the intense heat generated during combustion. These advancements demonstrate the versatility and continued evolution of solid rocket motors, making them indispensable in both military and space applications. Their robust, reliable, and increasingly adaptable designs have solidified their role as a cornerstone of rocket propulsion technology.

Solid propellants are simple and reliable, as they are pre-manufactured and stored directly in the rocket motor, reducing the complexity of on-site fuelling [187, 188]. They are highly dependable, as they lack the complex plumbing and machinery of liquid systems [187, 188]. Solid rocket motors are often more rugged and can operate in extreme environmental conditions [187, 188]. They are commonly used in military applications, missile systems, and as boosters for large launch vehicles [187, 188].

Once ignited, solid propellants cannot be shut off or throttled, limiting their flexibility during flight [187, 188]. Their performance, characterized by specific impulse, is generally lower than that of liquid propellants [187, 188]. Furthermore, the manufacturing and handling of solid propellants involve safety risks due to their sensitivity to impact and temperature [187, 188].

Solid propellants are ideal for applications requiring simplicity and high thrust, such as military missiles, first-stage boosters, and small satellite launch vehicles [187-190]. Examples include the Space Shuttle Solid Rocket Boosters (SRBs) and the Vega rocket [187-189].

Researchers are exploring new energetic materials and binders to improve the energy density and performance of solid propellants [191-194]. Composite propellants with trimodal oxidizer distributions and gel propellants are also being investigated to enhance specific impulse and safety [188, 195, 196].

The design of a solid rocket motor begins with determining the total impulse required, which directly influences the quantities of fuel and oxidizer needed. The impulse requirement sets the foundation for the overall mass of the propellant, and from there, the grain geometry and chemical composition are selected to meet the desired motor performance characteristics. This involves a delicate interplay of several factors, each influencing critical aspects of the motor's functionality and safety.

Solid Fuel Performance

The performance of solid rocket motors (SRMs) is a complex interplay of propellant chemistry, motor design, and the specific mission objectives. At their core, solid rockets provide a balance of simplicity, reliability, and high thrust, making them invaluable in many space and defence applications. Their specific impulse (I_{sp}), a critical measure of propulsion efficiency, typically ranges lower than that of liquid-fuelled rockets but remains competitive, particularly in applications where their unique advantages outweigh efficiency trade-offs.

A well-designed ammonium perchlorate composite propellant (APCP) motor, for instance, can achieve a vacuum-specific impulse as high as 285.6 seconds, as seen in the Titan IVB SRMU. While this is less than the Isp of liquid-fuelled engines like the RD-180 (339.3 s) or the hydrogen-fuelled RS-25 (452.3 s), SRMs offer other advantages. Their high thrust-to-weight ratio and compact design make them ideal for initial launch stages, where overcoming Earth's gravity and atmospheric drag are primary concerns. Solid rocket upper stages, like the Orbus 6E, slightly improve efficiency, achieving a specific impulse of 303.8 seconds in a vacuum, highlighting their ability to adapt to higher altitude and near-space conditions.

Solid rockets generally have high propellant fractions, often exceeding 90% of the total stage mass. For example, the Castor 120 first stage boasts a propellant mass fraction of 92.23%, while its counterpart, the Castor 30 upper stage, achieves 91.3%. These fractions reflect the efficient packing of propellant relative to structural and ancillary components, ensuring maximal performance per unit of stage mass. The Athena II, a four-stage solid rocket vehicle, effectively demonstrates the versatility of solid rockets, achieving milestone missions such as launching the Lunar Prospector probe in 1998.

The inherent simplicity and reliability of solid rockets make them especially well-suited for first-stage applications. Their ability to deliver massive thrust without the complexities of pumps, valves, or active cooling systems enables rapid and robust deployment. They also shine in satellite launches as final boost stages, where their compact size and minimal maintenance requirements provide cost-effective solutions for orbital insertion. Notably, spin-stabilized solid motors like Thiokol's Star series have been extensively employed, achieving high propellant fractions (up to 94.6%) and offering reliable performance in missions requiring additional velocity for interplanetary trajectories.

In the defence sector, solid rocket motors benefit from their ability to remain loaded for extended durations while maintaining reliability. Strategic missiles like the Peacekeeper ICBM have used advanced high-energy propellants such as HMX, providing increased energy density compared to APCP. The Peacekeeper's second stage, for example, achieved a specific impulse of 309 seconds using HMX-based propellant. Further advancements in propellant chemistry, such as CL-20 developed by the Naval Air Weapons Station, promise even greater performance, with a 14% increase in energy per mass compared to HMX. CL-20's enhanced energy density and compliance with insensitive munitions (IM) laws make it a potential candidate for future commercial launch applications, potentially pushing specific impulse values in the range of 320 seconds.

Solid rocket motors are particularly advantageous in military contexts due to their rapid response capability and long-term storage stability. Unlike liquid propellant systems, which often require complex handling and maintenance, solid motors can remain mission-ready for years, making them ideal for strategic deterrence and rapid deployment scenarios. As technology advances, solid rockets continue to play a critical role across commercial, scientific, and military applications, offering a reliable and efficient propulsion option tailored to specific mission needs.

Grain geometry

The grain's burning behaviour plays a pivotal role in defining motor performance. The burn rate is influenced by the surface area of the grain exposed to combustion and the chamber pressure within the motor. Chamber pressure, in turn, is governed by the size of the nozzle throat, which regulates the flow of exhaust gases, and the burn rate of the grain. This interdependence means that grain geometry, nozzle dimensions, and casing design must be solved simultaneously to achieve the desired performance while ensuring safety and structural integrity. The casing must be designed to withstand the internal pressures generated during combustion, with the allowable chamber pressure directly tied to the strength and material of the casing.

Another critical factor is the burn time, which depends on the "web thickness" of the grain. This term refers to the thickness of the grain in the direction perpendicular to the burning surface. As the grain burns, its thickness decreases until the propellant is depleted, determining the duration of the motor's operation. This relationship requires precise calculation to ensure that the motor provides thrust for the required time without over pressurization or premature burnout.

Grain design can also differ in its attachment to the casing. In some motors, the grain is bonded directly to the casing, which introduces additional complexities in the design process. The deformation characteristics of both the grain and the casing under operational stresses must be compatible to prevent structural failures. Non-bonded grains, while simpler in this regard, may introduce other challenges such as ensuring consistent combustion behaviour.

Solid rocket motors are susceptible to specific failure modes that designers must account for. One common failure is the fracturing of the grain, which can occur if internal stresses exceed the material's tolerance. Fractures or voids in the grain can lead to a sudden increase in the burn surface area, causing a rapid rise in exhaust gas production and chamber pressure. If the pressure exceeds the casing's structural limits, catastrophic failure can occur. Similarly, failure of the bond between the grain and the casing can result in uneven burning and pressure spikes, compromising the motor's integrity.

Casing seals also represent a critical point of vulnerability, particularly in motors where the casing must be opened to load the grain. A failure in these seals allows hot gases to escape, eroding the material around the leak and rapidly increasing the size of the escape path. This can result in the uncontrolled release of gas and destruction of the motor. Such a failure was famously responsible for the Space Shuttle Challenger disaster, where a faulty O-ring seal allowed hot gas to escape, leading to the destruction of the vehicle.

Designing a solid rocket motor is, therefore, a complex process that requires careful consideration of geometry, material behaviour, and operational stresses. Each component, from the grain to the casing and nozzle, must be optimized to work in harmony, while potential failure modes must be meticulously analysed and mitigated to ensure reliable and safe operation.

The geometry of the propellant grain in a solid rocket motor is a critical design parameter that significantly influences motor performance. This is because the propellant burns from its exposed surface in the combustion chamber, a process known as deflagration. The shape and configuration of the propellant dictate how the burn area evolves over time, which directly affects the thrust profile of the motor. The relationship between grain geometry and performance is a primary focus in the field of internal ballistics.

As the surface of the propellant burns, its geometry changes, altering the exposed surface area. This evolution of the burning surface area determines the rate at which combustion gases are generated. The instantaneous mass flow rate of the exhaust gases, which produce thrust, is calculated as the product of the propellant's density, the burning surface area, and the linear burn rate. This relationship can be expressed as:

$$\dot{m} = \rho \cdot A_s \cdot \dot{b}$$

Where:

- $\dot{m}$ is the mass flow rate of the combustion gases,
- ρ is the density of the propellant,
- A_s is the exposed surface area of the propellant,

- $\dot{b}$ is the linear burn rate.

Grain geometry configurations are chosen based on the desired thrust curve and mission requirements. Various designs are used to create specific performance profiles, including progressive, regressive, or neutral thrust. These configurations are described below:

Circular Bore: In a circular bore configuration, the propellant grain has a cylindrical hollow core that burns outward radially. When used in a BATES (Ballistic Test and Evaluation System) setup, this geometry produces a progressive-regressive thrust curve. Initially, the thrust increases as the burning surface area grows, then decreases as the surface area diminishes toward the end of the burn.

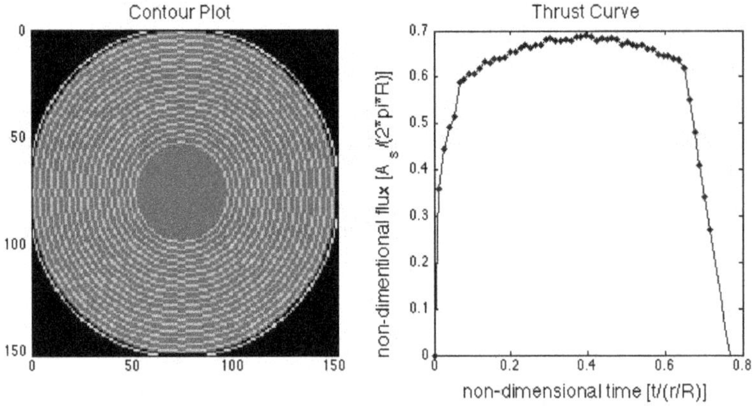

Figure 47: Low-res simulation of circular bore propellant burn profile. ThirdCritical, Graham Orr, CC BY-SA 3.0, via Wikimedia Commons.

End Burner: The end burner configuration involves propellant that burns from one axial end to the other, maintaining a steady and prolonged burn. This design is suited for applications requiring consistent, long-duration thrust. However, it poses challenges such as thermal management and significant shifts in the centre of gravity (CG) as the burn progresses.

C-Slot: In this configuration, a wedge-shaped cutout runs along the axial length of the propellant grain, creating a large burning surface. This results in a regressive thrust curve, where thrust decreases over time as the burning surface diminishes. While the design can provide a fairly long burn duration, it introduces challenges with asymmetric CG movement and thermal stresses.

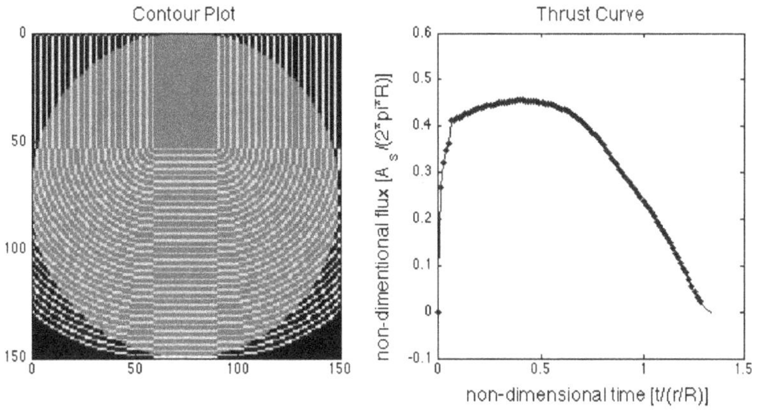

Figure 48: Low-res simulation of c-slot propellant burn profile. ThirdCritical, Graham Orr, CC BY-SA 3.0, via Wikimedia Commons.

Moon Burner: The moon burner uses an off-centre circular bore to achieve a progressive-regressive thrust curve with a longer burn duration. Although this design is efficient in some applications, the off-centre configuration introduces slight asymmetry in CG movement, which must be accounted for in stability calculations.

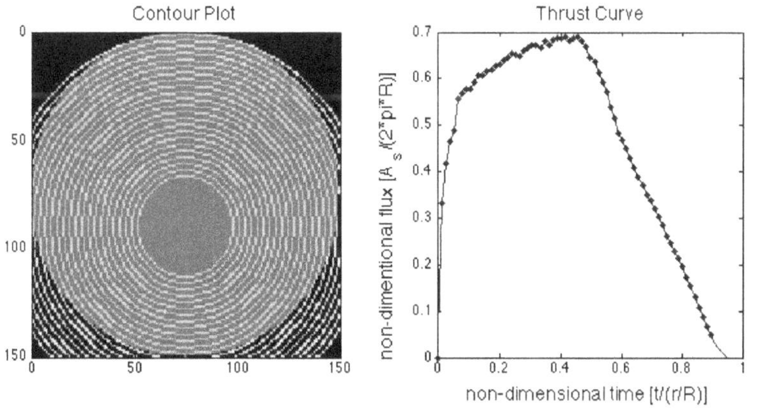

Figure 49: Low-res simulation of moon burner propellant burn profile. ThirdCritical, Graham Orr, CC BY-SA 3.0, via Wikimedia Commons.

Finocyl: A finocyl grain features a star-like cross-section, typically with five or six points. This design increases the initial burning surface area compared to a simple circular bore, resulting in a higher thrust at the beginning of the burn. The thrust profile is relatively stable, with a quicker burn time than the circular bore configuration due to the increased surface area. This geometry is often chosen for applications requiring rapid acceleration.

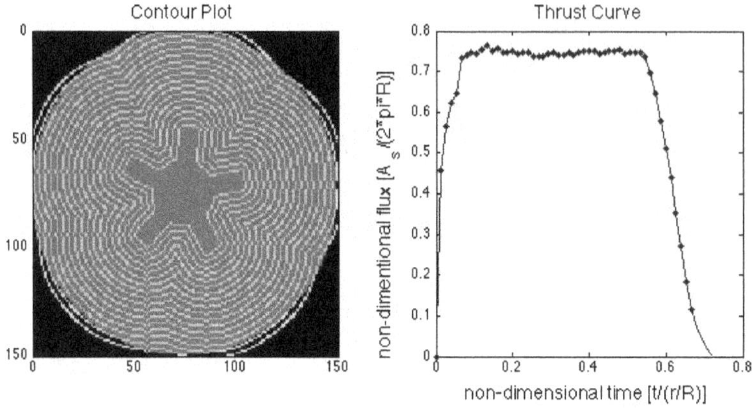

Figure 50: Low-res simulation of finocyl bore propellant burn profile. ThirdCritical, Graham Orr, CC BY-SA 3.0, via Wikimedia Commons.

The choice of grain geometry involves trade-offs between thrust profile, burn duration, and structural considerations. Designers must account for the impact of geometry on heat transfer, internal pressure distribution, and the structural integrity of the grain and casing. For instance, configurations like the end burner can lead to high thermal stresses and uneven CG movement, while more complex geometries like the finocyl or C-slot require precise manufacturing to ensure predictable performance. By tailoring grain geometry to the specific needs of a mission, engineers can optimize the efficiency and reliability of solid rocket motors.

Casing

The casing of a solid rocket motor is a critical structural component that functions as both the container for the propellant and a pressure vessel capable of withstanding the intense conditions within the motor during operation [197-199]. The choice of material and design for the casing depends on the scale of the rocket, the type of propellant used, and the intended performance requirements [197-199].

For small-scale rockets, such as black powder model motors, casings are often constructed from lightweight and inexpensive materials like cardboard [197-199]. These materials are sufficient to handle the modest pressures and thermal loads typical of such motors [197-199]. For larger hobbyist rockets using composite fuels, casings are typically made of aluminium, which offers a balance of strength, light weight, and manufacturability [197-199].

For larger, high-performance systems, more robust materials are required [197-199]. The steel casings used in the Space Shuttle Solid Rocket Boosters (SRBs) were designed to endure the significant pressures and thermal stresses associated with large-scale launches [197-199]. Steel provides exceptional strength and durability, ensuring the casing can safely contain the high pressures generated during combustion [197-199]. However, the weight of steel makes it less suitable for applications requiring maximum efficiency and payload capacity [197-199].

In cutting-edge and high-performance motors, filament-wound graphite epoxy casings are frequently used [197-200]. These composite materials offer superior strength-to-weight ratios, enabling high-pressure containment without the significant mass penalty of steel [197-200]. The filament-winding process allows for precise control of fibre orientation, optimizing the casing's ability to withstand the multi-axial stresses generated during motor operation [197-200]. Additionally, composite casings are more resistant to corrosion, an essential feature for prolonged storage or exposure to harsh environments [197-200].

The casing must be engineered to withstand not only the internal pressures of the motor but also the elevated temperatures resulting from combustion [197-200]. As a pressure vessel, it is designed with safety factors to account for these stresses [197-200]. The integrity of the casing is paramount, as failure can lead to catastrophic results [197-200]. The design process often involves extensive computational modelling and testing to ensure that the casing can maintain its structural integrity under operational conditions [197-200].

To protect the casing from the corrosive effects of the hot combustion gases, an internal sacrificial thermal liner is often employed [197-200]. This liner serves as a barrier, ablating during operation to absorb heat and reduce the direct impact of the hot gases on the casing material [197-200]. By sacrificially ablating, the liner prolongs the lifespan of the casing and ensures the motor can perform as designed [197-200]. The composition of the liner is typically tailored to the motor's specific requirements, balancing thermal resistance, ablation rate, and compatibility with the propellant [197-200].

Nozzle

The rocket nozzle is a critical component in solid rocket motor design, responsible for accelerating the combustion gases to generate thrust. A standard convergent-divergent nozzle configuration is used, which efficiently transforms the high-pressure and high-temperature gas from the combustion chamber into a high-velocity exhaust stream. This acceleration process not only produces thrust but also ensures the optimal use of energy stored in the propellant.

The nozzle must endure extreme thermal and mechanical stresses due to the hot, high-velocity exhaust gases. For this reason, nozzles are often made from materials with exceptional heat resistance. Carbon-based materials such as amorphous graphite or reinforced carbon–carbon (RCC) are common choices. These materials provide the necessary thermal stability and resistance to erosion, ensuring that the nozzle performs reliably over the motor's operational life.

In many advanced rocket systems, the nozzle design also incorporates directional control of the exhaust to aid in vehicle steering and stability. This is often achieved through several methods. Gimballing the nozzle, as seen in the Space Shuttle Solid Rocket Boosters (SRBs), allows the nozzle to pivot, directing the thrust vector and enabling precise control of pitch, yaw, and roll. Another method involves jet vanes, which are aerodynamic surfaces inserted into the exhaust stream to alter its direction. This technique was famously used in the V-2 rocket.

A more advanced control method is Liquid Injection Thrust Vectoring (LITV). In this approach, a liquid—often a reactive chemical—is injected into the exhaust stream after the nozzle throat. The injected liquid vaporizes upon contact with the hot gases, and in many cases, chemically reacts with the exhaust, increasing the mass flow on one side of the nozzle. This imbalance creates a control moment, effectively steering the rocket. For instance, the Titan IIIC solid boosters used nitrogen tetroxide for LITV, with tanks for this liquid visibly mounted between the main rocket stage and the boosters.

An alternative nozzle arrangement was employed in an early version of the Minuteman ICBM's first stage, which utilized a single motor with four gimballing nozzles. This design provided precise control over pitch, yaw, and roll without the need for external mechanisms, simplifying the thrust vectoring system.

Overall, the rocket nozzle serves not only as a propulsion component but also as a critical element in the vehicle's guidance and control system. Advances in materials, structural design, and thrust vectoring technologies have expanded the versatility and efficiency of nozzles, enabling them to meet the demands of modern rocketry while maintaining reliability under extreme conditions.

Liquid Propellants

Liquid propellants consist of a separate fuel and oxidizer stored in tanks and pumped into a combustion chamber where they mix and ignite [201]. They can be cryogenic, hypergolic, or storable, depending on the specific fuel-oxidizer combination [201].

Common liquid fuels include liquid hydrogen, kerosene (RP-1), and hydrazine [201]. Liquid oxygen (LOX) is a typical oxidizer for cryogenic systems, while nitrogen tetroxide (NTO) is used for hypergolic systems [201]. Liquid propellants require robust storage tanks and plumbing to handle their specific physical and chemical properties, such as cryogenic cooling for LOX and liquid hydrogen [201].

Liquid propellants offer high efficiency, with specific impulse values significantly higher than solid propellants [201]. They provide precise control, allowing for throttling, reignition, and shutdown during flight, which is essential for complex missions, including orbital insertion, interplanetary transfers, and crewed spaceflight [201].

However, liquid systems are mechanically complex, requiring pumps, valves, and pressurization systems to manage the fuel and oxidizer flow, increasing the risk of mechanical failure and demanding rigorous engineering and testing [201]. Cryogenic fuels require insulated tanks and careful handling to prevent boil-off or leakage, and liquid systems are more challenging to store and transport, making them less suited for rapid deployment or long-term readiness [201].

Liquid propellants are used in applications where high efficiency and precise control are necessary, such as upper stages, orbital manoeuvres, and deep-space missions [201]. Examples include the Saturn V's second and third stages (using liquid hydrogen and LOX) and SpaceX's Falcon 9 (using RP-1 and LOX) [201].

The principle of operation of a liquid rocket engine revolves around the careful management and combustion of liquid propellants to generate thrust. These engines consist of several critical components, including tanks and pipes for propellant storage and transfer, an injector system, combustion chambers, and nozzles. Each of these systems works together to enable efficient propulsion.

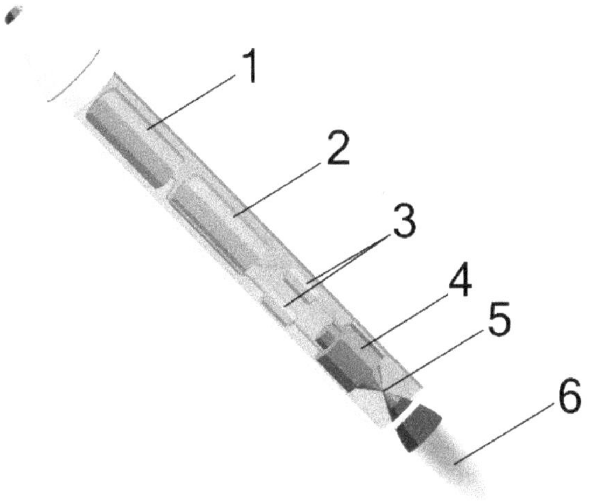

Figure 51: A simplified diagram of a liquid-propellant rocket. 1- Liquid rocket fuel; 2 – Oxidizer; 3 - Pumps carry the fuel and oxidizer; 4 - The combustion chamber mixes and burns the two liquids; 5 - Combustion product gasses enter the nozzle through a throat; 6 - Exhaust exits the rocket. Pbroks13, CC BY-SA 4.0, via Wikimedia Commons.

Figure 51 shows a liquid-fuel rocket diagram, which highlights the key components of a rocket engine that operates using liquid propellants. Liquid-fuel rockets rely on the controlled combustion of liquid fuel and oxidizer, which are stored separately and mixed inside the combustion chamber to produce thrust.

Nose Cone: At the top of the rocket, the nose cone is designed to minimize aerodynamic drag as the rocket moves through the atmosphere. It helps streamline the rocket and reduce air resistance. In some cases, the nose cone also houses payloads such as satellites, scientific instruments, or crew capsules, depending on the mission's objectives.

Fuel Tank: The fuel tank stores the liquid fuel used to power the rocket engine. Common liquid fuels include RP-1 (refined kerosene), liquid hydrogen, or other hydrocarbons, depending on the engine design and mission requirements. The fuel tank is pressurized to ensure the smooth flow of fuel into the combustion chamber.

Oxidizer Tank: The oxidizer tank contains the oxidizing agent required for combustion since there is no oxygen available in space. The most commonly used oxidizers are liquid oxygen (LOX) or other oxidizing agents like nitrogen tetroxide. The oxidizer is stored separately from the fuel to prevent premature combustion.

Pumps and Pipes: Liquid rockets use high-pressure pumps to transport both the liquid fuel and oxidizer from their respective tanks to the combustion chamber. These pumps are typically driven by turbines powered by a small amount of the propellant itself. The piping system ensures that the fuel and oxidizer reach the combustion chamber efficiently and in precise proportions.

Combustion Chamber: The combustion chamber is where the fuel and oxidizer are mixed and burned to produce high-temperature, high-pressure gases. Combustion is a highly energetic process that generates immense thrust by converting chemical energy into kinetic energy. The design of the combustion chamber must withstand extreme temperatures and pressures.

Nozzle: The rocket nozzle is a critical component that accelerates the combustion gases to high velocities to produce thrust. The nozzle follows a convergent-divergent design, where the gases are first compressed and then expanded as they exit. This expansion converts the thermal energy of the gases into kinetic energy, generating thrust in accordance with Newton's Third Law of Motion.

Exhaust Flame: The exhaust flame represents the high-speed stream of gases expelled from the rocket nozzle. The force generated by expelling the exhaust gases at high velocity creates thrust that propels the rocket upward. This is the fundamental principle of rocket propulsion.

In a liquid-fuel rocket, the process begins when the fuel and oxidizer are pumped from their respective tanks into the combustion chamber. Inside the chamber, they mix and ignite, producing an intense reaction that generates high-pressure gases. These gases are forced through the nozzle, where they are accelerated and expelled at tremendous speed. This action creates the thrust needed to propel the rocket into the atmosphere or space.

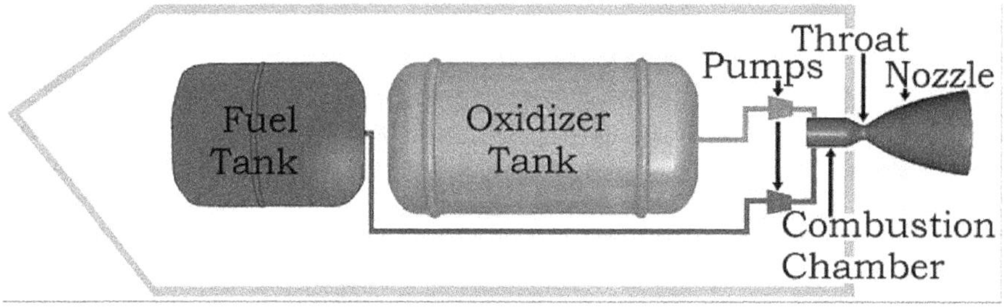

Figure 52: Liquid rocket schematic. Nwbeeson, CC BY-SA 3.0, via Wikimedia Commons.

Liquid propellants are typically dense, with densities similar to water, ranging from approximately 0.7 to 1.4 g/cm^3. This allows for lightweight tankage, as the tanks only need to contain the liquid without withstanding excessive pressures. For propellants like liquid hydrogen, however, the situation is different. Liquid hydrogen has a very low density, which necessitates larger tanks, heavier insulation, and more structural material to contain it. While hydrogen requires only modest pressure to remain in liquid form, the trade-off is a significant increase in tank mass. For dense propellants, the tank mass can be as low as 1% of the total propellant mass, but for liquid hydrogen, this increases to around 10%, primarily because of the additional insulation and the larger tank size required to store the same mass of fuel.

To deliver propellants into the combustion chamber, the propellant pressure at the injectors must exceed the chamber pressure, which is where combustion takes place. This pressure differential ensures a steady and controlled flow of propellants into the chamber. The method of achieving this pressure depends on the engine design. In most modern liquid rocket engines, this is accomplished using turbopumps, particularly centrifugal turbopumps. These pumps are highly efficient and lightweight, making them ideal for space applications. Turbopumps are powered by energy extracted from the propellant flow itself, typically using a gas generator or staged combustion process. The SpaceX Merlin 1D engine, for instance, demonstrates an outstanding thrust-to-weight ratio of up to 155:1, and its vacuum-optimized version achieves ratios as high as 180:1, showcasing the efficiency of pump-fed systems.

Alternatively, for simpler designs, pressure-fed systems are used. In these systems, an inert gas, such as helium, stored in a high-pressure tank, forces the propellants into the combustion chamber. While pressure-fed systems are mechanically simpler and more reliable, they are generally less efficient for large-scale rockets because the pressurant tank adds mass, reducing overall performance. However, for small rockets or satellite thrusters, where reliability is paramount and simplicity outweighs the performance loss, pressure-fed systems are ideal. These are often employed in orbit maintenance systems for satellites, where small engines are needed for adjustments over long durations.

Modern advancements have also introduced electric pumps powered by batteries as an alternative to turbopumps. This approach simplifies the engine design by removing the need for complex gas generators or combustion cycles to power the pumps. While electric pumps are not as power-dense as turbopumps, they offer excellent reliability and are increasingly being explored by companies for small launch vehicles and upper stages.

The major components of a liquid rocket engine include the combustion chamber (or thrust chamber), where the fuel and oxidizer are ignited and combusted to produce high-speed exhaust gases. A pyrotechnic igniter or electric igniter initiates combustion, ensuring a controlled and reliable ignition sequence. The propellant feed system consists of pumps, pressurization tanks, and valves, regulating the flow and pressure of propellants into the combustion chamber. Valves and regulators provide precision control over fuel delivery, ensuring consistent engine performance. Propellant tanks store the oxidizer and fuel under appropriate conditions, while the rocket engine nozzle expands and accelerates the exhaust gases to produce thrust efficiently.

Overall, liquid rocket engines operate using two primary feed methods—pressure-fed and pump-fed. Pump-fed systems are dominant in high-performance engines due to their ability to manage higher pressures and deliver more thrust with less tank mass. Pressure-fed engines, on the other hand, offer reliability and simplicity, making them suitable for smaller-scale propulsion systems. Together, these technologies form the foundation of modern rocket propulsion, enabling everything from satellite positioning to interplanetary travel.

Liquid-fuel rockets are advantageous because they allow precise control of thrust through the regulation of fuel and oxidizer flow rates. Additionally, they can be shut down and restarted, making them ideal for missions that require precise manoeuvring, such as satellite deployment, orbital adjustments, and interplanetary travel. Examples of rockets that use liquid-fuel engines include the Saturn V, SpaceX Falcon 9, and Blue Origin's New Shepard.

Figure 53 shows three bipropellant engines, originating from the Gemini and Apollo programs, were used for manoeuvring spacecraft in space. From left to right:

- **Gemini SE-6:** The SE-6 engines were used for attitude and manoeuvring control in the Gemini spacecraft. They operated in pairs to generate torque along the three axes of rotation. The Re-entry Control System (RCS) in the nose of the Gemini module consisted of sixteen SE-6 engines. Built from steel with an ablative nozzle by Rocketdyne, each engine produced 25 pounds of thrust.
- **Apollo Command Module SE-8:** The Apollo Command Module relied on twelve SE-8 engines for reaction control after separating from the Service Module. These engines performed functions such as rotation control, rate damping, and attitude control, each delivering 93 pounds of thrust.
- **Apollo Lunar Module R-4D:** The Lunar Module featured four clusters of four R-4D engines, with each cluster positioned at a corner of the module to provide attitude control. These engines used a molybdenum nozzle that was radiation-cooled and produced 100 pounds of thrust. Manufactured by Marquardt Corporation (part number 228686-501), this particular engine was test-fired four times during Lunar Module testing.

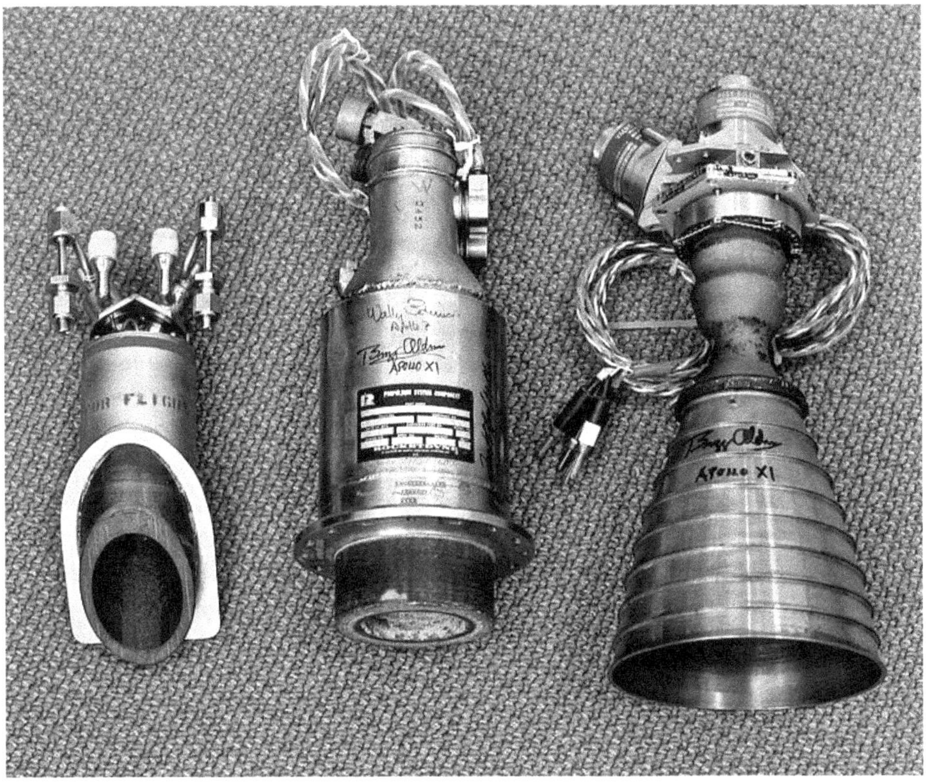

Figure 53: Three bipropellant engines, the first from the Gemini program and the second and third from the Apollo program. From left to right: Gemini SE-6, Apollo Command Module SE-8, Apollo Lunar Module R-4D. Steve Jurvetson, CC BY 2.0, via Wikimedia Commons.

Kerosene, particularly in its refined form known as RP-1, has played a crucial role in the evolution of liquid-fuelled rockets due to its practical advantages and compatibility with rocket design and operation [202, 203]. The initial development of kerosene as a rocket fuel stemmed from challenges faced with earlier fuels, such as ethyl alcohol, used in the V-2 rockets by Nazi Germany [202]. Ethyl alcohol, while sufficient for smaller rockets, lacked the energy density needed for larger rockets, and its water content, though beneficial for engine cooling, limited performance [202]. On the other hand, standard gasoline and kerosene provided more energy but presented new challenges, such as soot and combustion by-products that could clog rocket engines and plumbing systems [202].

By the 1950s, efforts in the United States chemical industry led to the formulation of RP-1 (Refined Petroleum-1), a highly purified form of kerosene [202, 203]. RP-1 addressed the challenges of traditional kerosene by burning cleaner, producing fewer residues, and reducing the risk of engine fouling [202, 203]. This made RP-1 the preferred choice for early American rockets, such as the Atlas, Titan I, and Thor missiles [202, 203]. The Soviet Union also adopted RP-1 for their R-7 rocket, laying the foundation for their space program, although later Soviet vehicles frequently transitioned to hypergolic propellants for long-term storage [202, 203].

RP-1 is favoured for first-stage rocket engines because of its energy density, storability, and safety [202, 203]. Unlike cryogenic fuels such as liquid hydrogen, kerosene remains in a liquid state at ambient temperatures, simplifying storage and handling [202, 203]. It does not require specialized insulated tanks, reducing weight and complexity [202, 203]. Additionally, kerosene poses fewer hazards to ground personnel since it does not produce explosive vapours like lighter hydrocarbons [202, 203]. These factors make kerosene particularly well-suited for rockets that require high thrust during the initial phase of launch when overcoming Earth's gravity and atmospheric drag is most critical [202, 203].

While liquid hydrogen (LH2) offers the highest specific impulse of all conventional fuels, it comes with significant challenges [202, 203]. Hydrogen's extremely low density necessitates large, insulated tanks, increasing the structural mass of the rocket and reducing the mass efficiency of the overall system [202, 203]. Kerosene, by contrast, has a much higher density, allowing for more compact and efficient fuel tanks [202, 203]. This makes RP-1 ideal for first stages where thrust is prioritized over efficiency [202, 203]. Hydrogen's advantages are more prominent in upper stages, where its higher specific impulse compensates for its bulkiness and the reduced atmospheric drag at high altitudes [202, 203].

Hydrogen fires, though dramatic, tend to dissipate quickly because hydrogen is buoyant and rises rapidly [202, 203]. In contrast, kerosene fires are far more damaging [202, 203]. Kerosene burns approximately 20% hotter than hydrogen and remains on the ground when spilled, causing prolonged fires that can damage launch infrastructure [202, 203]. This practical drawback requires additional safety precautions when testing and launching kerosene-fuelled rockets [202, 203].

As of 2024, RP-1 remains a dominant fuel for many modern rockets, especially for the first stages of orbital launch vehicles [202-204]. Rockets such as the Falcon 9 and Falcon Heavy by SpaceX use RP-1 and liquid oxygen (LOX) to produce high thrust and reliability [202-204]. The Electron rocket, a small-satellite launcher developed by Rocket Lab, also relies on RP-1 as its primary fuel [202-204]. These rockets demonstrate the continued relevance of kerosene in modern spaceflight, especially for cost-effective and reusable systems [202-204].

Despite advancements in cryogenic and methane-based fuels, RP-1 continues to be an optimal choice for many launch systems due to its storability, ease of handling, and ability to generate high thrust [202-204]. Its use in the rapidly expanding private space sector highlights its versatility and long-standing value in rocket propulsion [202-204].

Propellants

Liquid rocket engines operate with a variety of propellants, including cryogenic, semi-cryogenic, and storable hypergolic combinations. Each combination has unique advantages and limitations that determine their suitability for different missions.

Cryogenic Propellants: Cryogenic propellants, such as liquid oxygen (LOX) and liquid hydrogen (LH2), are among the most efficient fuel-oxidizer pairs used in space exploration. This combination is used in engines such as the Space Shuttle Main Engine, Space Launch System core stage, Ariane 5 main and second stages, Blue Origin's BE-3 engine, and upper stages like the Delta IV and Centaur. LOX and LH2 produce a clean burn with water vapor as the only byproduct and deliver high specific impulse. However, liquid hydrogen has significant challenges, including its extremely low storage temperature (20 K or −253°C) and very low density, which requires large, insulated tanks. The lightweight insulation used, such as in the Space Shuttle external tank, introduced risks, as seen in the Columbia disaster when a piece of insulation damaged the spacecraft during re-entry.

A more recent alternative to hydrogen is liquid methane (LNG) paired with LOX, used in SpaceX's Raptor and Blue Origin's BE-4 engines. LNG offers a balance between performance and practicality. While its specific impulse is lower than LH2, LNG has higher density, improving thrust-to-volume ratios and enabling smaller, lighter tanks. This characteristic makes it particularly suitable for reusable launch systems. LNG burns cleaner than kerosene (RP-1), producing less soot and minimizing engine coking, which improves reusability. It also operates at lower temperatures than LH2, reducing engine deformation and easing design challenges. Unlike RP-1, LNG does not require heaters in space to remain liquid, further simplifying long-duration missions. Additionally, LNG is cost-effective, widely available, and safer to handle than cryogenic hydrogen.

Semi-Cryogenic Propellants: Semi-cryogenic propellants typically involve liquid oxygen (LOX) combined with dense hydrocarbon fuels like RP-1 (kerosene). This combination powers many significant rocket systems, including the Saturn V's first stage, SpaceX's Falcon 9, the Soyuz rockets, and China's Long March family. LOX and kerosene offer a practical compromise between performance, density, and storability, making them ideal for high-thrust first stages. While RP-1 provides higher density than hydrogen, it produces soot, leading to challenges for reusability. Early rockets also utilized simpler fuels like alcohol (ethanol) and gasoline paired with LOX, as seen in World War II's V-2 rocket and Robert Goddard's pioneering designs.

More experimental approaches include LOX combined with carbon monoxide (CO), which has been proposed for Mars exploration. This combination offers a specific impulse around 250 seconds and can be produced on Mars through zirconia electrolysis, eliminating the need for transporting hydrogen and conserving local water resources.

Storable and Hypergolic Propellants: Storable propellants are non-cryogenic and can be stored for extended periods without special refrigeration, making them ideal for intercontinental ballistic missiles (ICBMs), spacecraft, and planetary probes. Many storable bipropellants are hypergolic, meaning they ignite on contact, simplifying engine design and eliminating the need for complex ignition systems.

Examples include combinations such as hydrazine (N_2H_4) with red fuming nitric acid (RFNA), or unsymmetrical dimethylhydrazine (UDMH) paired with dinitrogen tetroxide (N_2O_4). These fuels are used in rockets like Proton, Rokot, and Long March vehicles. The Titan II missile and Apollo's Lunar Module employed a mixture known as Aerozine-50, composed of equal parts hydrazine and UDMH, paired with N_2O_4. Monomethylhydrazine (MMH), combined with N_2O_4, powered systems like the Space Shuttle Orbital Maneuvering System (OMS) engines and the Reaction Control System (RCS).

SpaceX's Dragon spacecraft also utilizes hypergolic MMH and N_2O_4 for its Draco and SuperDraco engines.

Although highly reliable and storable, hypergolic fuels are toxic and carcinogenic, creating safety and environmental concerns. This has driven interest in safer alternatives. For example, hybrid rocket systems using non-toxic fuels and oxidizers are planned for vehicles like Dream Chaser and SpaceShipTwo, offering cleaner and more manageable propulsion systems.

The choice of propellants depends on the specific mission requirements, including performance, storability, density, and reusability. Cryogenic combinations like LOX-LH2 and LOX-LNG are preferred for high efficiency and reusability. Semi-cryogenic options, such as LOX-RP1, provide dense, high-thrust solutions ideal for initial rocket stages. Storable and hypergolic propellants, while toxic, remain essential for long-term missions where refrigeration is impractical. Advances in hybrid systems and safer alternatives continue to drive innovation in rocket propulsion technologies.

Injectors

The injector is a critical component in liquid rocket engines, directly influencing combustion efficiency, engine performance, and thermal management. The injector determines how effectively propellants (fuel and oxidizer) are mixed and combusted, which dictates the percentage of the nozzle's theoretical performance that can be achieved. Inefficient injectors allow unburnt propellants to leave the engine, leading to lower efficiency and reduced thrust. Additionally, injectors play a significant role in managing the thermal loads on the nozzle walls. By enriching the fuel concentration around the chamber's edges, the injector lowers the temperatures at the nozzle walls, reducing thermal stress and enabling better cooling.

Liquid rocket injectors range from simple designs to highly sophisticated configurations tailored for optimal propellant mixing and combustion stability. Early designs, such as those in the V-2 rocket, featured parallel fuel and oxidizer jets that combusted upon entering the chamber. While functional, these injectors delivered poor mixing efficiency and reduced overall performance.

Modern injectors typically consist of small holes drilled in carefully constructed patterns. These holes aim the propellant jets to collide at specific points just downstream of the injector plate. The collision of the fuel and oxidizer breaks the propellants into fine droplets, increasing the surface area and promoting efficient combustion.

Several common injector designs are in use today, including:

- **Showerhead injectors**: These consist of multiple holes that spray propellants like a showerhead. While simple, they are less efficient at atomizing propellants.
- **Self-impinging doublet injectors**: Jets of fuel and oxidizer are directed toward each other at an angle, ensuring they collide and atomize effectively.

- **Cross-impinging triplet injectors**: Three jets (two oxidizer and one fuel, or vice versa) intersect, creating improved droplet formation and mixing.
- **Centripetal or swirling injectors**: Rotational motion is imparted to the liquid propellants as they exit through small holes, creating cone-shaped sheets that rapidly atomize into fine droplets. Developed by Valentin Glushko in the early 1930s, this design has been widely used in Russian rocket engines due to its efficiency.
- **Pintle injectors**: These use a central pintle (a rod-like element) surrounded by propellant jets, offering excellent control of fuel-oxidizer mixing across a wide range of flow rates. Pintle injectors are notable for their use in the Apollo Lunar Module's Descent Propulsion System, the Kestrel engine, and SpaceX's Merlin engines powering the Falcon 9 and Falcon Heavy rockets.

Advanced injectors can also incorporate special designs to enhance combustion stability and efficiency. For example, the RS-25 engine, used in the Space Shuttle, employs a system of fluted posts. Heated hydrogen vaporizes the liquid oxygen as it flows through the centre of the posts, improving combustion stability and preventing oscillations. This design solved the high-frequency oscillation problems that plagued earlier engines, such as the F-1 engines on the Saturn V.

Figure 54: The Rocketdyne RS-25 is a liquid-fuel cryogenic rocket engine, burning LOX and hydrogen. Steve Jurvetson, CC BY 2.0, via Flickr.

Combustion stability is a major concern in liquid rocket engines, as instabilities can severely damage or destroy the engine. One common instability is chugging, a low-frequency oscillation caused by pressure variations in the chamber and propellant feed system. To mitigate this, the injector design must ensure sufficient pressure drop across the injectors—typically at least 20% of the chamber pressure. This pressure drop ensures that the flow rate remains stable, regardless of fluctuations in the chamber pressure.

However, larger engines often experience high-frequency combustion oscillations, which are far more destructive. These oscillations disrupt the gas boundary layer on the chamber walls, leading to excessive heat transfer that can overwhelm cooling systems and cause engine failure. These issues were particularly prevalent during the development of the Saturn V engines but were eventually resolved through rigorous testing and redesign.

To combat these instabilities, some engines, like the RS-25, use Helmholtz resonators. These damping mechanisms suppress specific resonant frequencies within the combustion chamber, preventing oscillations from growing uncontrollably.

Helmholtz resonators are devices used to suppress acoustic oscillations or vibrations at specific frequencies in systems such as rocket engine combustion chambers. They are named after Hermann von Helmholtz, a 19th-century physicist who studied the resonance of sound waves. In liquid rocket engines, Helmholtz resonators are implemented to stabilize combustion and prevent destructive high-frequency pressure oscillations.

The Helmholtz resonator works by exploiting the natural behaviour of air or gases within a cavity. The resonator consists of two main components:

1. A chamber or cavity that holds gas.
2. A narrow neck or opening through which gas can move in and out.

When a sound wave or pressure oscillation strikes the resonator, it causes the gas in the neck to oscillate like a piston. The gas inside the chamber acts as a spring, compressing and expanding in response to these oscillations. This interaction creates a specific resonant frequency at which the resonator absorbs and dissipates energy from the incoming oscillations.

The resonant frequency *f* of a Helmholtz resonator is determined by the volume of the cavity, the cross-sectional area of the neck, and the length of the neck, and can be expressed as:

$$f = \frac{c}{2\pi}\sqrt{\frac{A}{V \cdot L_{\text{eff}}}}$$

Where:

- f = Resonant frequency
- c = Speed of sound in the gas

- A = Cross-sectional area of the neck
- V = Volume of the cavity
- L_{eff} = Effective length of the neck (including corrections for the oscillating gas at the ends).

The resonator's natural frequency can be carefully tuned to match the dominant pressure oscillation in the rocket engine's combustion chamber. By doing so, the resonator absorbs the oscillation energy, effectively reducing its amplitude and dampening the instability.

In rocket engines, combustion instabilities often arise from pressure oscillations that can resonate within the combustion chamber. These oscillations disrupt the combustion process and may lead to catastrophic failure if not controlled. Helmholtz resonators are mounted in or around the combustion chamber to suppress these oscillations. They work by:

1. Absorbing Energy: When oscillations at the resonator's tuned frequency occur, the gas inside the resonator oscillates. This motion dissipates energy, reducing the amplitude of the pressure wave.
2. Breaking Resonance: The resonator prevents feedback loops that could reinforce oscillations, effectively stabilizing the combustion process.

For a Helmholtz resonator to work effectively in a rocket engine:

- Its frequency must match the problematic oscillation frequency in the combustion chamber. This requires careful tuning of the cavity's volume and neck dimensions.
- Multiple resonators may be placed around the chamber to address multiple modes of oscillations or complex wave patterns.
- The resonators must withstand the extreme temperatures and pressures in the combustion chamber environment.

In addition, the RS-25 incorporates pre-vaporization of propellants before injection, ensuring gas-phase combustion and enhanced stability. Further studies revealed that vaporizing the propellants prior to injection was sufficient to eliminate instability issues without requiring additional features.

Ensuring combustion stability involves rigorous testing. A common approach is the use of impulsive excitation tests, where small explosives are detonated inside the combustion chamber during operation. The pressure response of the chamber is analysed to assess how quickly the disturbances die out. If the pressure trace reveals sustained oscillations, the injector and chamber designs are adjusted to improve stability.

Engine Cycles

Engine cycles in liquid-propellant rockets refer to the methods by which fuel and oxidizer are pressurized and delivered into the combustion chamber. The chosen engine cycle affects the rocket's performance, complexity, and efficiency. Each cycle balances trade-offs between specific impulse, thrust, and system simplicity, depending on the mission requirements.

Pressure-Fed Cycle: In a pressure-fed cycle, the propellants are stored in pressurized tanks, and the pressure forces the fuel and oxidizer into the combustion chamber. This cycle eliminates the need for turbopumps, reducing system complexity and weight. Helium is often used as the pressurant because it is inert and lightweight, minimizing the risk of chemical reactions. However, the tanks must withstand high pressures, making them relatively heavy. As a result, this cycle works best for engines operating at lower chamber pressures, which limits the engine's thrust and power.

The pressure-fed cycle is highly reliable and efficient because all the fuel is used for thrust generation, and no energy is wasted in auxiliary systems. It is particularly suitable for upper stages or orbital manoeuvring systems, where lower thrust is acceptable. Notable examples include the AJ-10 engine used in the Space Shuttle's Orbital Maneuvering System (OMS) and Apollo's Service Propulsion System (SPS).

Electric Pump-Fed Cycle: The electric pump-fed cycle uses electric motors, typically powered by battery packs, to drive the pumps that deliver propellant to the combustion chamber. This design simplifies the engine architecture by removing turbomachinery, which reduces mechanical complexity and improves reliability. However, the inclusion of batteries adds significant dry mass to the system, reducing the vehicle's overall performance.

Despite this drawback, the electric pump-fed cycle is particularly useful for small launch vehicles, where simplicity and cost-effectiveness are critical. A notable example is Rocket Lab's Rutherford engine, which powers the Electron rocket and demonstrates the practicality of electric pump-fed engines for lightweight applications.

Gas-Generator Cycle: In a gas-generator cycle, a small amount of propellant is burned in a separate preburner to drive the turbopumps. The exhaust from this preburner, which provides no meaningful thrust, is expelled through a separate nozzle or low on the main nozzle. While this design allows for very powerful turbopumps, enabling high thrust levels, it sacrifices efficiency because not all the propellant contributes to the main thrust generation.

The gas-generator cycle strikes a balance between simplicity and high thrust, making it one of the most widely used engine cycles for booster stages. Examples include the F-1 engine on the Saturn V, which powered the first stage of the Apollo missions, and the Merlin engine used in SpaceX's Falcon 9 and Falcon Heavy rockets.

Tap-Off Cycle: The tap-off cycle takes hot gases directly from the main combustion chamber to drive the turbopumps. These gases are then exhausted, making this an open-cycle design similar to the gas-generator cycle. However, the tap-off cycle is more efficient because it uses gases already present in the main combustion chamber rather than burning extra propellant. This approach simplifies the engine design but can be challenging to implement because the gases must be carefully controlled to avoid performance losses or instability.

The BE-3 engine by Blue Origin, used on the New Shepard rocket, and the experimental J-2S engine are prominent examples of tap-off cycle engines.

Expander Cycle: The expander cycle leverages cryogenic fuels like liquid hydrogen or methane to cool the combustion chamber and nozzle walls. As the fuel absorbs heat, it vaporizes and expands, driving the turbopumps before entering the combustion chamber. This cycle is extremely efficient because it uses thermal energy that would otherwise be lost.

However, the expander cycle's power is limited by the amount of heat transferred to the fuel, restricting the engine's thrust. To increase power, a bleed version of the cycle can exhaust some vaporized fuel overboard, enabling higher turbopump performance at the cost of minor efficiency losses. Examples include the RL10 engine, used on the Atlas V and Delta IV second stages, and Japan's LE-5 engine, which employs a bleed cycle.

Staged Combustion Cycle: The staged combustion cycle burns a fuel-rich or oxidizer-rich mixture in a preburner to drive the turbopumps. The resulting high-pressure exhaust is then fed directly into the main combustion chamber, where the remaining fuel or oxidizer completes combustion. This design allows for extremely high chamber pressures and efficiency because no propellant is wasted.

The increased performance comes at the cost of higher system complexity and mass, as the engine must handle the corrosive and hot exhaust gases from the preburner. The Space Shuttle Main Engine (SSME) and Russia's RD-191 engine are notable examples of staged combustion engines.

Full-Flow Staged Combustion Cycle: The full-flow staged combustion cycle improves on the standard staged combustion cycle by burning fuel-rich and oxidizer-rich mixtures in separate preburners to drive their respective turbopumps. Both high-pressure exhaust streams are then injected into the main combustion chamber, where they combine and combust. This cycle achieves extremely high efficiency, allows for very high combustion pressures, and reduces engine wear because the fuel-rich and oxidizer-rich gases are less corrosive at lower temperatures.

Although this design is even more complex than standard staged combustion, it delivers the highest performance among engine cycles. SpaceX's Raptor engine, used on the Starship rocket, is a prominent example.

Selecting an engine cycle involves trade-offs between performance, complexity, and cost. The gas-generator cycle offers high thrust and simplicity but sacrifices efficiency. The expander cycle delivers high efficiency but has limited power due to thermal constraints. The staged combustion cycle and full-flow staged combustion cycle maximize efficiency and thrust but require complex systems and higher costs. The pressure-fed cycle and electric pump-fed cycle simplify design and are ideal for low-thrust applications but result in lower performance for large rockets.

Ultimately, the choice of engine cycle depends on mission requirements, such as thrust, efficiency, and reliability, as well as practical considerations like cost and technical constraints.

Engine Cooling

Cooling in liquid rocket engines is a critical aspect of design, ensuring that the extreme temperatures produced during combustion do not damage or destroy the engine components. One of the primary methods to achieve this is through the strategic layout of injectors. By creating a fuel-rich layer along the walls of the combustion chamber, the temperature near these surfaces is significantly reduced. This fuel-rich boundary layer insulates the chamber wall, throat, and nozzle from the full heat of combustion, which can exceed thousands of degrees Celsius. This design not only protects the engine materials but also allows for higher chamber pressures, enabling the use of more efficient, high-expansion ratio nozzles that improve specific impulse (ISP) and overall system performance.

A common cooling method used in liquid rocket engines is regenerative cooling. This technique involves circulating the liquid propellant—usually the fuel but occasionally the oxidizer—through cooling channels around the combustion chamber and nozzle. As the propellant flows through these channels, it absorbs heat from the engine walls, keeping the structural components below their thermal limits. The absorbed heat also preheats the propellant, enhancing its energy content and improving combustion efficiency once it is injected into the chamber.

Regenerative cooling is particularly effective for cryogenic propellants, such as liquid hydrogen, which have a high capacity for absorbing heat. By combining the protective benefits of fuel-rich boundary layers and the efficient heat transfer properties of regenerative cooling, rocket engines can achieve greater performance while maintaining structural integrity under the extreme conditions of operation. This integration of cooling methods is vital for the success of modern high-pressure, high-performance rocket engines.

Ignition

Ignition is a crucial aspect of liquid rocket engine operation, as it requires a consistent and reliable source to initiate combustion [205]. The precise timing and method of ignition are vital because even a slight delay on the scale of tens of milliseconds can result in excessive propellant buildup within the combustion chamber, leading to over pressurization and a phenomenon known as a "hard start" [206]. In extreme cases, this can destroy the engine through an explosion [206].

To ensure reliable ignition, systems are designed to provide a flame or heat source across the injector surface at the start of the engine cycle [207]. This flame typically interacts with a mass flow of propellant equivalent to about 1% of the engine's full mass flow rate, creating the conditions necessary for stable combustion to follow [207]. Safety mechanisms, such as interlocks, are sometimes integrated to confirm the presence of an ignition source before the main propellant valves open [207]. However, these interlocks introduce their own reliability challenges, and for critical systems like the RS-25 engine used on the Space Shuttle, interlocks are employed to shut down the engine in the event of ignition failure, ensuring safety during crewed missions [207].

Ignition methods vary widely and include pyrotechnic, electrical, and chemical approaches [208]. Pyrotechnic systems rely on small explosive charges to generate the initial flame, while electrical systems use sparks or hot wires to ignite the propellant [208]. Chemical methods leverage highly reactive substances to initiate combustion, such as the use of hypergolic propellants, which ignite spontaneously upon contact with each other, minimizing the risk of hard starts and improving reliability [209-211].

One particularly effective chemical ignition method involves pyrophoric agents such as triethylaluminium (TEA) [212]. TEA is highly reactive, igniting spontaneously upon exposure to air or cryogenic liquid oxygen, and even decomposing in water or other oxidizers [212]. Its enthalpy of combustion is exceptionally high, making it a powerful ignition source [212]. Often combined with triethylborane (TEB) to form TEA-TEB mixtures, this method provides a reliable ignition system, particularly in engines that require precise and rapid ignition under extreme conditions [212].

Hybrid Propulsion Systems

Hybrid propulsion systems are a form of rocket propulsion that combine elements of both solid and liquid propellants. These systems utilize a solid fuel and a liquid or gaseous oxidizer. This combination allows hybrid rockets to leverage the simplicity of solid rockets while incorporating some of the controllability and higher specific impulse of liquid rockets. The defining characteristic of hybrid propulsion is that the fuel and oxidizer exist in different physical states, separated until ignition [127].

Hybrid propulsion systems offer a unique balance of safety, cost-effectiveness, and controllability, positioning them as an attractive option for a range of rocket applications. One of their key advantages lies in safety and storage. Unlike traditional solid or liquid systems, the separation of fuel and oxidizer in hybrid systems significantly reduces the risk of accidental detonation. Additionally, these systems are non-hypergolic, meaning they require an external ignition source, which adds an extra layer of safety during handling and storage.

Another advantage is their throttleability and shutdown capabilities. By controlling the oxidizer flow rate, hybrid systems can be throttled, restarted, or even shut down mid-flight, offering operational flexibility that is unavailable in traditional solid rockets. This feature makes hybrids especially suitable for missions requiring precise thrust control. The simplicity of hybrid systems further enhances their appeal. Unlike liquid engines, hybrids do not require complex turbopumps or plumbing systems, simplifying the propulsion design and lowering production costs. Cost efficiency is another significant benefit, as hybrid systems often use inexpensive materials such as HTPB (hydroxyl-terminated polybutadiene) for fuel and nitrous oxide for oxidizers.

Despite these advantages, hybrid propulsion systems face several challenges. Combustion instabilities, for instance, can arise from uneven mixing of oxidizer and fuel pyrolysis gases, leading to localized hotspots or oscillations. Achieving uniform combustion remains a technical hurdle. Limited regression rates, or the rate at which the solid fuel burns, can also constrain the system's thrust capabilities. To address this, innovations such as swirl injectors or advanced grain geometries are employed. Oxidizer handling presents another challenge, particularly when using cryogenic or high-pressure oxidizers, which require meticulous engineering to manage thermal and pressure conditions. Scaling up hybrid engines for high-thrust applications also introduces difficulties, as maintaining consistent combustion across larger fuel grains becomes more complex.

Recent innovations in hybrid rocket technology are addressing these challenges. Additive manufacturing techniques, such as 3D printing, are now used to optimize fuel grain geometry and improve regression rates. Swirl injection technology enhances combustion efficiency by increasing turbulence and improving the mixing of oxidizer with fuel gases. Paraffin-based fuels, which exhibit higher regression rates due to their ability to form a thin liquid layer during combustion, are gaining popularity. Additionally, metal additives such as aluminum or magnesium can be incorporated into the solid fuel to increase energy density and combustion temperatures.

Hybrid propulsion systems are used in various applications. They are popular in suborbital and educational rockets due to their safety and simplicity, making them ideal for research vehicles and student-designed projects. In space tourism, Virgin Galactic's SpaceShipTwo employs a hybrid system that combines rubber-based solid fuel with a nitrous oxide oxidizer. Military applications are also exploring hybrid propulsion for tactical missiles, where safety and controllability are crucial. Furthermore, the throttleability and restart capability of hybrids make them an appealing option for reusable launch systems under development.

The basic components of a hybrid propulsion system include:

Combustion Chamber and Solid Fuel Grain: The combustion chamber contains the solid fuel, typically cast or molded into a specific geometry (e.g., cylindrical or star-shaped) to optimize surface area and combustion characteristics. Common solid fuels include synthetic rubbers like hydroxyl-terminated polybutadiene (HTPB), polyethylene, or paraffin wax [125, 213, 214].

Oxidizer Supply System: The oxidizer is stored in a separate tank, typically in liquid or gaseous form. Examples of oxidizers include liquid oxygen (LOX), nitrous oxide (N_2O), or hydrogen peroxide (H_2O_2) [213, 215, 216]. The oxidizer is introduced into the combustion chamber via injectors, which atomize the oxidizer to enhance mixing and combustion efficiency [127].

Injector System: The injectors deliver and distribute the oxidizer into the combustion chamber. Their design influences the mixing rate between the oxidizer and the pyrolyzed fuel gases from the solid grain [127].

Ignition System: Ignition initiates the combustion process, often using pyrotechnic devices or hypergolic igniters to provide the heat necessary to start the reaction [217, 218].

Nozzle: The combustion gases exit the chamber through a convergent-divergent (C-D) nozzle, accelerating the exhaust to produce thrust. The nozzle is typically designed to handle a wide range of flow conditions due to the variability in hybrid combustion dynamics [127].

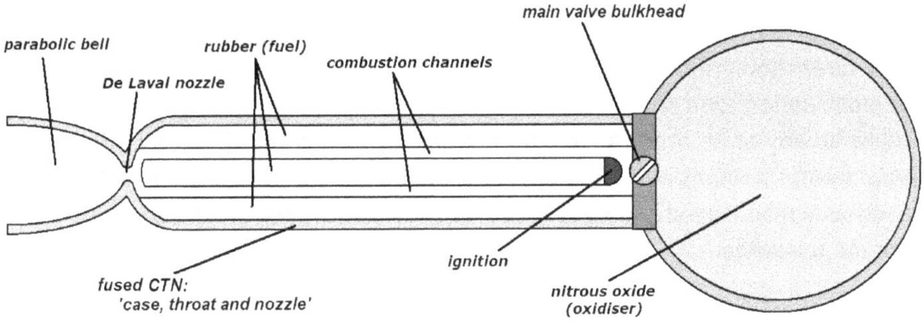

Figure 55: Hybrid rocket motor detail of SpaceShipOne. Jack, Public domain, via Wikimedia Commons.

Hybrid rocket engines combine solid fuel with a liquid or gas oxidizer, providing many advantages over conventional liquid or solid rocket propulsion systems, such as low cost, fuel handling simplicity, durability, throttling capability, and eco-friendly attributes [127]. Researchers have recognized the potential of hybrid propulsion and have investigated various aspects, including combustion modelling [127], new propellant formulations [213], ignition systems [217, 218], and applications for space missions [214, 219].

The principle of operation of a hybrid propulsion system begins with ignition, which provides the energy required to pyrolyze, or thermally decompose, the surface of the solid fuel. This process creates a layer of combustible gases near the fuel surface, forming the foundation for sustained combustion.

Once ignition is initiated, the oxidizer, in either liquid or gaseous form, is introduced into the combustion chamber through the injector system. The oxidizer interacts with the pyrolyzed fuel gases, and this interaction triggers combustion. The reaction between the oxidizer and the fuel gases releases significant heat, which maintains the pyrolysis of the solid fuel. This cycle creates a self-sustaining combustion process that continues as long as the oxidizer flow is maintained.

Combustion takes place at the interface of the solid fuel surface and the incoming oxidizer, where the reaction zone generates heat and energy. The burn rate of the fuel is influenced by several factors, including the oxidizer flow rate, the chamber pressure, and the geometry of the solid fuel grain. These parameters determine the overall efficiency and performance of the hybrid engine.

The combustion process generates high-temperature, high-pressure gases, which are expelled through a convergent-divergent nozzle. As the gases exit the nozzle at high velocity, they produce thrust according to Newton's third law of motion. This expulsion of exhaust gases provides the forward momentum necessary for the rocket to achieve lift and propulsion. The combination of controlled combustion and efficient expulsion of gases makes hybrid propulsion systems a versatile and effective solution in rocket engineering.

Transient Operation Model for Hybrid Rockets

The **transient operation model** for hybrid rockets refers to the theoretical and computational framework used to understand and predict the time-dependent behaviour of hybrid rocket propulsion systems. Unlike steady-state models, transient models focus on the dynamic processes occurring during ignition, ramp-up, combustion, shutdown, and any operational fluctuations. These transient phases are critical for understanding the rocket's performance, ensuring system stability, and mitigating potential failures.

Key Phases of Transient Operation:

1. Ignition Phase: During ignition, the solid fuel grain's surface begins to pyrolyze due to an external energy source, such as a spark, pyrotechnic device, or hypergolic reaction. This process initiates the production of combustible gases, creating a localized region of combustion. The transient model must account for:

- The delay in ignition, which could lead to unburnt oxidizer accumulation and potential overpressure (hard start).
- The thermal and chemical interactions between the igniter, solid fuel, and incoming oxidizer flow.
- Initial chamber pressure build-up and the development of a stable combustion front.

2. Ramp-Up Phase: After ignition, the combustion chamber reaches operational conditions as the chamber pressure, temperature, and mass flow rates stabilize. The transient model addresses:

- The rate of pressure increase, which is influenced by oxidizer injection rate and fuel pyrolysis rate.
- Heat transfer mechanisms, including thermal conduction into the solid fuel grain and heat loss to the combustion chamber walls.
- The time-dependent changes in oxidizer-to-fuel ratio (O/F ratio), which are critical for optimizing performance and preventing instability.

3. Steady-State Combustion: During steady-state operation, the hybrid rocket achieves a stable thrust output. While primarily steady, slight transient effects may still occur due to:

- Variations in oxidizer flow rate caused by injector or tank pressure fluctuations.
- Changes in regression rate of the fuel grain as the burn progresses, altering the surface geometry and combustion characteristics.
- Erosive burning effects, particularly in regions of high oxidizer velocity near the grain surface.

4. Shutdown Phase: During shutdown, the oxidizer flow is terminated, and combustion ceases. Transient modelling during this phase focuses on:

- Residual combustion and heat dissipation in the combustion chamber.
- The impact of thermal stress on the solid fuel grain and chamber walls as they cool.
- Potential afterburn effects if unburnt oxidizer remains in the chamber.

Mathematical Framework

Transient operation models are typically governed by a set of coupled partial differential equations (PDEs) that describe the conservation of mass, momentum, and energy:

Mass Conservation:

$$\frac{\partial \rho}{\partial t} + \nabla \cdot (\rho \vec{v}) = \dot{m}_{\text{fuel}} + \dot{m}_{\text{oxidizer}}$$

Here, ρ is the gas density, $\vec{v}$ is the velocity field, and $\dot{m}_{\text{fuel}}$ and $\dot{m}_{\text{oxidizer}}$ represent the mass flow rates of fuel and oxidizer, respectively.

Momentum Conservation:

$$\frac{\partial (\rho \vec{v})}{\partial t} + \nabla \cdot (\rho \vec{v} \vec{v}) = -\nabla P + \mu \nabla^2 \vec{v} + \vec{F}$$

This equation accounts for pressure (P), viscous effects (μ), and external forces ($\vec{F}$).

Energy Conservation:

$$\frac{\partial e}{\partial t} + \nabla \cdot ((e + P)\vec{v}) = \dot{q} - \nabla \cdot \vec{q}_{\text{cond}} - \nabla \cdot \vec{q}_{\text{rad}}$$

Here, e is the internal energy, $\dot{q}$ is the heat release due to combustion, and $\vec{q}_{\text{cond}}$ and $\vec{q}_{\text{rad}}$ are conductive and radiative heat losses, respectively.

Fuel Regression Rate: The fuel regression rate $(\dot{r})$ is modelled using empirical or semi-empirical correlations:

$$\dot{r} = aG_{\text{ox}}^{n}$$

where a and n are constants derived from experimental data, and G_{ox} is the oxidizer mass flux.

Chamber Pressure Dynamics: Chamber pressure (P_c) is dynamically linked to mass flow rates and combustion properties:

$$P_c = \frac{\dot{m}_{\text{total}} C^*}{A_t}$$

where $\dot{m}_{\text{total}}$ is the total mass flow rate, C^* is the characteristic velocity, and A_t is the throat area.

Rocket design involves dynamic changes in various physical properties such as mass, momentum, energy, and combustion characteristics. Transient operation models rely on partial differential equations (PDEs) to describe these changes, providing a framework for analysing and optimizing hybrid and liquid rocket systems. Each formula within these models serves a specific purpose in ensuring the stable and efficient operation of the engine.

The mass conservation equation ensures that mass is preserved within the system. It tracks how the gas density changes over time and space due to the addition of mass from fuel and oxidizer. This equation is particularly useful during the ignition phase for modelling the initial filling of the combustion chamber, during the combustion phase to match reactant flow rates with exhaust gas rates, and during shutdown to simulate reactant depletion. Accurate mass conservation predictions are essential for maintaining the correct oxidizer-to-fuel ratio, preventing inefficiencies or overpressures that could lead to system failure.

The momentum conservation equation describes changes in momentum within the combustion chamber, accounting for pressure gradients, viscous forces, and external forces such as gravity or turbulence. This equation is vital for analysing combustion stability, optimizing thrust by ensuring smooth acceleration of exhaust gases, and designing injectors to model the interaction of fuel and oxidizer streams entering the chamber. Understanding momentum distribution prevents instabilities, such as chugging or pogo oscillations, that could reduce performance or damage the engine.

The energy conservation equation tracks energy flow in the system, considering internal energy, heat release from combustion, and energy losses through conduction and radiation. This equation is essential for optimizing heat transfer in combustion chamber design, predicting heat flux in cooling systems, and modelling the energy release characteristics of various propellants. Efficient energy management ensures high specific impulse and prevents overheating of critical engine components, maintaining thermal stability and prolonging engine lifespan.

The fuel regression rate equation models the rate at which the solid fuel grain burns, influenced by the oxidizer mass flux and constants specific to the fuel and oxidizer. This equation is crucial in grain geometry design to ensure even burning and prevent hotspots, predicting thrust curves over time, and simulating changes in the fuel surface during operation. Accurate regression rate modelling maintains a stable oxidizer-to-fuel ratio and avoids combustion instabilities.

The chamber pressure dynamics equation links the chamber pressure to total mass flow rate, characteristic velocity, and nozzle throat area. This equation is vital for nozzle design to achieve the desired thrust and pressure, engine cycle analysis to balance pressure-fed or pump-fed systems, and modelling transient behaviours during startup or shutdown phases. Managing chamber pressure effectively ensures optimal thrust and prevents overpressures that could damage the combustion chamber or nozzle.

These equations collectively play a crucial role in rocket engine design, enabling engineers to simulate and predict dynamic behaviours during ignition, steady combustion, and shutdown phases. By integrating mass, momentum, and energy conservation with regression rate and chamber pressure models, designers can ensure stable combustion, maximize efficiency, and prevent catastrophic failures. These models are particularly critical for scaling engines, managing thermal stresses, and achieving high-performance propulsion in both hybrid and liquid rocket systems.

Hybrid Rocket Chamber Pressure Equation

The chamber pressure equation for hybrid rockets is expressed as:

$$P_c = \frac{\dot{m}_{\text{total}} \cdot C^*}{A_t}$$

where:

- P_c is the chamber pressure,
- $\dot{m}_{\text{total}}$ is the total mass flow rate (fuel + oxidizer),
- C^* is the characteristic velocity, and
- A_t is the nozzle throat area.

This equation provides a relationship between the key parameters that govern the pressure inside the combustion chamber of a hybrid rocket. Understanding and applying this equation is critical for the design and analysis of hybrid rocket engines.

The chamber pressure equation is an essential tool in rocket design, used alongside other engine design equations to calculate or verify chamber pressure under specific operating conditions. The process begins with determining the total mass flow rate, which involves calculating the oxidizer mass flow rate based on the injector design and supply system. The fuel regression rate is then derived using the regression rate equation and combined with the grain geometry to calculate the fuel mass flow rate. These two components are added together to find the total mass flow rate. Next, the characteristic velocity is evaluated, which depends on the propellant combination and combustion properties and is typically derived from empirical data or thermochemical simulations.

The next step involves selecting the nozzle throat area, determined by the desired throat diameter, as this significantly influences nozzle design and exhaust flow characteristics. With these parameters in place, the chamber pressure is calculated by substituting the total mass flow rate, characteristic velocity, and throat area into the equation. If the calculated chamber pressure does not meet the design criteria, iterative adjustments are made to variables such as throat area, oxidizer flow rate, or fuel grain geometry, and the calculations are repeated.

The chamber pressure equation is applied during various stages of hybrid rocket development. During conceptual design, it estimates the chamber pressure for a specific combination of fuel, oxidizer, and geometry, ensuring alignment with material and structural constraints. It is also used for performance optimization, allowing engineers to assess how design changes impact chamber pressure and engine performance. For combustion stability, the equation ensures the chamber pressure stays within a stable range, preventing instabilities like chugging or hard starts. In transient analysis, it models the dynamic response of chamber pressure during ignition, steady operation, and shutdown phases.

This equation serves multiple purposes in hybrid rocket engineering. It helps determine structural requirements by influencing the strength and material properties needed for the combustion chamber and nozzle, balancing between overdesign and catastrophic failure. In thrust prediction, the chamber pressure directly correlates with thrust, requiring careful management to avoid structural and thermal constraints. It also guides nozzle optimization by affecting the expansion ratio, which influences exhaust velocity and specific impulse. Moreover, it aids in fuel grain and oxidizer injector design, ensuring an appropriate oxidizer-to-fuel ratio is maintained. Lastly, the equation is critical for system safety and reliability, ensuring chamber pressure remains within safe limits to mitigate risks of overpressure and maintain combustion stability.

Fuel Grain Regression Model

The Fuel Grain Regression Model is a critical framework in the design and analysis of hybrid rocket propulsion systems. It provides a mathematical representation of how the solid fuel burns over time in response to the interaction with the oxidizer flow. This model is essential for predicting the fuel mass flow rate, maintaining the oxidizer-to-fuel ratio, and ensuring consistent thrust generation throughout the rocket's operation.

In a hybrid rocket, the solid fuel grain does not burn uniformly across its surface. Instead, the regression, or burn rate, of the fuel surface depends on multiple factors, including the oxidizer mass flux, fuel geometry, and combustion conditions. The regression model captures these dependencies and allows for the accurate simulation of the combustion process.

The regression rate of the fuel is typically expressed as:

$$\dot{r} = aG_{\mathrm{ox}}^{n}$$

Here:

- $\dot{r}$ is the regression rate (m/s), representing the rate at which the fuel surface recedes due to combustion.
- G_{ox} is the oxidizer mass flux (kg/m^2·s), defined as the mass flow rate of the oxidizer per unit area of the fuel surface.
- *a* is an empirical constant dependent on the fuel and oxidizer combination and the combustion chamber conditions.

- n is a regression rate exponent, typically between 0.5 and 0.8 for most hybrid systems, determined experimentally.

The regression model describes how the interaction between the oxidizer and the fuel surface drives the combustion process. The oxidizer flows into the combustion chamber, and as it comes into contact with the solid fuel grain, it causes pyrolysis of the fuel. The heat from the combustion process sustains this pyrolysis, creating a layer of combustible gases that mix with the oxidizer, driving the reaction forward.

The oxidizer mass flux, G_{ox}, plays a dominant role in determining the burn rate. A higher oxidizer flow rate results in increased regression due to more energetic combustion. Additionally, the geometry of the fuel grain influences the surface area exposed to the oxidizer, directly impacting the fuel mass flow rate.

The fuel grain regression model is applied throughout the design, testing, and operational phases of a hybrid rocket system. It is particularly useful for:

1. **Fuel Grain Geometry Design**: By using the regression model, engineers can design grain geometries that optimize the surface area and ensure even burning throughout the rocket's operation. Common geometries include cylindrical bores, star-shaped patterns, or helical grooves.
2. **Thrust Prediction**: Since the regression rate determines the fuel mass flow rate, it directly impacts the thrust produced by the engine. The model allows engineers to predict and optimize thrust profiles over the burn time.
3. **O/F Ratio Management**: The oxidizer-to-fuel ratio is critical for achieving efficient combustion. The regression model helps ensure this ratio remains within the desired range by predicting how the fuel burn evolves over time.
4. **Transient Analysis**: During transient phases such as ignition or shutdown, the regression model helps simulate dynamic changes in the burn rate and assess stability.
5. **Scaling and Performance Evaluation**: For larger hybrid rockets, the regression model aids in scaling designs and evaluating performance under various operating conditions.

The fuel regression model is crucial because it addresses some of the inherent challenges in hybrid propulsion systems, such as limited regression rates compared to liquid or solid rockets. Innovations like swirl injectors, paraffin-based fuels, or metal additives are often evaluated using this model to enhance regression rates and improve performance.

Accurate modelling ensures stable and efficient combustion, prevents localized hotspots or uneven burning, and allows engineers to predict how changes in operating conditions (e.g., oxidizer flow rate, chamber pressure) will affect performance. By integrating the regression model with other rocket design equations, hybrid systems can achieve higher reliability, scalability, and thrust consistency, making them suitable for various applications in suborbital flights, space tourism, and small satellite launches.

Enthalpy Balance Regression Model History

The enthalpy balance approach to modelling fuel grain regression in hybrid rockets emerged as an extension of classic heat transfer principles, building upon earlier empirical models by incorporating energy conservation laws at the solid-liquid-gas interface [220, 221]. The aim was to improve accuracy in predicting regression rates by considering the detailed thermochemical interactions occurring during combustion [221].

Early models, such as the Marxman and Gilbert relation, primarily relied on semi-empirical correlations derived from experimental observations [221, 222]. While effective for simple geometries and limited conditions, these models struggled to capture the complexities of modern hybrid rocket systems, such as swirl injection, alternative fuels, and high-pressure oxidizer flows [223, 224]. Enthalpy-balance models were developed to address these limitations by explicitly accounting for the heat transfer between the reacting oxidizer-fuel mixture, the pyrolyzing fuel grain, and the surrounding environment [220, 221].

This approach allows for more precise predictions of fuel regression rates, especially in systems with advanced grain geometries or varying operating conditions [225, 226]. Enthalpy-balance models are now widely used in hybrid rocket research and development [227].

The Marxman and Gilbert relation, developed in the early 1960s, was one of the first comprehensive attempts to model fuel regression in hybrid rockets [221, 222]. It is based on the concept of convective heat transfer from the oxidizer flow to the fuel surface, which drives pyrolysis and regression [221, 222].

The regression rate, $\dot{r}$, is expressed as:

$$\dot{r} = aG_{\text{ox}}^{n}$$

Here:

- G_{ox} is the oxidizer mass flux (kg/m^2·s), defined as the oxidizer flow rate per unit surface area of the fuel grain.
- *a* and *n* are empirically determined constants that depend on the specific oxidizer-fuel pair and operating conditions.

The Marxman and Gilbert model assumes that:

1. The boundary layer over the fuel grain is turbulent.
2. The fuel regression is dominated by the heat flux from the gas phase to the solid fuel surface.
3. The regression rate is proportional to the oxidizer flux raised to a power (*n*), typically between 0.5 and 0.8.

This model provides a simple yet effective means of predicting regression rates for basic hybrid rocket designs. However, it does not account for complexities like multidimensional heat transfer, varying grain geometries, or advanced injection techniques.

Enthalpy-Balance Fuel Grain Regression Model for Hybrid Rockets

The enthalpy-balance regression model improves upon the Marxman and Gilbert relation by explicitly considering the energy balance at the fuel surface. It calculates the regression rate by equating the heat flux from the gas phase to the surface with the heat required to pyrolyze the solid fuel.

The governing equation is:

$$q_{\text{conv}} = \dot{r} \cdot \Delta h$$

Here:

- q_{conv} is the convective heat flux from the oxidizer flow to the fuel surface.
- $\dot{r}$ is the regression rate (m/s).
- Δh is the enthalpy of pyrolysis, representing the energy required to decompose the solid fuel into gaseous products.

In this model:

1. The heat flux q_{conv} is derived from boundary layer theory and depends on the oxidizer velocity, temperature, and composition.
2. Δh incorporates both the latent heat of pyrolysis and any additional heat absorbed by the fuel during its phase change.
3. The regression rate is directly tied to the energy balance, making the model more robust for varying conditions.

This approach is particularly effective for modern hybrid rockets that use complex grain geometries, swirl injectors, or paraffin-based fuels. It also provides a better framework for analysing transient effects, such as ignition or shutdown phases.

Wall Blowing Correction

The wall blowing correction is an adjustment applied to account for the effect of pyrolyzed fuel gases escaping from the surface of the fuel grain. This phenomenon alters the boundary layer dynamics by introducing additional momentum and thermal energy into the flow.

Wall blowing modifies the convective heat transfer coefficient, *h*, reducing the heat flux to the surface and consequently affecting the regression rate. The corrected heat flux is given by:

$$q_{\text{conv}}^{\text{corrected}} = q_{\text{conv}} \cdot (1 - B)$$

Here:

- B is the blowing parameter, which quantifies the influence of the escaping fuel gases on the boundary layer.

The wall blowing correction is significant for fuels with high pyrolysis rates or systems with high oxidizer flux. It ensures that the regression model remains accurate even under conditions where mass addition from the fuel surface substantially alters the boundary layer.

Skin Friction Coefficient Model

The skin friction coefficient model relates the shear stress at the surface of the fuel grain to the flow properties of the oxidizer. This is important because the convective heat flux, which drives the regression rate, depends on the turbulent boundary layer, which in turn is influenced by skin friction.

The skin friction coefficient, C_f, is typically expressed as:

$$C_f = \frac{\tau_w}{\frac{1}{2}\rho U^2}$$

Here:

- τ_w is the wall shear stress.
- ρ is the gas density.
- U is the freestream velocity.

In hybrid rockets, the skin friction coefficient is influenced by:

1. The geometry of the fuel grain.
2. The Reynolds number of the flow.
3. The blowing effects from pyrolyzed fuel gases.

A higher skin friction coefficient generally corresponds to increased heat transfer to the fuel surface, leading to higher regression rates. Accurate modelling of C_f is essential for predicting how variations in flow conditions or grain design will affect combustion performance.

Together, these models provide a comprehensive framework for understanding and optimizing the combustion process in hybrid rockets. They enable engineers to:

1. Predict regression rates for various fuel-oxidizer combinations and operating conditions.
2. Design fuel grain geometries that maximize surface area and combustion efficiency.
3. Account for complex boundary layer interactions, such as wall blowing or turbulent heat transfer.
4. Ensure stable combustion and prevent localized hotspots that could lead to failure.

By integrating these models into computational simulations and experimental testing, hybrid rocket systems can achieve greater reliability, scalability, and performance, making them suitable for applications ranging from suborbital research to space tourism and military systems.

Example to Demonstrate Transient Operation Models in Hybrid Rocket Design

Let's design a hybrid rocket for a student-led suborbital research mission, with a goal of achieving a maximum altitude of 50 km. The hybrid rocket will use paraffin wax as the solid fuel and liquid nitrous oxide (N_2O) as the oxidizer. The design requires integrating transient operation models for mass conservation, momentum conservation, energy conservation, fuel regression rate, and chamber pressure dynamics.

Step 1: Mass Conservation

The conservation of mass equation is applied to track the flow of fuel gases and oxidizer into the combustion chamber. Assume:

- The oxidizer mass flow rate, $\dot{m}_{\text{oxidizer}}$, is regulated by a valve that delivers 2 kg/s of N_2O.
- The fuel regression rate is modelled to calculate $\dot{m}_{\text{fuel}}$, derived from the regression rate formula,

$$\dot{r} = aG_{\text{ox}}^n.$$

We find $\dot{m}_{\text{fuel}}$ based on the grain geometry (a cylindrical core of 0.1 m diameter and 1.0 m length).

The oxidizer flux, G_{ox}, is:

$$G_{\text{ox}} = \frac{\dot{m}_{\text{oxidizer}}}{A_{\text{fuel}}}$$

where A_{fuel} is the exposed surface area. The mass conservation equation ensures that:

$$\frac{\partial \rho}{\partial t} + \nabla \cdot (\rho \vec{v}) = \dot{m}_{\text{fuel}} + \dot{m}_{\text{oxidizer}}$$

By solving this equation, we validate that the total mass flow rate equals the exhaust flow rate.

Step 2: Momentum Conservation

To ensure stable combustion and thrust generation, the momentum conservation equation is used:

$$\frac{\partial (\rho \vec{v})}{\partial t} + \nabla \cdot (\rho \vec{v} \vec{v}) = -\nabla P + \mu \nabla^2 \vec{v} + \vec{F}$$

Here:

- P is the pressure gradient in the chamber.
- μ is the viscosity of the exhaust gases.
- $\vec{F}$ includes turbulence-induced forces.

Using this, we design the injector to provide uniform oxidizer distribution and calculate the thrust generated at the nozzle exit. Numerical simulations confirm no adverse instabilities like chugging or oscillations.

Step 3: Energy Conservation

The energy conservation equation is crucial for ensuring that the combustion chamber operates within thermal limits:

$$\frac{\partial e}{\partial t} + \nabla \cdot \big((e + P)\vec{v}\big) = \dot{q} - \nabla \cdot \vec{q}_{\text{cond}} - \nabla \cdot \vec{q}_{\text{rad}}$$

Here:

- $\dot{q}$ is the heat released from combustion.
- $\vec{q}_{\text{cond}}$ and $\vec{q}_{\text{rad}}$ represent heat losses.

Given the enthalpy of pyrolysis for paraffin wax (3×10^5 J/kg^3), we calculate the required heat flux, ensuring that the chamber materials withstand thermal stress. Regenerative cooling with liquid N_2O is modelled to absorb excess heat, protecting the chamber walls.

Step 4: Fuel Regression Rate

The regression rate is modelled using:

$$\dot{r} = aG_{\text{ox}}^n$$

For paraffin wax, empirical data provides a=0.002 and n=0.8. With G_{ox} =25 kg/m²·s, the regression rate is:

$$\dot{r} = 0.002 \times 25^{0.8} \approx 0.018\,\text{m/s}$$

Using the regression rate, we calculate the fuel burn area over time to maintain a stable oxidizer-to-fuel ratio (*O/F*≈3).

Step 5: Chamber Pressure Dynamics

The chamber pressure, P_c, is modelled using:

$$P_c = \frac{\dot{m}_{\text{total}} C^*}{A_t}$$

Here:

- $\dot{m}_{\text{total}} = \dot{m}_{\text{oxidizer}} + \dot{m}_{\text{fuel}}$,
- C^* is the characteristic velocity (1500 m/s),
- A_t is the nozzle throat area (0.001 m²).

Substituting $\dot{m}_{\text{total}}$ =2.2 kg/s, the chamber pressure is:

$$P_c = \frac{2.2 \times 1500}{0.001} = 3.3\,\text{MPa}$$

This value is within material limits for the chamber and ensures efficient thrust generation.

Using these equations, the hybrid rocket is optimized for safe operation and consistent thrust. The system achieves a specific impulse of 250 s, propelling the rocket to the target altitude of 50 km. These transient models guide design choices, from injector geometry to cooling system layout, ensuring the rocket performs reliably under dynamic conditions.

Emerging Technologies: Ion Thrusters and Plasma Drives

Space Propulsion Systems: Thruster Types

Electric propulsion (EP) systems have advanced significantly, becoming a cornerstone of modern spacecraft propulsion due to their efficiency and adaptability. Numerous types of electric thrusters have been developed, each suited to specific mission scenarios or manoeuvres. These propulsion systems can be broadly categorized into three main types: electrothermal, electrostatic, and electromagnetic thrusters. Each category represents a distinct method of converting electrical energy into kinetic energy for propulsion [228].

Electrothermal Thrusters:

Electrothermal thrusters generate thrust by heating a gaseous propellant electrically and expanding it through a nozzle. This category includes resistojets and arcjets, which are relatively simple in design. In resistojets, the propellant is heated through an electric heater element, relying on Ohmic heating. This increases the propellant's exhaust velocity, though the velocities achieved are generally lower compared to other EP systems. Arcjets, on the other hand, use an arc discharge to heat the propellant. Collisions between the arc's particles and the propellant result in higher temperatures and exhaust velocities than resistojets. While electrothermal thrusters are cost-effective and straightforward, their specific impulse (Isp)—a measure of propulsion efficiency—is comparatively low, making them suitable for missions requiring modest thrust over shorter durations [228].

Electrostatic Thrusters: Electrostatic thrusters generate thrust by accelerating ions using electrostatic fields. The ionization of the propellant and the subsequent acceleration of ions occur in two distinct stages. For instance, gridded ion engines (GIEs) ionize the propellant within a discharge chamber and extract ions through a system of multi-aperture grids. The ions are then accelerated to high velocities by the potential difference between the grids, achieving a specific impulse that can reach up to 10,000 seconds. Another type, the Hall-effect thruster (HET), employs crossed electric and magnetic fields to generate and accelerate plasma. HETs, along with high-efficiency multistage plasma thrusters (HEMPTs), ionize the propellant using electrons emitted by a neutralizer cathode. The ion acceleration occurs within a discharge channel, where the plasma is confined by magnetic fields. While HEMPTs offer higher plasma thrust densities and reduced channel erosion compared to HETs, both systems are known for their efficiency and adaptability to a wide range of missions [228].

Electrostatic thrusters also include colloid emitters and field emission electric propulsion (FEEP) systems. These devices use electrostatic fields to extract charged droplets or ions from a liquid propellant, offering precise control and high efficiency for specialized applications like station-keeping or small satellite manoeuvres [228].

Electromagnetic Thrusters: Electromagnetic thrusters rely on Lorentz forces, generated by the interaction of electric currents and magnetic fields, to accelerate ionized propellant. In pulsed plasma thrusters (PPTs), the magnetic field is created by the arc discharge itself, combining ionization and acceleration into a single process. Magnetoplasmadynamic thrusters (MPDTs) represent a more powerful variant, where intense arc discharges ionize and accelerate the propellant to high velocities. MPDTs are particularly efficient at high power levels, with applied-field MPDTs (AF-MPDs) incorporating external magnetic fields to enhance performance. Conversely, self-field MPDTs (SF-MPDs) rely solely on fields generated by the discharge [228].

Another notable type is the electrodeless magnetic nozzle electron cyclotron resonance (ECR) thruster, which uses complex plasma interactions to accelerate the propellant. Unlike other electromagnetic thrusters, ECR systems lack electrodes and achieve thrust through ambipolar fields, which remain an area of active research due to the intricate physical processes involved [228].

All electric thrusters share the fundamental principle of converting electrical energy into kinetic energy, but their efficiencies, thrust capabilities, and operational complexities vary significantly. Electrothermal thrusters are valued for their simplicity but provide lower specific impulse.

Electrostatic thrusters, particularly GIEs and HETs, offer high efficiency and specific impulse, making them ideal for long-duration missions, though their thrust densities are modest. Electromagnetic thrusters excel at high thrust densities and are suitable for missions demanding powerful acceleration, though they often require higher power levels and exhibit shorter operational lifetimes due to erosion and complexity [228].

The choice of a propulsion system depends on mission requirements. For example, resistojets and arcjets are often used for station-keeping and orbital manoeuvres in satellites, while GIEs and HETs are commonly employed in deep-space exploration due to their high specific impulse. MPDTs and ECR thrusters, with their high thrust densities, are being explored for future missions requiring rapid transit or heavy payloads [228].

As electric propulsion technologies continue to evolve, their adaptability and efficiency are transforming spacecraft design and mission planning, paving the way for more ambitious exploration and commercial endeavours in space [228].

Basics of Electric Propulsion

Electric propulsion (EP) systems are gaining prominence in space exploration due to their efficiency and adaptability. The choice of a propulsion system depends on two critical criteria: the electrical power available on the satellite and the specific impulse (I_{sp}). The electrical power directly correlates with the thrust-to-power ratio, a measure of how efficiently a propulsion system converts electrical energy into thrust. Higher thrust-to-power ratios enable faster travel, making this parameter vital for time-critical missions. Meanwhile, I_{sp}, defined as the change in momentum (Δp) per unit of propellant mass (Δm), measures propulsion efficiency. Expressed mathematically as $I_{sp} = \frac{v_{ex}}{g}$ is the exhaust velocity and g is the gravitational acceleration, I_{sp} is often represented in seconds across all unit systems to avoid discrepancies, such as those that caused the Mars Climate Orbiter loss.

High I_{sp} values indicate better mass efficiency, requiring less propellant for the same thrust. However, achieving high I_{sp} often demands greater electrical power, reducing the thrust-to-power ratio. Selecting a propulsion system, therefore, involves a trade-off between available electrical power and the amount of propellant that can be carried. The jet power (P_{jet}), representing the energy associated with the exhaust ion beam, becomes a key factor. The relationship $T = \dot{m}v_{ex}$ ties thrust to the mass flow rate $(\dot{m})$ and exhaust velocity, linking propulsion performance to fundamental mission parameters.

For example, consider a satellite with a dry mass of one ton carrying 500 kg of propellant and equipped with a 20 kW engine operating at an I_{sp} of 5000 seconds. This system could achieve a Δv of approximately 20.3 km/s, sufficient for outer planetary missions, delivering a thrust of around 0.8 N. However, if the mission extends to more distant destinations, increasing propellant alone would yield diminishing returns due to the logarithmic dependence of Δv on mass in the Tsiolkovsky equation. Instead, increasing I_{sp} becomes more effective, although this requires higher power, as jet power scales with I_{sp}^2. Advanced technologies like gridded ion engines (GIEs) with four-grid ion optics can achieve these higher I_{sp} values.

Efficiency Metrics and Characteristics

Ion thrusters further refine efficiency through parameters like mass utilization efficiency (η_m) and electrical efficiency (η_e). Mass utilization efficiency quantifies the proportion of ions emitted relative to the propellant admitted, where singly charged ions satisfy $\eta_m = \frac{I_b}{\dot{m}_p e/M}$, with I_b representing ion beam current, *e* the elementary charge, and MMM the atomic mass. Electrical efficiency, defined as $\eta_e = \frac{P_{jet}}{P_T}$, evaluates the ratio of jet power to total input power (P_T), accounting for power used in ion production (P_d). For ion thrusters, the exhaust velocity depends on the acceleration voltage (U), with $v_{ex} = \sqrt{\frac{2Ue}{M}}$, incorporates beam divergence, acceleration voltage, and contributions from multi-charged ions, allowing for meaningful comparisons across different EP systems.

Mission-Specific Considerations

The relationship between I_{sp}, thrust, and power influences the choice of EP system for various missions. High I_{sp} values, associated with greater mass efficiency, are ideal for applications requiring minimal propellant usage over long durations, such as interplanetary missions. Conversely, systems emphasizing thrust-to-power ratios suit time-sensitive manoeuvres or orbital transfers.

Electric propulsion systems offer unparalleled efficiency and adaptability, making them indispensable in modern space missions. By balancing thrust, specific impulse, and power, engineers can tailor propulsion systems to meet the demands of exploration, satellite operation, and deep-space ventures. Through continued innovation, EP technologies promise to push the boundaries of human and robotic presence in space.

Ion Thrusters

Ion thrusters are highly efficient propulsion systems designed for spacecraft, primarily used in deep space missions and orbital adjustments. They operate by expelling ions, or charged particles, at extremely high velocities using electromagnetic forces. Unlike chemical propulsion, which relies on combustion, ion thrusters utilize electric and magnetic fields to ionize and accelerate a propellant, making them a fundamentally different and highly efficient propulsion method.

The primary propellant used in ion thrusters is xenon, an inert gas chosen for its high atomic mass, low ionization energy, and inert chemical properties. Xenon's high mass ensures greater momentum transfer when expelled from the thruster, enhancing propulsion efficiency. To operate, the xenon propellant must first be ionized. Ionization typically occurs through electron bombardment, where a hollow cathode emits electrons that collide with xenon atoms, stripping away electrons to create positively charged xenon ions (Xe^+). Alternatively, radio frequency (RF) ionization uses an oscillating electromagnetic field to ionize the xenon gas without direct electron bombardment, a method often employed in gridded ion thrusters. The resulting plasma consists of positive xenon ions and free electrons.

Ion acceleration occurs after ionization. In gridded ion thrusters, positively charged ions are directed through two electrostatic grids. The screen grid is positively charged, permitting ions to pass through, while the negatively charged acceleration grid creates a strong electric field that propels the ions to velocities as high as 30 km/s. The force acting on the ions is defined by $F=qE$, where q is the ion's charge and E is the electric field strength. Hall Effect thrusters, in contrast, use a radial magnetic field to trap electrons in a Hall current, which induces a potential difference to accelerate the ions. Unlike gridded thrusters, Hall Effect thrusters operate without grids, relying instead on the self-induced electric field.

To maintain electrical neutrality, a cathode called the neutralizer emits electrons into the ion stream. These electrons combine with the expelled ions, forming neutral xenon atoms and preventing the spacecraft from accumulating a net charge that could interfere with operations.

Gridded electrostatic ion thrusters are advanced propulsion systems that have been under development since the 1960s. These thrusters have been used for both commercial satellite propulsion and scientific space missions due to their high efficiency and precision. Their defining characteristic is the physical separation of the ionization process from the ion acceleration process, which allows for controlled and efficient operation.

In these thrusters, ionization occurs within a discharge chamber where the propellant, typically xenon, is bombarded with energetic electrons. The energy from these electrons ejects valence electrons from the xenon atoms, creating positively charged ions. The electrons required for this process can be produced by a hot cathode filament that emits electrons, which are then accelerated towards an anode by a potential difference. Alternatively, ionization can be achieved using a radio frequency (RF) ionization method, where an oscillating electric field created by an alternating electromagnet induces a self-sustaining discharge, eliminating the need for a cathode.

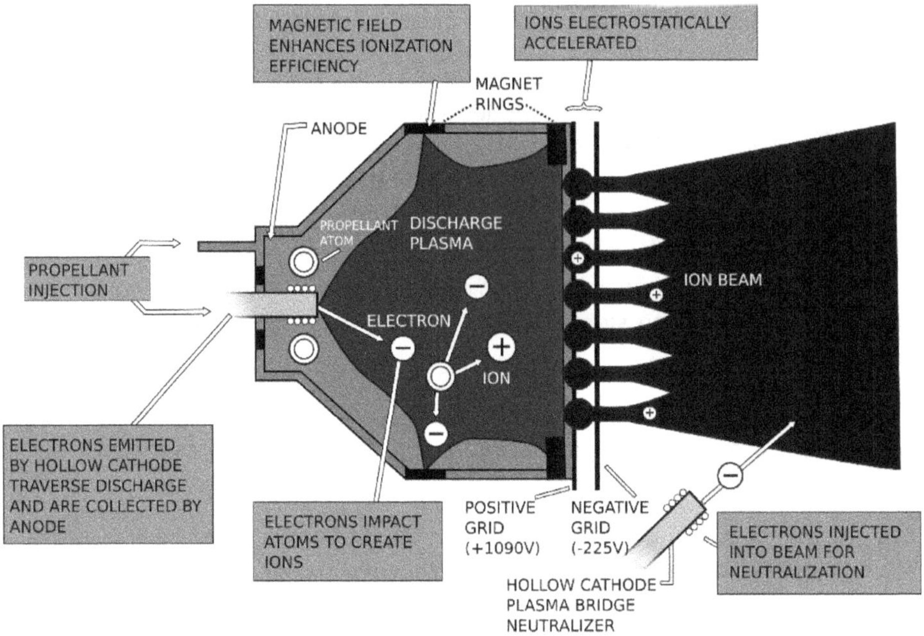

Figure 56: Gridded electrostatic ion engine (multipole magnetic cusp type). Vectorization: Chabacano, Public domain, via Wikimedia Commons.

The positively charged ions are extracted from the discharge chamber through a system of two or three multi-aperture grids. The ions enter the grid system near the plasma sheath and are accelerated by the potential difference between the screen grid (positively charged) and the accelerator grid (negatively charged). This acceleration imparts the ions with significant energy, typically in the range of 1–2 keV, resulting in thrust. The ions are expelled from the thruster at high velocities, generating a reactive force that propels the spacecraft forward.

To prevent the spacecraft from accumulating a positive charge due to the ion beam, a neutralizer cathode emits electrons into the ion stream. These electrons combine with the expelled ions to form neutral xenon atoms, ensuring the spacecraft remains electrically neutral. This step is crucial to prevent the ions from being attracted back to the spacecraft, which would negate the thrust produced.

Numerous research and development efforts have focused on improving gridded ion thruster technology. Notable examples include NASA's Solar Technology Application Readiness (NSTAR) thruster, used on two successful missions, and the NASA Evolutionary Xenon Thruster (NEXT), which achieved flight qualification and was used on the DART mission. Other projects include the Nuclear Electric Xenon Ion System (NEXIS), the High Power Electric Propulsion (HiPEP) system capable of 25 kW, and the European EADS Radio-frequency Ion Thruster (RIT). Additionally, the Dual-Stage 4-Grid (DS4G) thruster represents an innovative approach with increased efficiency and thrust capabilities.

Hall-effect thrusters are another form of electric propulsion that uses an electric potential and a radial magnetic field to accelerate ions. These thrusters are commonly used for station-keeping and orbital adjustments due to their simplicity and reliability. Unlike gridded ion thrusters, Hall-effect thrusters do not have physically separated ionization and acceleration processes.

In a Hall-effect thruster, most of the xenon propellant is introduced near a cylindrical anode located at one end of the thruster. The propellant is ionized as it flows toward a negatively charged plasma, which acts as the cathode. The resulting positively charged ions are accelerated by the electric potential between the anode and the plasma cathode. As the ions are expelled at high velocities, they pick up electrons from the cathode's emitted electron stream, neutralizing the ion beam to maintain spacecraft charge neutrality.

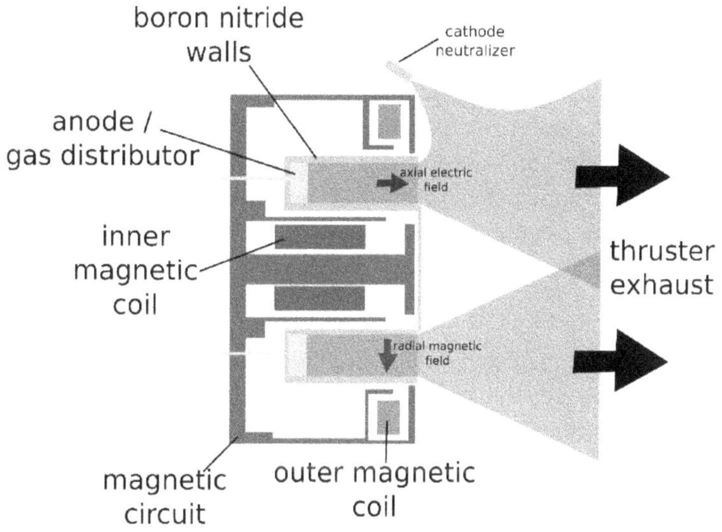

Figure 57: Schematic of a Hall effect thruster electric propulsion device. This is a cross-section of a radially-symmetric device. Finlay McWalter, Public domain, via Wikimedia Commons.

The thruster's design includes a cylindrical tube with the anode at one end and a central spike wound with coils to create a radial magnetic field. The heavy ions are largely unaffected by this magnetic field due to their mass, but the lighter electrons are trapped. These electrons spiral along the magnetic field lines in a Hall current, circulating around the spike. As they spiral, they ionize neutral xenon atoms near the anode. The electrons eventually reach the anode, completing the circuit.

Hall-effect thrusters are simpler in design compared to gridded ion thrusters and offer higher thrust at the expense of slightly lower efficiency. Their operational simplicity and robustness make them a popular choice for long-duration missions and commercial satellite operations. Both Hall-effect and gridded ion thrusters continue to be pivotal in advancing space propulsion technologies, offering complementary capabilities tailored to different mission requirements.

Ion thrusters are characterized by their exceptional efficiency and performance metrics. They achieve specific impulses between 2,000 and 10,000 seconds, significantly exceeding the efficiency of chemical rockets, which typically reach about 450 seconds. This high specific impulse is a result of the thrusters' high exhaust velocity, calculated using $I_{sp}=v_e/g_0$ is the exhaust velocity and g_0 is the gravitational constant. Although thrust is relatively low, usually between 1 and 250 millinewtons, the high exhaust velocity compensates for the limited thrust over long durations, making ion thrusters ideal for extended missions. Additionally, they convert between 60% and 90% of input electrical energy into ion acceleration, highlighting their operational efficiency.

Ion thrusters require substantial electrical power, which is typically supplied by solar panels or nuclear reactors, depending on the mission's location and duration. The power demands range from a few hundred watts to several kilowatts, making their application heavily dependent on the spacecraft's energy generation capabilities.

The advantages of ion thrusters include their high efficiency, which makes them well-suited for long-duration missions by minimizing propellant use. Their ability to finely control thrust enables precise orbital adjustments, and the compact storage requirements for xenon propellant reduce payload mass compared to chemical propulsion systems. However, ion thrusters also have limitations. Their low thrust output renders them unsuitable for launches or rapid manoeuvres. Additionally, the significant electrical power requirement can be a constraint for missions with limited energy resources, and their precision engineering adds complexity and cost.

Ion thrusters are widely employed in space applications. They are commonly used for station-keeping, maintaining the position of geostationary satellites. In deep space exploration, missions like NASA's Dawn spacecraft utilized ion propulsion to explore celestial bodies such as Vesta and Ceres. They are also used for orbital transfers, enabling satellites to move efficiently from one orbit to another.

As a cutting-edge propulsion technology, ion thrusters have revolutionized spacecraft design. Their ability to sustain long-duration, high-efficiency thrust has opened new possibilities in space exploration and satellite operations, enabling missions that would be impractical with traditional chemical propulsion systems.

The operation of an ion thruster relies on the principles of plasma generation and ion acceleration to produce thrust [229]. The process begins with the injection of a propellant, typically xenon, into a discharge chamber where it is ionized to form a plasma [230]. Plasma consists of positively charged ions and free electrons, which are manipulated to generate thrust [231].

There are multiple methods for producing the electrostatic ions required within the discharge chamber. One common approach, used in Kaufman-type thrusters, employs electron bombardment [232]. In this method, a potential difference between a hollow cathode and an anode creates an electron stream that bombards the neutral xenon atoms, stripping electrons and producing positively charged xenon ions [233]. Examples of thrusters utilizing this method include the NSTAR, NEXT, T5, and T6 thrusters [234].

Alternatively, radio frequency (RF) oscillation can induce an electric field in the discharge chamber using an alternating electromagnet [235]. This RF-induced field sustains a self-sufficient discharge, ionizing the xenon atoms without the need for a cathode [236]. Thrusters such as the RIT 10, RIT 22, and μN-RIT employ this technique [237]. A third method involves microwave heating, where microwaves energize the xenon gas, ionizing it through dielectric heating [238]. This approach is used in thrusters like the μ10 and μ20 [235].

The choice of ionization method affects the power supply requirements and thruster design [239]. Electron bombardment systems require power supplies for the cathode, anode, and chamber, while RF and microwave systems need additional power supplies for their respective generators but omit anode or cathode power connections [240].

Once ions are generated, they diffuse toward the chamber's extraction system, which consists of two or three multi-aperture grids [241]. The ions enter the plasma sheath at the grid holes, where they experience a strong electric field created by the potential difference between the screen grid (positively charged) and the accelerator grid (negatively charged) [242]. This electric field accelerates the ions through the extraction holes, imparting high velocity to the ions [243]. The energy of the ions is primarily determined by the plasma potential, which is generally slightly higher than the voltage of the screen grid [244].

The negatively charged accelerator grid plays a critical role in preventing backstreaming electrons from the beam plasma outside the thruster from re-entering the discharge plasma [245]. However, if the negative potential of the accelerator grid is insufficient, this backstreaming can occur, leading to the end of the thruster's operational life [246].

The expelled ions generate thrust by propelling the spacecraft in the opposite direction, as described by Newton's Third Law of Motion [247]. To ensure charge neutrality, a separate cathode, called the neutralizer, emits low-energy electrons into the ion beam [237]. These electrons neutralize the positively charged ions, forming neutral xenon atoms [248]. Neutralization is essential to prevent the spacecraft from accumulating a net negative charge, which would attract ions back toward the spacecraft, reducing or cancelling the thrust [249].

Overall, the ion thruster's operation is a sophisticated balance of plasma generation, ion acceleration, and charge neutrality management [52]. By carefully controlling these processes, ion thrusters achieve the high efficiency and precise thrust control needed for long-duration space missions and orbital adjustments [52].

Plasma Drives

Plasma propulsion engines are an advanced form of electric propulsion that generate thrust by accelerating a quasi-neutral plasma [58, 250, 251]. Unlike ion thrusters that rely on external high-voltage grids or anodes/cathodes to accelerate ions, plasma thrusters internally generate the currents and potentials needed for ion acceleration [58, 250, 251]. This internal mechanism often results in a lower exhaust velocity compared to ion thrusters, but it provides several operational and design advantages [58, 250, 251].

One of the primary advantages of plasma thrusters is the elimination of high-voltage grids or anodes, which are prone to ion erosion over time and can limit the operational life of the thruster [58, 250, 251]. Additionally, the quasi-neutral nature of the plasma exhaust ensures that the expelled ions and electrons neutralize each other naturally in the exhaust plume, negating the need for an external neutralizing electron gun or hollow cathode [58, 250, 251]. This simplifies the overall system design and improves reliability [58, 250, 251].

Plasma thrusters also offer versatility in propellant choice, as they can utilize a wide variety of propellants, including inexpensive or readily available gases like argon, carbon dioxide, and even unconventional sources such as astronaut urine [58, 250, 251]. This ability to operate with diverse propellants makes plasma thrusters particularly attractive for interplanetary missions where refuelling opportunities are limited or where in-situ resource utilization (ISRU) might be necessary [58, 250, 251].

Plasma thrusters typically generate plasma using radio frequency (RF) or microwave energy [252-254]. External antennas create electromagnetic fields that ionize the propellant, producing a plasma without the need for high-voltage discharge mechanisms [252-254]. This method of plasma generation is efficient and minimizes wear on critical components, further extending the operational life of the system [252-254].

Due to their high specific impulse, plasma propulsion engines are well-suited for interplanetary missions [255-257]. Specific impulse is a measure of propulsion efficiency, and the high values achieved by plasma engines enable long-duration missions with minimal propellant consumption [255-257]. This efficiency makes them ideal for deep-space exploration, satellite station-keeping, and potentially crewed interplanetary journeys [255-257].

Numerous space agencies and research institutions have contributed to the development and advancement of plasma propulsion systems, including the European Space Agency (ESA), Iranian Space Agency, and Australian National University (ANU) [258-261]. For instance, ANU co-developed a double-layer thruster, which is a notable variant of plasma propulsion that offers enhanced performance characteristics [258-261].

Helicon plasma thrusters utilize low-frequency electromagnetic waves, known as Helicon waves, to ionize a neutral gas and generate plasma in the presence of a static magnetic field. An RF antenna, which surrounds the gas chamber, emits electromagnetic waves that excite the neutral gas, converting it into plasma. Once the plasma is formed, it is expelled at high velocity to produce thrust. The acceleration of the plasma involves the use of carefully configured electric and magnetic fields, which together form an optimal topology for efficient thrust generation. These thrusters fall into the category of electrodeless propulsion systems, meaning they do not require electrodes or grids for ion acceleration, thus eliminating potential issues related to erosion and component degradation.

A significant advantage of Helicon plasma thrusters is their ability to operate with multiple types of propellants, including gases such as argon, krypton, and hydrogen. This flexibility makes them particularly suitable for extended missions or interplanetary travel, where resupply of a specific propellant may not be feasible. Additionally, the simplicity of their design allows for their construction from easily obtainable materials, such as glass, making them a cost-effective solution for experimental or small-scale applications.

Magnetoplasmadynamic thrusters operate on the principle of the Lorentz force, which is generated when an electric current flows through a plasma in the presence of a magnetic field. The interaction between the magnetic field and the electric current creates a force that accelerates the plasma, producing thrust. This mechanism makes MPD thrusters particularly effective at high power levels, where the generated electromagnetic fields can efficiently ionize and accelerate the propellant.

Figure 58: NASA Pulsed Magnetoplasmadynamic Thruster. Defense Visual Information Distribution Service, Public Domain, via GetArchive.

MPD thrusters are well-suited for missions requiring high thrust-to-power ratios, making them ideal for deep-space propulsion and large spacecraft. However, their efficiency depends heavily on the magnitude of the electric and magnetic fields, requiring significant power input to achieve optimal performance. Their reliance on the Lorentz force also means that they share operational principles with other plasma-based propulsion systems, such as pulsed plasma thrusters.

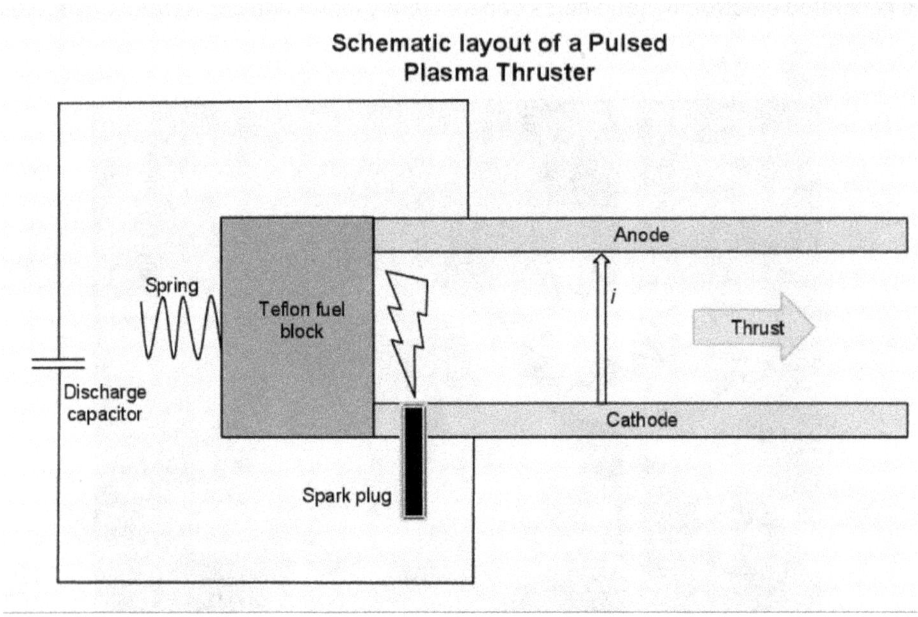

Figure 59: Schematic layout of a Pulsed Plasma Thruster. Ulrich Walach, CC BY-SA 3.0, via Wikimedia Commons.

Pulsed inductive thrusters also leverage the Lorentz force to generate thrust, but they use a unique method that avoids direct contact with electrodes. Instead of relying on physical electrodes, PITs induce ionization and electric currents within the plasma by applying a rapidly oscillating magnetic field. This design eliminates the erosion problems commonly associated with electrode-based systems, thereby extending the operational life of the thruster.

PITs are particularly valuable for applications requiring low maintenance and high durability, as they are not subject to wear and tear from direct plasma interactions. This makes them an attractive option for long-duration missions or spacecraft with limited opportunities for maintenance or repair.

Electrodeless plasma thrusters employ the ponderomotive force, a unique phenomenon that arises when plasma or charged particles are exposed to a strong gradient in electromagnetic energy density. This force accelerates plasma electrons and ions in the same direction, eliminating the need for a neutralizer. By operating without electrodes or cathodes, electrodeless plasma thrusters overcome many of the mechanical and material challenges faced by traditional electric propulsion systems.

The simplicity and versatility of electrodeless plasma thrusters make them a promising candidate for diverse mission profiles, particularly in scenarios where robust and low-maintenance systems are required. Their ability to operate without a neutralizer simplifies the system architecture and reduces the potential for mechanical failures.

VASIMR, or Variable Specific Impulse Magnetoplasma Rocket, represents a cutting-edge plasma propulsion system that uses radio waves to ionize a propellant into plasma. A strong magnetic field then accelerates the plasma out of the engine, generating thrust. VASIMR's unique design allows for the adjustment of specific impulse, enabling it to switch between high-thrust, low-efficiency modes and low-thrust, high-efficiency modes depending on mission requirements.

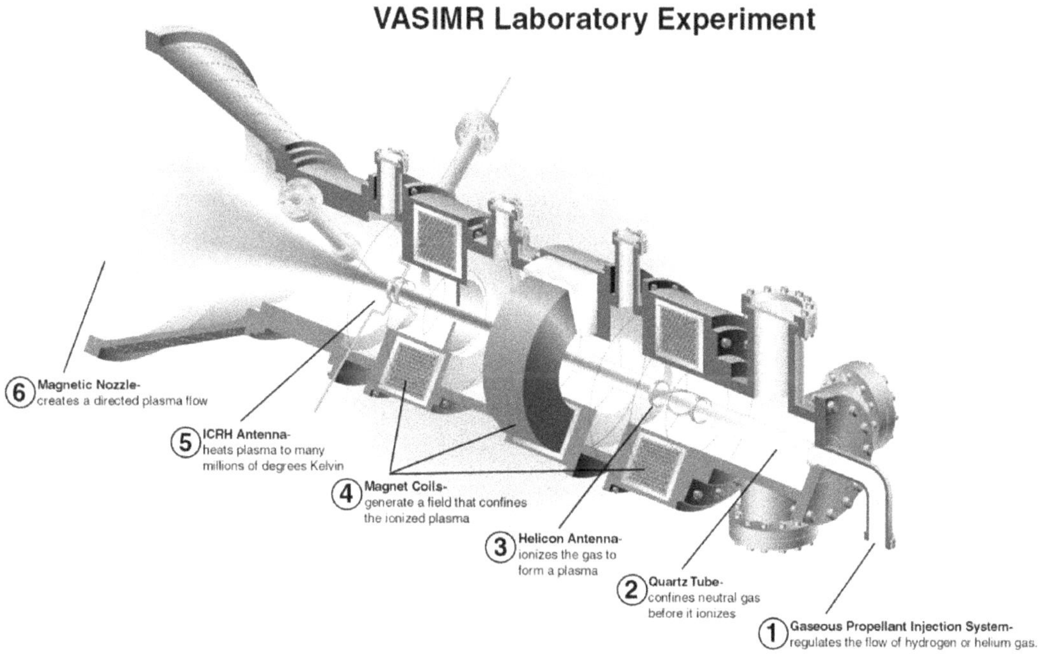

Figure 60: The Variable Specific Impulse Magnetoplasma Rocket (VASIMR) is an electro-magnetic thruster for spacecraft propulsion. It uses radio waves to ionize a propellant and magnetic fields to accelerate the resulting plasma to generate thrust. NASA, Public domain, via Wikimedia Commons.

One of VASIMR's most remarkable potential applications is drastically reducing travel times for interplanetary missions. For instance, a 200-megawatt VASIMR engine could reduce the travel time from Earth to Mars from six months to just 39 days, and journeys to Jupiter or Saturn from six years to approximately 14 months. This capability makes VASIMR an attractive option for future crewed missions to distant planets, where reducing travel time is critical for the safety and well-being of astronauts.

Calculations Used in Plasma Thruster Design

Designing a plasma thruster involves a combination of theoretical physics, empirical models, and engineering principles. The goal is to achieve efficient propulsion by optimizing parameters such as thrust, specific impulse, power consumption, and system lifetime. Below is an in-depth explanation of the key calculations involved in plasma thruster design.

1. Thrust Calculation

Thrust (*T*) in a plasma thruster is the product of the mass flow rate ($\dot{m}$) of the propellant and its exhaust velocity (v_{ex}):

$$T = \dot{m} \cdot v_{\text{ex}}$$

Components:

- $\dot{m}$: The mass flow rate is determined by the amount of propellant ionized and expelled. It can be calculated using the density of the propellant (ρ) and the flow area (*A*):

$$\dot{m} = \rho \cdot v_{\text{flow}} \cdot A$$

- v_{ex}: The exhaust velocity is given by:

$$v_{\text{ex}} = \sqrt{\frac{2qU}{M}}$$

Where:

- *q*: Charge of the ion (Coulombs).
- *U*: Acceleration voltage applied to the ions (Volts).
- *M*: Mass of the ion (kg).

This equation assumes ideal electrostatic acceleration and is central to determining thrust efficiency.

2. Specific Impulse (I_{sp})

Specific impulse (I_{sp}) measures the efficiency of the thruster and is defined as the thrust per unit propellant mass flow rate. It can be expressed as:

$$I_{\text{sp}} = \frac{T}{\dot{m} \cdot g_0} = \frac{v_{\text{ex}}}{g_0}$$

Where:

- g_0: Gravitational acceleration on Earth (9.807 m/s^2).
- I_{sp}: Typically measured in seconds, it indicates how efficiently the thruster uses its propellant.

3. Power Requirements

The total electrical power required for the thruster (P_{total}) includes:

- The power used to ionize the propellant.
- The power used to accelerate the ions.

The power to accelerate ions (jet power, P_{jet}) is given by:

$$P_{\text{jet}} = \frac{1}{2}\dot{m}v_{\text{ex}}^2$$

The total power is often higher due to inefficiencies in the system, such as ionization energy (P_{ion}) and other losses (P_{losses}):

$$P_{\text{total}} = P_{\text{jet}} + P_{\text{ion}} + P_{\text{losses}}$$

The ionization energy is calculated as:

$$P_{\text{ion}} = \frac{\dot{m}_{\text{ionized}}}{M} \cdot E_{\text{ion}}$$

Where E_{ion} is the ionization energy of the propellant.

4. Plasma Density and Ionization

The density of the plasma (n_e) and ionization rate are critical for efficient operation. Plasma density is often related to the input power (P_{input}) and the chamber dimensions:

$$n_e \propto \frac{P_{\text{input}}}{V_{\text{chamber}} \cdot \langle E \rangle}$$

Where:

- $V_{chamber}$: Volume of the ionization chamber.
- $\langle E \rangle$: Average energy of electrons in the plasma.

Ionization rate (Γ_{ion}) is calculated using the cross-section of ionization (σ_{ion}), electron density (n_e), and neutral propellant density (n_{neutral}):

$$\Gamma_{\text{ion}} = n_e \cdot n_{\text{neutral}} \cdot \sigma_{\text{ion}} \cdot v_{\text{thermal}}$$

5. Magnetic Field Calculations

For thrusters such as Hall-effect thrusters or magnetoplasmadynamic thrusters, magnetic field strength (*B*) is critical. The Lorentz force (*FL*) is calculated as:

$$F_L = q \cdot v \cdot B$$

The magnetic field must be optimized to confine electrons and allow efficient ion acceleration. The field is usually generated by electromagnets, with calculations based on Ampère's law and the coil geometry.

6. Beam Divergence and Efficiency

Beam divergence reduces thrust efficiency and must be minimized. The thrust efficiency (η_T) is defined as the ratio of the jet power to the total input power:

$$\eta_T = \frac{P_{\text{jet}}}{P_{\text{total}}}$$

Beam divergence angle (θ) affects the effective thrust and is calculated using the plasma exit velocity components:

$$\text{Effective thrust} = T \cdot \cos(\theta)$$

7. Thermal and Structural Design

Thermal considerations include calculating the heat flux from the plasma to the thruster walls and ensuring materials can withstand these conditions. The heat flux (q) is given by:

$$q = \sigma \cdot T^4$$

Where:

- σ: Stefan-Boltzmann constant.
- *T*: Temperature of the plasma.

Structural integrity must account for the forces exerted by the magnetic field, plasma pressure, and thermal stresses.

8. Lifetime and Erosion

Thruster lifetime is affected by erosion of grids, cathodes, or other components. Erosion rate is often estimated using sputtering yield (*Y*) and ion impact energy (E_{impact}):

$$\text{Erosion rate} = Y \cdot \frac{\text{ion flux}}{A}$$

Where *A* is the area of the eroded component.

Plasma thruster design calculations are interconnected and require balancing multiple parameters, such as thrust, efficiency, power input, and structural considerations. The process involves iterative optimization to meet mission requirements, making plasma thrusters highly effective for applications requiring efficient propulsion over long durations, such as satellite station-keeping and deep-space exploration.

Practical Example of Plasma Thruster Design Using Calculations

Scenario: Designing a Plasma Thruster for a Deep-Space Mission

Objective: Design a plasma thruster to propel a 500 kg spacecraft on a deep-space mission. The system should achieve a thrust of 0.5 N, with a specific impulse (I_{sp}) of 3000 s, using xenon as the propellant. The electrical power available is 10 kW. The aim is to calculate the exhaust velocity, mass flow rate, power requirements, and evaluate whether the design meets the mission criteria.

Step 1: Calculate Exhaust Velocity

The specific impulse (I_{sp}) is related to the exhaust velocity (v_{ex}) by:

$$v_{\text{ex}} = I_{\text{sp}} \cdot g_0$$

Where:

- g_0=9.807 m/s2 (gravitational acceleration).
- I_{sp} =3000 s.

Substituting the values:

$$v_{\text{ex}} = 3000 \cdot 9.807 = 29,421\,\text{m/s}$$

The exhaust velocity is 29,421 m/s.

Step 2: Calculate Mass Flow Rate

The thrust equation relates the thrust (*T*), exhaust velocity (v_e), and mass flow rate $(\dot{m})$:

$$T = \dot{m} \cdot v_{\text{ex}}$$

Rearranging for $\dot{m}$:

$$\dot{m} = \frac{T}{v_{\text{ex}}}$$

Where:

- T=0.5 N,
- v_{ex}=29,421 m/s

Substituting the values:

$$\dot{m} = \frac{0.5}{29,421} = 1.7 \times 10^{-5}\,\text{kg/s}$$

The mass flow rate is $1.7 \times 10^{-5}\,\text{kg/s}$.

Step 3: Calculate Jet Power

Jet power (P_{jet}) is the kinetic energy of the exhaust stream per unit time:

$$P_{\text{jet}} = \frac{1}{2}\dot{m}v_{\text{ex}}^2$$

Substituting:

- $\dot{m} = 1.7 \times 10^{-5}\,\text{kg/s}$,
- $v_{\text{ex}} = 29,421\,\text{m/s}$.

$$P_{\text{jet}} = \frac{1}{2} \cdot (1.7 \times 10^{-5}) \cdot (29,421)^2$$

$$P_{\text{jet}} = 7.3\,\text{kW}$$

The jet power is 7.3 kW.

Step 4: Evaluate Total Power Requirements

The total power required (P_{total}) includes inefficiencies due to ionization and other losses. Assuming an efficiency (η_T) of 70%:

$$P_{\text{total}} = \frac{P_{\text{jet}}}{\eta_T}$$

Substituting:

- $P_{\text{jet}} = 7.3\,\text{kW}$,
- $\eta_T = 0.7$.

$$P_{\text{total}} = \frac{7.3}{0.7} = 10.4\,\text{kW}$$

The total power requirement is 10.4 kW. This exceeds the available power slightly (10 kW), indicating a need for design optimization, such as improving efficiency or reducing thrust.

Step 5: Ion Acceleration Voltage

The exhaust velocity is related to the ion acceleration voltage (*U*):

$$v_{\text{ex}} = \sqrt{\frac{2qU}{M}}$$

Rearranging for *U*:

$$U = \frac{Mv_{\text{ex}}^2}{2q}$$

For xenon $(M = 2.18 \times 10^{-25}\,\text{kg}, q = 1.6 \times 10^{-19}\,\text{C})$:

$$U = \frac{(2.18 \times 10^{-25}) \cdot (29,421)^2}{2 \cdot (1.6 \times 10^{-19})}$$

$$U = 592\,\text{V}$$

The required acceleration voltage is 592 V.

Step 6: Check Beam Divergence Efficiency

Assume a beam divergence angle (θ) of 15°. The effective thrust is:

$$T_{\text{effective}} = T \cdot \cos(\theta)$$

Substituting:

- T=0.5 N,
- θ=15°.

$$T_{\text{effective}} = 0.5 \cdot \cos(15^\circ) \approx 0.48\,\text{N}$$

The effective thrust after accounting for beam divergence is 0.48 N.

Step 7: Evaluate Design Feasibility

- The calculated thrust meets the mission requirements, though efficiency improvements could reduce power consumption slightly below the 10 kW limit.
- Specific impulse (I_{sp} =3000 s) ensures efficient propellant usage.
- Beam divergence is minimal, maintaining effective thrust close to the design target.

This example demonstrates how plasma thruster design calculations guide parameter selection and iterative optimization to meet mission criteria.

Chapter 7
Avionics and Guidance Systems

Navigational Systems for Space Travel

Rocket navigational systems are fundamental to space travel, ensuring that spacecraft can reach their destinations, conduct scientific missions, and return safely to Earth. These systems combine advanced hardware, sophisticated software, and mathematical models to determine position, velocity, and orientation, enabling continuous trajectory adjustments for precision navigation.

Guidance, Navigation, and Control (GNC) is a specialized branch of engineering that focuses on the development and implementation of systems used to direct and control the movement of vehicles [262, 263]. These vehicles include automobiles, ships, aircraft, spacecraft, and autonomous systems like drones [262]. GNC systems are critical for ensuring that a vehicle follows a desired path, maintains stability, and reaches its target efficiently [262].

Guidance refers to the process of determining the desired trajectory or path for a vehicle to follow [262, 263]. It involves calculating the necessary changes in velocity, acceleration, and rotational motion to maintain the desired trajectory [262]. Algorithms like proportional navigation and predictive models play a critical role in modern guidance systems [262, 264, 265].

Navigation focuses on determining the vehicle's state vector, including its position, velocity, and orientation (attitude) [262]. This information is critical for the guidance system to plan an appropriate trajectory [262]. Navigation systems integrate data from a variety of sensors, such as GPS, inertial measurement units (IMUs), and star trackers, to provide real-time feedback [262, 266, 267].

Control refers to the manipulation of the forces acting on the vehicle to ensure it follows the guidance commands while maintaining stability [262, 268]. Control systems adjust steering mechanisms, thrusters, reaction wheels, or other actuators to generate the desired motion [262]. Closed-loop control systems, which rely on real-time feedback from sensors, are the standard in GNC [262, 269].

The integration of GNC systems is particularly critical for space missions, where the systems work together to enable precise station-keeping, trajectory corrections, and attitude alignment [262, 269]. Advancements in sensors, artificial intelligence, and computational power are enhancing GNC capabilities, enabling greater autonomy, improved accuracy, and lower operational costs [262, 266].

GNC systems are widely applied across multiple domains, including aerospace, automotive, and maritime industries [262]. They are essential for autonomous vehicles, spacecraft, drones, and aircraft, where precision and reliability are paramount [262, 270]. Emerging technologies, such as quantum sensors, X-ray navigation, and AI-driven guidance systems, promise to revolutionize GNC capabilities in the future [262, 266, 271].

The primary functions of navigational systems include determining the spacecraft's location relative to celestial bodies or reference frames, calculating optimal trajectories to minimize time and fuel usage, managing the spacecraft's orientation for proper alignment of instruments or propulsion systems, and adjusting the path using thrusters, reaction wheels, or gimbaled engines to stay on course. Each function is critical for ensuring mission success.

Inertial navigation systems (INS) form the backbone of many navigational setups. These systems utilize accelerometers to measure linear acceleration along three axes and gyroscopes to track angular rotation and changes in orientation. By integrating these measurements, INS provides real-time data on position, velocity, and orientation. However, INS can experience drift over time, necessitating periodic corrections from external sources. In low Earth orbit (LEO), global navigation satellite systems (GNSS), such as GPS, are used extensively for this purpose. GNSS offers highly accurate position and velocity data by communicating with satellites, although it is unavailable beyond LEO, requiring alternative methods for deep-space navigation.

Star trackers are critical for deep-space missions, where GNSS signals cannot be used. These optical devices capture images of stars, comparing observed patterns to onboard catalogues to determine spacecraft orientation with high precision. Similarly, sun sensors measure the spacecraft's position relative to the Sun and are often used alongside star trackers for orientation corrections. Ground-based tracking systems like NASA's Deep Space Network (DSN) provide additional support by using Doppler shift measurements, two-way ranging, and very long baseline interferometry (VLBI) to track spacecraft.

Optical navigation, which uses cameras to capture images of celestial bodies or landmarks, is another vital technique for calculating relative positions. Autonomous navigation (AutoNav) systems further enhance capabilities by enabling spacecraft to calculate and adjust their trajectories independently using onboard sensors and algorithms. This autonomy is particularly useful for missions like NASA's Deep Space 1, where real-time decision-making is crucial.

Navigational strategies vary depending on the mission phase. During launch and ascent, the focus is on maintaining stability and ensuring accurate orbital insertion using INS, GNSS, and ground-based radar tracking. In LEO operations, precise orbit maintenance, rendezvous manoeuvres, and debris avoidance rely on GNSS and ground tracking systems. For deep-space navigation, the emphasis shifts to interplanetary transfers, gravity assists, and orbital insertions at distant destinations, employing star trackers, DSN, AutoNav, and optical navigation. Landing and descent phases require precision tools like radar altimeters and lidar systems to execute safe landings on planets or moons.

Despite their sophistication, rocket navigational systems face significant challenges. GNSS signals are unavailable in deep space, necessitating reliance on autonomous systems. Spacecraft must also contend with electromagnetic interference from cosmic radiation and solar activity. Time delays in ground-based commands increase with distance, complicating real-time adjustments. Additionally, inertial navigation systems can drift over time, requiring external corrections, and complex gravitational fields add further complications to trajectory calculations.

Recent advances in navigational technology address many of these challenges. Quantum accelerometers offer drift-free inertial navigation by leveraging quantum mechanics. AI-powered AutoNav systems enable faster and more accurate decision-making, while laser ranging systems improve distance measurement precision. X-ray navigation (XNAV) is an emerging technology that uses X-ray pulsar signals as deep-space beacons for precise navigation.

The applications of rocket navigation systems are vast. Earth-orbiting satellites rely on GNSS and INS for precise orbits in communication, weather forecasting, and reconnaissance. Interplanetary missions, such as Mars Perseverance, utilize AutoNav and star trackers to reach distant planets, moons, and asteroids. Optical navigation and AutoNav systems guide spacecraft in asteroid and comet rendezvous missions, as demonstrated by ESA's Rosetta mission. Human spaceflight missions, like NASA's Artemis program, depend on advanced navigational systems to ensure safe crewed trajectories.

Spacecraft utilize onboard sensors, star trackers, and inertial measurement units to autonomously determine their position and velocity [272-274]. These systems rely on measurements from cameras, accelerometers, and gyroscopes to estimate the spacecraft's state without external assistance [272-274].

Optical measurement devices that are resistant to electromagnetic interference can be used for celestial navigation, providing good navigation accuracy [275, 276]. Star trackers calculate a spacecraft's attitude by determining its orientation in space based on the positions of stars in its field of view [276].

Global navigation satellite systems (GNSS) like GPS, Galileo, GLONASS, and Beidou provide autonomous geopositioning by using signals from a network of satellites [277].

Optical navigation systems use cameras to compute a body's position in the camera reference frame and derive its target location in space [278]. These navigation systems can be used in combination, with optical navigation and celestial navigation augmenting inertial navigation systems to improve accuracy, especially during critical mission phases like planetary approach and landing [278].

Looking to the future, rocket navigation systems will continue evolving to address the challenges of autonomous interstellar travel and long-duration missions. Technologies like quantum navigation, XNAV, and advanced AI are likely to become foundational, enabling spacecraft to navigate with unprecedented precision and reliability in the vastness of space. These advancements promise to revolutionize space exploration, expanding the frontiers of human knowledge and capability.

The Guidance, Navigation, and Control (GNC) subsystem plays a central role in spacecraft operations, ensuring precise position determination and attitude management for successful mission execution. The GNC subsystem integrates components for determining the spacecraft's position and orientation while also enabling control functions for trajectory corrections and stability [279].

For spacecraft in Earth orbit, onboard position determination is typically achieved through the use of a Global Positioning System (GPS) receiver. This allows real-time tracking and navigation with high accuracy. Ground-based radar tracking systems serve as an alternative for determining spacecraft position. In cases where onboard knowledge is required but GPS is unavailable, radar observations can be paired with suitable orbit propagators. The U.S. Air Force (USAF) provides Two-Line Element (TLE) sets, which are commonly processed with the SGP4 propagator to predict orbital positions. For deep-space missions, position determination relies on the Deep Space Network (DSN) in combination with an onboard radio transponder. Additionally, emerging technologies use optical detection of celestial bodies or X-ray pulsars to calculate spacecraft position autonomously in deep space [279].

Navigating SmallSats in cislunar space and beyond presents unique challenges compared to low Earth orbit. For instance, reliance on Earth's magnetic field for attitude determination and control is not feasible in interplanetary missions, requiring alternative ADCS designs. Demonstrations like NASA's Mars Cube One (MarCO) CubeSats during the 2018 Insight mission to Mars showcased the capability of SmallSats for interplanetary missions, including precise pointing for communication in deep space [279].

The Attitude Determination and Control System (ADCS) within the GNC subsystem includes sensors and actuators that manage the spacecraft's orientation and trajectory corrections. Sensors such as star trackers, sun sensors, horizon sensors, magnetometers, and gyros are used to measure the spacecraft's attitude and spin rate. During trajectory correction maneuvers, accelerometers play a key role in terminating burns when the desired velocity change is achieved. Actuators like magnetic torquers, reaction wheels, and thrusters are employed to adjust spacecraft attitude and impart velocity changes. These components work in tandem to ensure that the spacecraft remains aligned with mission objectives, whether pointing antennas for communication or orienting instruments for scientific observations [279].

Miniaturization of GNC technologies has been a significant trend, particularly for micro- and nano-class spacecraft. While three-axis stabilized spacecraft equipped with GPS and weighing around 100 kg have been operational for decades, advancements in miniaturization have made these capabilities accessible to much smaller spacecraft. Advanced sensors such as gyros and accelerometers also provide enhanced stability and precision, critical for maintaining spacecraft orientation in dynamic environments [279].

Deep space navigation systems have expanded capabilities with bands like X, Ka, S, and UHF, which facilitate long-distance communication and tracking. Altimeters, achieving altitude measurements with centimetre-level accuracy, contribute to precise landing and proximity operations. Additionally, atomic clocks with high-frequency ranges are emerging as critical components for improving onboard timing accuracy and enabling autonomous navigation [279].

The performance and Technology Readiness Level (TRL) of these GNC components vary based on mission-specific requirements, payload constraints, and environmental conditions.

Technology Readiness Level (TRL) is a structured framework developed by NASA to evaluate and communicate the maturity of a technology. It provides a common language for assessing the development progress of a technology, from its initial conceptualization to full deployment in operational settings. This scale has been widely adopted in industries such as aerospace, defence, and energy, serving as a tool to manage risks and make informed decisions during the technology development process.

The TRL scale spans nine levels, starting at TRL 1, which represents the observation of basic scientific principles, and culminating at TRL 9, where the technology is fully operational and proven in real-world applications. Each level corresponds to a specific stage in the technology's lifecycle, requiring progressively rigorous testing and validation. At TRL 1, foundational research focuses on understanding the underlying principles of a concept, such as studying a physical phenomenon in a laboratory setting. TRL 2 involves formulating the basic application of the research, where a discovery is conceptualized to solve a particular problem. TRL 3 advances to experimental proof of concept through small-scale laboratory experiments that validate the feasibility of the technology under controlled conditions.

As the technology matures, TRL 4 sees the development and testing of a prototype in a laboratory environment, such as evaluating the core functionality of a system within a vacuum chamber. At TRL 5, the prototype is tested under conditions that simulate its intended operational environment, ensuring its relevance and robustness. TRL 6 involves demonstrating the system or subsystem in an environment representative of its actual application, such as flight testing on a high-altitude balloon or suborbital rocket. TRL 7 integrates the system into its operational setting for initial testing, such as deploying and evaluating a satellite subsystem in orbit. At TRL 8, the technology undergoes comprehensive validation during operational use, such as a communication system functioning successfully on multiple deployed satellites. Finally, TRL 9 signifies a fully operational and proven technology, having demonstrated its reliability in real-world missions.

Progressing through the TRL levels requires iterative refinement, comprehensive documentation, and validation through rigorous testing in environments that progressively mimic real-world conditions. Higher TRL levels involve greater integration, moving from individual components to complete systems functioning under realistic circumstances. As a technology advances, testing environments transition from controlled laboratories to operational settings, ensuring its performance in the intended application.

TRL is used extensively for various purposes. It serves as a tool for risk assessment, allowing organizations to identify and mitigate technical risks associated with developing new technologies. It also informs funding decisions, guiding investments based on a technology's maturity and potential. TRL facilitates project planning by providing a clear roadmap for transitioning a technology from concept to operational use. Additionally, it enhances stakeholder communication by offering a standardized scale to convey the current status and future milestones of a technology.

For instance, a small satellite project might begin with a novel propulsion concept at TRL 1, where basic principles are explored. Through TRLs 2 to 4, the system is validated in laboratory settings. At TRL 5 and 6, a prototype undergoes testing in simulated space environments. By TRLs 7 to 9, the technology is flight-tested on experimental missions and eventually deployed on operational satellites.

The TRL framework brings several advantages. It standardizes the evaluation of technology readiness across industries, ensuring consistency and clarity. It enhances transparency by providing a clear understanding of development progress and gaps. Moreover, it aids in managing risks by identifying readiness levels and associated uncertainties, enabling developers to rigorously test and validate technologies, thus minimizing risks and maximizing success in real-world applications.

While most state-of-the-art technologies have achieved TRLs of 7-9, indicating high maturity and flight readiness, continued research and development are essential for expanding capabilities, especially for small spacecraft. NASA and other space agencies encourage the exploration of emerging technologies and collaboration with industry partners to enhance the performance and reliability of GNC subsystems [279].

Integrated Units

Integrated units are compact systems that consolidate multiple components required for spacecraft Guidance, Navigation, and Control (GNC). These systems streamline spacecraft design by offering an all-in-one solution tailored to meet the specific requirements of attitude determination and control. Typically, integrated units include essential components such as reaction wheels, magnetometers, magnetic torquers, and star trackers, as well as onboard processors and software capable of executing sophisticated attitude control algorithms [279].

A significant advantage of integrated units is their ability to provide precise pointing and stabilization capabilities in a lightweight, compact package. For example, Blue Canyon Technologies' XACT units have demonstrated outstanding performance in missions like NASA's MarCO and ASTERIA, which utilized 6U platforms, and in smaller 3U missions like MinXSS. These units have flown successfully in various space environments, showcasing their reliability and versatility for both nanosatellites and microsatellites [279].

Integrated units come with a range of performance metrics, tailored to the size and requirements of the spacecraft. For instance, Blue Canyon Technologies' XACT series boasts remarkable pointing accuracy, with models like XACT-15, XACT-50, and XACT-100 achieving precision as fine as 0.003° to 0.007°. The systems typically include three reaction wheels for attitude control, three magnetorquers for magnetic stabilization, and star trackers for accurate orientation determination. Additionally, many integrated units, such as the XACT series, feature three-axis magnetometers to provide comprehensive attitude sensing [279].

Other manufacturers, such as AAC Clyde Space and Berlin Space Technologies, offer integrated units with comparable capabilities but cater to different mission profiles. For example, AAC Clyde Space's iADCS-200 and iADCS-400 systems include inertial measurement units (IMUs), high-precision magnetometers, and sun sensors, enabling robust performance with pointing accuracy better than

1°. Berlin Space Technologies' IADCS-100 system provides a lightweight option (0.4 kg) with similar functionality, making it suitable for weight-sensitive missions [279].

CubeSpace Satellite Systems also offers a variety of integrated solutions under their CubeADCS series, catering to small satellites with different payload and performance requirements. For instance, their CubeADCS 3-Axis Small unit includes reaction wheels, magnetorquers, sun sensors, and magnetometers, achieving pointing accuracies of less than 1°. Advanced configurations with star trackers improve this precision to better than 0.1°. These units are designed to balance mass and power constraints while maintaining reliable performance across diverse mission scenarios [279].

Integrated units cater to various mission types, from low Earth orbit (LEO) operations to interplanetary missions. The ability to combine sensors, actuators, and processing capabilities into a single module simplifies integration and reduces the complexity of spacecraft GNC systems. These units are particularly beneficial for small spacecraft, where mass, volume, and power budgets are tightly constrained.

For interplanetary missions, fine pointing accuracy is critical for maintaining communication and executing precise trajectory corrections. The successful deployment of Blue Canyon Technologies' XACT units in the MarCO mission, which provided real-time relay support for NASA's Insight Mars lander, exemplifies the adaptability and performance of integrated units in challenging deep-space environments.

The performance of integrated units depends on the size and mission objectives of the spacecraft. The units vary in mass, actuator and sensor configurations, and pointing accuracy. For example, CubeSpace's CubeADCS Y-Momentum system is a minimalistic solution designed for less demanding missions, with a single momentum wheel and coarse sun sensors providing pointing accuracy under 5°. On the other hand, the Flexcore system by Blue Canyon Technologies offers unparalleled accuracy of 0.002°, suitable for missions requiring extreme precision [279].

Advances in miniaturization and manufacturing technologies have significantly enhanced the capabilities of integrated units. These systems are no longer limited to larger spacecraft; they are now accessible for nano- and microsatellites, expanding their applications across a broader range of missions. With technology readiness levels (TRLs) between 7 and 9, these integrated units are highly mature and have been extensively validated through flight heritage [279].

Reaction Wheels

Reaction wheels are critical components in spacecraft attitude control systems, providing precise three-axis pointing capability for satellites and other spacecraft. Their role is rooted in the conservation of angular momentum. By spinning a wheel in one direction, the spacecraft counter-rotates around its centre of mass in the opposite direction, allowing for controlled adjustments in orientation. This mechanism is fundamental for tasks such as aligning antennas, cameras, or scientific instruments with specific targets [279].

The design and selection of reaction wheels depend on several key factors, including the spacecraft's mass, its required rotation performance rates, and mission-specific constraints. The torque and momentum storage capacity of reaction wheels determine their ability to manage the spacecraft's angular motion. Typically, three wheels are mounted orthogonally for full three-axis control, but a four-wheel configuration is often employed to provide fault tolerance. In such configurations, the fourth wheel acts as a redundant backup, ensuring functionality even if one wheel fails.

To enhance reliability and operational flexibility, multiple reaction wheels are often arranged in a "skewed" or angled configuration. This design allows for cross-coupling of torques among the wheels, ensuring a reduced but functional torque capability in multiple axes if one wheel becomes inoperable. While this approach slightly compromises torque performance along individual axes, it significantly boosts overall system robustness, which is particularly valuable for long-duration or high-risk missions.

Reaction wheels require periodic desaturation to manage the buildup of angular momentum caused by external forces, such as gravitational gradients or solar radiation pressure. Desaturation involves using external actuators, such as thrusters or magnetic torquers, to offload excess momentum and restore the wheels' operational range. This process ensures the continuous functionality of the attitude control system [279].

The advent of miniaturized reaction wheels has made it possible to equip small spacecraft, including CubeSats and nanosatellites, with high-precision pointing capabilities. These compact systems are designed to meet the unique challenges of small spacecraft, such as limited power budgets, tight volume constraints, and reduced payload capacities. Modern reaction wheels are optimized for low mass, minimal power consumption, and high radiation tolerance to withstand the harsh conditions of space.

There are a range of high-heritage miniature reaction wheels currently available on the market. These models vary in mass, power requirements, torque capacity, and momentum storage, providing options for different spacecraft sizes and mission profiles. For instance, Berlin Space Technologies' RWA05 wheel offers a peak torque of 0.016 Nm and a momentum capacity of 0.5 Nms, making it suitable for small to medium-sized satellites. On the other hand, Blue Canyon Technologies' XACT series provides state-of-the-art solutions with exceptional pointing accuracy, down to 0.003° in certain models, and high momentum capacities for demanding missions [279].

Reaction wheels are essential for spacecraft requiring precise attitude control, such as Earth observation satellites, astronomical observatories, and interplanetary missions. Their performance metrics, including torque and momentum capacity, are carefully matched to mission requirements. The choice of reaction wheels must also account for environmental factors, such as radiation levels, which can impact long-term reliability.

In addition to their functional role, reaction wheels are integral to advancing the miniaturization of spacecraft technologies. By combining low mass, compact designs, and high performance, these systems enable the deployment of highly capable small spacecraft for diverse applications, from scientific research to commercial operations. As reaction wheel technology continues to evolve, it plays a pivotal role in expanding the capabilities and accessibility of space exploration [279].

Magnetic Torquers

Magnetic torquers, also known as magnetorquers, are essential spacecraft components used for attitude control and momentum management. These devices generate control torques by interacting with the local external magnetic field, providing a simple and effective way to adjust a spacecraft's orientation. The generated torque is always perpendicular to the magnetic field, making magnetic torquers a valuable tool for specific applications but limiting their ability to provide full three-axis stabilization without supplemental systems.

Magnetic torquers are frequently employed to desaturate reaction wheels. Over time, reaction wheels accumulate momentum due to external forces such as solar radiation pressure or gravitational interactions. Magnetic torquers help offload this excess momentum by creating an external torque that counteracts the buildup, allowing reaction wheels to return to their operational range and maintain precise attitude control. This synergy between magnetic torquers and reaction wheels is vital for spacecraft with extended mission durations [279].

Magnetic torquers operate by generating a magnetic dipole moment, which interacts with the Earth's magnetic field to produce a torque. The magnitude of this torque depends on the dipole strength of the torquer and the local magnetic field intensity. However, this reliance on the ambient magnetic field restricts their use to environments where a significant magnetic field is present. In low Earth orbit (LEO), where Earth's magnetic field is strong, magnetic torquers are highly effective. For missions beyond LEO or in interplanetary space, their utility diminishes unless a significant local magnetic field exists, necessitating alternative control methods.

Since magnetic torquers can only generate torque in the plane perpendicular to the magnetic field, they cannot achieve full three-axis stabilization independently. For comprehensive attitude control, magnetic torquers are typically used alongside other actuators, such as reaction wheels or thrusters, to ensure complete control across all axes.

Magnetic torquers are available in various designs, including rod-shaped, coil-based, and planar configurations, tailored to different spacecraft requirements. As an example, CubeSpace Satellite Systems offers a range of CubeTorquer models with different dipole strengths, catering to small satellites with varying control needs. ZARM Technik's offerings span a broad spectrum of dipole moments, making them suitable for both nanosatellites and larger spacecraft [279].

These torquers are often integrated into spacecraft in single or multiple-axis configurations, depending on mission requirements. Single-axis torquers are used for specific applications, such as desaturating a reaction wheel along a particular axis, while multi-axis setups provide more versatile control. Some advanced designs incorporate skewed arrangements to enhance redundancy and ensure continued functionality in case of partial system failure.

The successful application of magnetic torquers requires careful consideration of the mission environment. For interplanetary missions or those in regions with weak or non-existent magnetic fields, such as deep space or around non-magnetic celestial bodies, alternative attitude control methods must supplement or replace magnetic torquers. Additionally, the torque generated by these devices is proportional to the local magnetic field strength, which varies significantly with altitude and geographic location in Earth orbit. This variability necessitates precise modelling and calibration during mission planning.

Magnetic torquers are also limited by their power requirements and radiation tolerance. While they are generally energy-efficient, higher dipole moments may require increased power, which can strain small spacecraft with limited energy budgets. Radiation tolerance is another critical factor, especially for missions in high-radiation environments, where electronics and materials must withstand prolonged exposure [279].

Recent advancements in magnetic torquer technology have focused on miniaturization and increased efficiency, enabling their integration into nanosatellites and CubeSats. These developments have expanded the utility of magnetic torquers for a wide range of missions, from Earth observation to deep-space exploration. Their simplicity, reliability, and low-cost design make them a popular choice for attitude control in small satellite platforms.

Magnetic torquers are particularly valuable for missions requiring long-term stability and precision, such as scientific observation satellites, communication platforms, and constellations. Their ability to work in tandem with reaction wheels and other actuators ensures robust and flexible attitude control, making them a cornerstone of modern spacecraft design. As technology continues to evolve, magnetic torquers will likely remain an integral component of space missions, providing efficient and dependable solutions for attitude and momentum management [279].

Star Trackers

Star trackers are critical components in spacecraft navigation systems, providing precise three-axis attitude determination. These devices use onboard cameras and sophisticated algorithms to capture and analyse images of the stars. By comparing these images to an onboard star catalog, star trackers can calculate the absolute orientation of the spacecraft in real-time. This capability ensures accurate alignment of communication antennas, scientific instruments, and propulsion systems.

Star trackers are particularly advantageous because they provide autonomous, high-precision attitude determination without the need for continuous ground intervention. This makes them invaluable for missions requiring precise pointing, such as Earth observation satellites, interplanetary probes, and space telescopes.

A star tracker consists of an optical system, image sensors, and a processing unit. The optical system captures an image of the star field visible from the spacecraft's location. The image sensor converts this visual data into electronic signals, which are then processed to identify individual stars. The star tracker compares the detected stars to an onboard catalogue, calculating the spacecraft's orientation in terms of roll, pitch, and yaw.

Star trackers typically update attitude information several times per second, allowing for dynamic control in response to changes in the spacecraft's orientation. They can also track multiple stars simultaneously, increasing the accuracy and robustness of the attitude determination.

Star trackers are widely used across various types of missions. They are particularly effective in deep-space and low-Earth orbit (LEO) missions where high-precision attitude control is essential. For example, the Arcsec Sagitta Star Tracker demonstrated its effectiveness aboard the SIMBA CubeSat launched in 2020, highlighting the feasibility of using star trackers on small spacecraft.

There are a variety of star tracker models suitable for small spacecraft. For instance, the CubeSpace CubeStar offers a lightweight and low-power option with a wide FOV, making it ideal for nanosatellites. On the other hand, high-precision models like the Blue Canyon Technologies Extended NST provide exceptional accuracy for more demanding missions [279].

Star trackers are often integrated with other components in the spacecraft's Guidance, Navigation, and Control (GNC) system. They work in conjunction with sun sensors, gyroscopes, and reaction wheels to provide comprehensive attitude determination and control. For optimal performance, star trackers must be placed in positions with minimal obstructions to their view of the star field and away from potential sources of thermal or electromagnetic interference.

The performance of a star tracker depends on several factors, including its field of view, the sensitivity of its sensors, and the processing speed of its onboard software. Models with wider fields of view can detect more stars at once, which improves redundancy and robustness in environments with limited visible star density. Additionally, high radiation tolerance is crucial for missions in harsh environments such as geostationary orbits or deep space.

Star trackers face challenges such as stray light from the Sun, Moon, or Earth, which can interfere with the detection of stars. To mitigate this, advanced optical designs and shielding are employed. Another limitation is the potential for sensor saturation in high-radiation environments, which can degrade performance. Some models are specifically designed to withstand such conditions [279].

Star trackers also require significant computational power to process images and compare them to star catalogues. As a result, they are often paired with efficient processors to minimize power consumption while maintaining high processing speeds.

Advances in sensor technology, algorithms, and miniaturization are expanding the capabilities of star trackers. Innovations in radiation-hardened components and artificial intelligence-based image processing are expected to enhance their reliability and accuracy. These developments will enable broader applications, including use on smaller platforms like CubeSats and in more challenging environments such as interstellar missions.

In summary, star trackers are indispensable for modern spacecraft, offering unmatched accuracy and autonomy in attitude determination. By integrating high-performance star trackers into GNC systems, spacecraft can achieve precise navigation and control, ensuring mission success across a wide range of applications [279].

Magnetometers

Magnetometers are crucial instruments in spacecraft attitude determination systems, providing measurements of the local magnetic field. These measurements are used to estimate two-axis attitude information by comparing the observed magnetic field vectors to predicted values based on magnetic field models like the International Geomagnetic Reference Field (IGRF). Magnetometers are lightweight, consume minimal power, and are integral to many small spacecraft, particularly those operating in low Earth orbit (LEO), where Earth's magnetic field is strong and stable [279].

Magnetometers measure the strength and direction of the local magnetic field using magnetic sensors. Common types of magnetometers include fluxgate, anisotropic magnetoresistive (AMR), and optically pumped magnetometers. These devices sense changes in magnetic fields and convert them into electrical signals, which are then processed to determine the field's magnitude and direction. For spacecraft, three-axis magnetometers are used to measure magnetic fields along three perpendicular axes, allowing for a complete vector representation of the field.

In attitude determination, magnetometers rely on magnetic field models to compare the measured field with expected values based on the spacecraft's position. By analysing the differences, magnetometers can estimate the spacecraft's orientation relative to Earth's magnetic field. This method is particularly useful for determining pitch and roll axes but requires additional sensors, such as sun sensors or star trackers, for full three-axis attitude determination.

Magnetometers are commonly used in conjunction with magnetorquers for attitude control and momentum management. By aligning or opposing the spacecraft's magnetic dipole moment with the local magnetic field, magnetorquers create torque, enabling adjustments in spacecraft orientation. Magnetometers provide real-time feedback for this process, ensuring precise control.

For small satellites, magnetometers offer a cost-effective and reliable solution for attitude determination and control. Their lightweight and low-power design make them ideal for CubeSats and nanosatellites, where size and power constraints are critical.

Models such as the AAC Clyde Space MM200 and the NewSpace Systems NMRM-Bn25o485 are lightweight, consume less than 1 watt of power, and offer high resolution, making them suitable for small spacecraft operating in challenging environments.

For high-precision applications, models like the ZARM Technik Fluxgate Magnetometer FGM-A-75 provide exceptional sensitivity and reliability, with a radiation tolerance of 50 krad and a resolution of ±75,000 nT. This makes them suitable for long-duration missions in higher radiation environments.

Magnetometers are often integrated into spacecraft alongside other attitude determination sensors like gyroscopes, star trackers, and sun sensors. While magnetometers excel in providing real-time measurements of the magnetic field, their accuracy depends on the strength and predictability of the local field. This limits their effectiveness in interplanetary missions or deep-space environments, where Earth's magnetic field is weak or non-existent. In such cases, alternate attitude determination methods must be employed.

Additionally, spacecraft-generated magnetic noise from onboard electronics and actuators can interfere with magnetometer readings. To mitigate this, careful placement and shielding are employed during spacecraft design. For example, magnetometers are often placed on booms or extended structures away from sources of interference.

Advancements in magnetometer technology aim to improve sensitivity, reduce power consumption, and enhance radiation tolerance. Digital magnetometers, such as the ZARM Technik AMR-D-100-EFRS485, integrate advanced processing capabilities to minimize noise and provide higher resolution, enabling their use in more demanding missions [279].

Sun Sensors

Sun sensors are critical components in spacecraft attitude determination systems, providing measurements of the Sun's position relative to the spacecraft body frame. By estimating the Sun's direction, these sensors contribute valuable data for determining spacecraft orientation, though they often require complementary attitude information from other sources (e.g., Earth nadir vector or a star tracker) to achieve full three-axis attitude determination. The Sun's brightness and ease of identification make sun sensors reliable tools for fault detection and recovery, especially in scenarios requiring rapid re-establishment of spacecraft orientation. However, their measurements can be perturbed by reflected light from the Earth (albedo) or the Moon, necessitating careful calibration and placement.

Sun sensors come in various designs, each with distinct operating principles and performance characteristics.

Cosine detectors are simple, low-cost photocells that generate an electrical current proportional to the cosine of the angle between the sensor's boresight and the Sun. These sensors are typically arranged in multiple orientations to provide full-sky coverage. Despite their simplicity, cosine detectors are limited in accuracy, offering estimates typically within a few degrees of precision. They require analogue-to-digital converters to process the generated current, making them straightforward but less precise compared to advanced alternatives. Figure 5.5 illustrates a typical cosine detector design.

Quadrant sun sensors consist of a 2x2 array of photodiodes, with sunlight entering through a square window and striking the array. The intensity of light measured by each photodiode depends on the Sun's direction relative to the sensor boresight. By combining the currents generated by the four photodiodes mathematically, the Sun's angles can be calculated. Quadrant detectors provide better accuracy than cosine detectors and are commonly used in spacecraft attitude systems [279].

Digital sun sensors operate by projecting sunlight through a narrow slit onto a coded geometric bit mask. Beneath the mask is an array of photodiodes that generate photocurrents based on the illumination pattern. These currents are processed and converted into digital bit outputs, which are then used to calculate the Sun's angle based on the known slit geometry. Digital sun sensors are highly accurate and compact, making them suitable for modern spacecraft requiring precise attitude data.

Sun cameras represent the most advanced type of sun sensors, using a small camera to capture an image of the Sun. The optical system includes elements to reduce light throughput, ensuring the Sun's brightness does not saturate the imaging sensor. The onboard computer processes the image, identifies the Sun's position, and calculates its centroid to determine the Sun's direction with high precision. Some sun cameras incorporate multiple apertures to further enhance accuracy. These systems are more complex but offer unparalleled precision, making them ideal for missions with stringent attitude requirements.

Sun sensors are widely used across various spacecraft applications, from fault detection to precise attitude determination. They are especially critical in fault recovery scenarios where re-establishing orientation relative to the Sun is essential for ensuring solar panel alignment and power generation. In such cases, sun sensors provide a reliable initial reference, even in the absence of other attitude determination systems [279].

For missions requiring high accuracy, such as scientific observations or interplanetary navigation, digital sun sensors and sun cameras are preferred due to their precision and ability to handle complex scenarios. On the other hand, simpler cosine and quadrant detectors are cost-effective solutions for smaller satellites or missions with less stringent accuracy demands.

While sun sensors are generally reliable, several challenges must be addressed to maximize their performance. Perturbations from reflected light (e.g., Earth's albedo or lunar reflections) can introduce errors in measurements. Proper placement of the sensors on the spacecraft and shielding techniques can mitigate these effects. Regular calibration is also necessary to account for potential degradation of sensor components over time, especially in harsh space environments [279].

Horizon Sensors

Horizon sensors are critical tools in spacecraft attitude determination systems, providing data on the spacecraft's orientation relative to a planetary horizon. They leverage the stark temperature contrast between a planet and the surrounding space to identify the horizon line, offering reliable information for aligning and stabilizing the spacecraft. These sensors come in various forms, ranging from basic infrared horizon crossing indicators (HCIs) to advanced thermopile sensors capable of detecting nuanced thermal variations. While commonly used in terrestrial (Earth-based) applications, horizon sensors can also be adapted for other planetary bodies, making them versatile for interplanetary missions [279].

Infrared Horizon Crossing Indicators (HCIs): Infrared horizon crossing indicators are among the simplest and most widely used horizon sensors. These devices measure the infrared radiation emitted by the Earth or another planetary body and compare it to the cold background of space. As the spacecraft's motion causes the planet's horizon to cross the sensor's field of view, the infrared detector registers a sharp change in radiation levels, marking the horizon. HCIs are typically lightweight, compact, and energy-efficient, making them ideal for smaller spacecraft and missions with limited budgets [279].

Thermopile Horizon Sensors: Thermopile sensors represent a more advanced type of horizon sensor, capable of detecting subtle temperature differences across a planetary surface. These sensors consist of multiple thermocouples arranged to measure thermal radiation over a broad field of view. By analysing the temperature gradient, thermopile sensors can distinguish between the warmer equatorial regions and the cooler poles of a planet. This capability allows for more precise attitude determination and is particularly useful for missions requiring detailed orientation data, such as remote sensing or scientific observation [279].

Horizon sensors are most commonly referred to as Earth sensors when used in terrestrial applications, where they help spacecraft maintain proper alignment with the Earth's horizon. They are indispensable for Earth-orbiting satellites, ensuring that instruments, antennas, and solar panels are correctly oriented. Beyond Earth, these sensors can be adapted for use on other planets by calibrating them to account for the specific thermal characteristics of the target body. For instance, a mission to Mars or Venus could use horizon sensors to determine the spacecraft's position relative to the planetary horizon, aiding in orbital insertion or surface observations.

Horizon sensors offer several advantages, including reliability, simplicity, and the ability to operate autonomously without requiring external signals. They are particularly useful in low-Earth orbit (LEO), where the Earth's infrared signature is strong and consistent. Additionally, their ability to work independently of GPS or other external navigation aids makes them valuable for deep-space missions and in environments where external references are unavailable.

However, these sensors also face certain challenges. For example, they rely on a clear thermal contrast between the planetary surface and space, which may be reduced during certain conditions, such as atmospheric interference or solar radiation. Furthermore, the accuracy of horizon sensors can be affected by spacecraft motion, thermal noise, and the specific characteristics of the planetary body being observed. Advanced designs, such as thermopile sensors, address some of these limitations by offering greater sensitivity and the ability to measure temperature gradients [279].

Inertial Sensing

Inertial sensing is a cornerstone of modern spacecraft navigation, relying on gyroscopes and accelerometers to measure angular changes and velocity changes, respectively. These sensors are crucial for propagating a spacecraft's state, especially during periods when updates from non-inertial sensors, such as star trackers, are unavailable or infrequent. The inertial sensing system bridges the gap between these updates, providing continuous and accurate state information essential for maintaining control and stability during a mission.

Inertial sensors are available in various configurations, ranging from single-axis devices to comprehensive multi-axis packages. A single-axis gyroscope or accelerometer measures changes along one axis, while more advanced systems integrate multiple sensors for three-dimensional sensing. An Inertial Reference Unit (IRU) typically contains three orthogonal gyroscopes to measure angular changes in all three spatial dimensions. In contrast, an Inertial Measurement Unit (IMU) combines three orthogonal gyroscopes with three orthogonal accelerometers, enabling the measurement of both angular and linear changes in a fully integrated system.

IMUs play a vital role in spacecraft navigation by propagating the vehicle's state between updates from external sensors. For instance, star trackers, which determine spacecraft attitude by observing celestial bodies, often provide updates at only a few Hertz. During the intervals between these updates, an IMU is used to estimate the spacecraft's attitude, ensuring that the control system has the accurate, real-time information needed for stability and guidance [279].

Gyroscopes are at the heart of inertial sensing, with several technologies available for spacecraft applications. Among the most commonly used in small spacecraft are fiber optic gyroscopes (FOGs) and microelectromechanical systems (MEMS) gyros. FOGs operate by exploiting the Sagnac effect, where a light beam traversing a fiber loop experiences a phase shift proportional to angular rotation. This technology offers high precision and stability, making FOGs ideal for missions requiring exceptional accuracy. However, their superior performance comes with a trade-off in terms of higher mass, cost, and power consumption.

MEMS gyros, on the other hand, use vibrating structures to detect angular changes. These sensors are compact, lightweight, and cost-effective, making them well-suited for small satellites and CubeSats. Although MEMS gyros typically offer lower performance compared to FOGs, their size, weight, and power (SWaP) advantages make them an attractive choice for many missions.

Other gyroscope technologies, such as resonator gyros and ring laser gyros, are less commonly used in the SmallSat and CubeSat domain due to their larger size, higher power requirements, and cost. However, these technologies may be employed in larger spacecraft or missions with specialized requirements [279].

Gyroscope performance is a complex topic, characterized by numerous parameters that determine their accuracy and reliability. Key performance metrics include bias stability, angular random walk, and scale factor errors. Bias stability refers to the consistency of the gyroscope's output when no angular motion is present, while angular random walk measures the noise in the output over time. Scale factor errors describe deviations in the gyroscope's output relative to the actual angular velocity. These parameters, along with environmental factors such as temperature sensitivity and radiation tolerance, are critical for evaluating the suitability of a gyroscope for a particular mission.

Inertial sensors are indispensable for a wide range of spacecraft applications. They enable precise attitude control during manoeuvres, support autonomous navigation, and provide stability in dynamic environments. For example, during deep-space missions, where external navigation signals such as GPS are unavailable, inertial sensors are essential for maintaining accurate state estimates. Additionally, in low-Earth orbit, IMUs are used in conjunction with star trackers, magnetometers, and sun sensors to deliver robust and redundant navigation solutions.

GPS Receivers in Spacecraft Navigation

For spacecraft in low-Earth orbit (LEO), GPS receivers have become the standard technology for orbit determination, largely replacing traditional ground-based tracking methods. This shift has been driven by the availability of onboard GPS systems, which now provide real-time, accurate position and velocity data, making them an integral part of small spacecraft navigation. These receivers are considered a mature and reliable technology, and advancements in commercial-off-the-shelf (COTS) GPS solutions have further enhanced their accessibility and performance [279].

Modern GPS receivers used in spacecraft are highly miniaturized and efficient, allowing them to fit within the stringent size, weight, and power (SWaP) constraints of small satellites and CubeSats. For example, next-generation chip-sized COTS GPS boards, such as the NovaTel OEM 719, have replaced older models like the OEMV1, offering enhanced performance in smaller form factors. These advancements have made GPS receivers a practical and cost-effective solution for a wide range of missions [279].

The accuracy of GPS receivers in space is influenced by several factors. One significant limitation is the propagation variance of GPS signals as they pass through the exosphere, where the ionosphere and atmospheric layers can introduce errors. Additionally, civilian-use GPS relies on the Coarse Acquisition (C/A) code, which has lower precision compared to military GPS signals. These limitations can affect the overall accuracy of position and velocity data provided by the receivers.

Another critical factor is the export control of GPS units under the Export Administration Regulations (EAR). To access unrestricted GPS capabilities, such as higher velocity thresholds and altitudes, operators must secure licenses to remove Coordinating Committee for Multilateral Export Control (COCOM) limits. These limits are intended to prevent the misuse of GPS technology but can impose constraints on certain mission profiles.

While GPS technology is primarily designed for use within LEO, recent experiments have demonstrated the feasibility of using weak GPS signals at geostationary orbit (GSO) and even in cislunar space. These developments suggest that GPS receivers may soon be able to support missions beyond traditional LEO applications. Researchers are actively exploring methods to enhance signal processing and receiver sensitivity, which could enable reliable GPS functionality in regions farther from Earth.

The widespread adoption of GPS receivers for LEO missions has streamlined orbit determination and reduced reliance on ground-based tracking infrastructure. The ability to integrate GPS receivers into small spacecraft has enabled autonomous navigation and improved mission flexibility. Moreover, the potential expansion of GPS usability to higher altitudes and cislunar distances represents a significant advancement in navigation technology, promising to extend its benefits to interplanetary missions and deep-space exploration [279].

GPS receivers have revolutionized spacecraft navigation, offering real-time, precise orbit determination for LEO missions. Advances in COTS technology and ongoing research into signal processing are poised to expand their applicability to GSO and cislunar missions, making GPS an increasingly versatile tool in space exploration. As this technology continues to evolve, it will play a pivotal role in enhancing the autonomy, efficiency, and scope of space missions across a variety of domains.

Deep Space Navigation

Navigating spacecraft in deep space presents unique challenges that require specialized systems and infrastructure. Unlike low-Earth orbit missions, where GPS and ground-based systems dominate, deep space navigation relies on advanced radio transponders working in coordination with the Deep Space Network (DSN). These systems provide precise tracking, enabling spacecraft to determine their position and trajectory as they explore distant regions of the solar system.

The DSN is a global network of large antennas designed to maintain communication with spacecraft beyond Earth's orbit. By measuring the time delay and Doppler shift of signals sent between the spacecraft and the DSN, navigation teams can calculate the spacecraft's distance, velocity, and trajectory. This data is critical for trajectory corrections, scientific observations, and mission planning [279].

At the heart of deep space navigation systems are radio transponders that manage communication with the DSN. As of 2020, the Small Deep Space Transponder (SDST) was the only flight-proven transponder suitable for small spacecraft. Designed by NASA's Jet Propulsion Laboratory (JPL) and manufactured by General Dynamics, the SDST is highly reliable for deep-space operations. It is capable of handling the precise timing and signal processing required for DSN-based navigation [279].

To accommodate the growing use of CubeSats in deep-space missions, JPL developed the IRIS V2 transponder, a compact and efficient alternative to the SDST. Derived from the Low Mass Radio Science Transponder (LMRST), IRIS V2 is specifically designed for the size, weight, and power constraints of CubeSats. This transponder represents a significant advancement in making deep-space navigation accessible to smaller spacecraft.

The IRIS V2 transponder has been successfully deployed on several high-profile missions, demonstrating its capabilities and reliability. It first flew aboard the MarCO CubeSats during the 2018 Mars InSight mission, where it provided critical navigation support for the lander's descent. More recently, it was used on the LICIACube mission, which performed an asteroid flyby in September 2022, and the 12U CAPSTONE spacecraft, which entered a unique lunar orbit in November 2022. Additionally, IRIS V2 was included in six secondary CubeSat payloads on NASA's Artemis I mission, including Lunar Flashlight, LunaH-Map, ArgoMoon, CubeSat for Solar Particles, BioSentinel, and NEA Scout [279].

The IRIS V2 is also set to fly on upcoming missions such as INSPIRE, further solidifying its role in deep-space navigation for small spacecraft. Its compact design and proven performance make it a cornerstone technology for enabling cost-effective exploration of deep-space targets, from lunar orbits to asteroid flybys and interplanetary missions [279].

Atomic Clocks

Atomic clocks, long a cornerstone of precision timing for larger spacecraft in low-Earth orbit (LEO), are now becoming integral to small spacecraft systems. These highly accurate timekeeping devices enable precise navigation, improving spacecraft autonomy and reducing reliance on ground-based infrastructure. While their integration into smaller spacecraft is a relatively new development, atomic clocks and oscillators are emerging as critical tools for advanced space missions, particularly those requiring precise timing and navigation in deep space.

Historically, spacecraft navigation has relied on two-way tracking systems that involve ground-based antennas equipped with atomic clocks. The ground station sends a signal to the spacecraft, which then responds, allowing the ground station to measure the time delay and calculate the spacecraft's position, velocity, and trajectory. However, this process is inherently limited. The spacecraft must wait for instructions from the ground, reducing real-time decision-making capabilities. Additionally, ground stations can only track one spacecraft at a time, as they must wait for a return signal before proceeding with other tracking tasks. These limitations become more pronounced in deep-space missions, where the vast distances necessitate measuring radio signal accuracy within a few nanoseconds to maintain precision.

For deep-space navigation, the integration of atomic clocks onboard spacecraft represents a transformative step. By carrying highly accurate and stable atomic clocks, small spacecraft can independently measure signal timings and calculate their own position and velocity in real time. This capability reduces the need for continuous ground-based support, allowing spacecraft to make autonomous navigation decisions and freeing up ground stations to handle other tasks. Atomic clocks are particularly valuable in deep space, where the time delay in communication with ground stations can span several minutes or even hours, depending on the mission's distance from Earth [279].

While atomic clocks have been used on larger spacecraft for years, adapting them for small spacecraft presents unique challenges. Small spacecraft, such as CubeSats, have stringent constraints on power consumption, volume, and weight. Engineers are now developing compact and power-efficient atomic clocks and oscillators specifically designed for these limitations. These new systems are also tailored for missions that may require multiple radios, ensuring synchronization across various communication and navigation subsystems.

The integration of atomic clocks into small spacecraft has far-reaching implications for mission design. By enabling autonomous navigation, spacecraft can operate with greater independence, reducing reliance on continuous ground support. This autonomy is particularly critical for missions involving multiple spacecraft, where managing simultaneous communication with ground stations would otherwise be infeasible. Furthermore, the precision offered by atomic clocks enhances scientific observations, enabling better coordination for missions requiring synchronized measurements or observations, such as planetary flybys or asteroid rendezvous.

Light Detection and Ranging (LiDAR)

LiDAR, or Light Detection and Ranging, is an advanced sensor technology that is increasingly finding applications in spacecraft navigation. Originally developed for terrestrial use and widely adopted in automotive and other industries, LiDAR has undergone significant technological maturation over the past decade. This evolution has paved the way for its integration into spacecraft, particularly those involved in proximity operations or missions requiring precise spatial measurements [279].

LiDAR operates by emitting laser pulses and measuring the time it takes for these pulses to return after reflecting off surfaces. This information is used to create detailed 3D maps of the surrounding environment or to calculate distances with high precision. In the context of space missions, LiDAR sensors are valuable for tasks requiring detailed situational awareness, such as proximity operations, landing, and relative navigation between spacecraft.

LiDAR technology is already being deployed on larger spacecraft like NASA's Orion, where it supports critical proximity operations. For small spacecraft, LiDAR offers potential applications in altimetry, relative navigation, and formation flying. It is particularly useful for scenarios such as:

- **Mars Helicopter Operations:** LiDAR can provide precise altitude data and terrain mapping to aid navigation and obstacle avoidance for aerial vehicles operating in extraterrestrial environments.
- **Rendezvous and Docking:** LiDAR enables small spacecraft to approach and dock with larger platforms or other small spacecraft by offering highly accurate distance and orientation measurements.
- **Formation Flying:** For missions involving multiple spacecraft flying in tight formation, LiDAR can provide real-time relative position and orientation data, enabling coordinated manoeuvres.

LiDAR sensors are compact, lightweight, and capable of functioning in a variety of environmental conditions, making them ideal for small spacecraft with stringent size, weight, and power (SWaP) constraints. The precision offered by LiDAR systems is unmatched by many other sensor types, particularly in scenarios requiring real-time three-dimensional mapping or highly accurate distance measurements [279].

Fitting Navigation Systems on Rockets

The successful operation of rocket navigation systems relies on their precise placement and integration into the vehicle. These systems ensure accurate positioning, orientation, and trajectory corrections throughout various mission phases, from launch to deep space operations. Fitting navigation systems onto a rocket involves a carefully designed layout where each subsystem is placed according to its function, environmental requirements, and data flow within the Guidance, Navigation, and Control (GNC) framework.

Inertial navigation systems (INS), including gyroscopes and accelerometers, are typically housed in the rocket's central avionics bay or inertial measurement unit (IMU). This section is centrally located to minimize vibration and ensure accurate measurements of angular and linear changes during flight. INS components must be placed as close to the spacecraft's center of mass as possible to reduce mechanical noise and rotational disturbances. The IMU integrates multiple gyroscopes and accelerometers to track position, velocity, and orientation by calculating changes in the rocket's motion. Placement in the avionics bay ensures they are shielded from harsh environmental conditions such as extreme temperatures and radiation, while still maintaining connectivity with the spacecraft's computing system for real-time state propagation.

Star trackers, which provide precise three-axis attitude determination, are mounted externally on the rocket or spacecraft structure. They are positioned where they have an unobstructed field of view of the star field to capture clear images. Typically, star trackers are placed on panels away from large heat sources, such as propulsion systems, to avoid thermal distortion that could affect imaging accuracy. For small spacecraft or CubeSats, star trackers may be located on deployable arms or booms to achieve a wider field of view. Sun sensors, on the other hand, are distributed strategically across the spacecraft body, ensuring continuous detection of the Sun's position regardless of the vehicle's orientation. This redundancy enables fault recovery by realigning the spacecraft's solar panels for power generation during critical phases of the mission.

Magnetometers are placed on booms or extended structures away from the rocket's main body to minimize interference from onboard electronics and actuators, which could distort magnetic field measurements. They measure the local magnetic field to estimate spacecraft orientation relative to Earth's magnetic field. Magnetic torquers, used to create control torques for momentum management, are embedded within the spacecraft body or mounted near the structure's edges. They are oriented to produce torque perpendicular to the Earth's magnetic field, ensuring effective interaction for attitude control. Both systems work in tandem to adjust the spacecraft's orientation and desaturate reaction wheels, particularly in low Earth orbit (LEO) missions.

Reaction wheels, which provide fine pointing control through angular momentum exchange, are integrated inside the spacecraft body in an orthogonal or skewed configuration. This configuration ensures full three-axis control with redundancy in case of wheel failure. Reaction wheels are mounted close to the spacecraft's centre of mass, where their angular rotation results in controlled counter-rotation of the spacecraft around its axes. To offset the momentum buildup caused by external forces, desaturation mechanisms like magnetic torquers or thrusters are fitted alongside reaction wheels, often within the same attitude control framework.

GPS receivers, used primarily for orbit determination in LEO missions, are fitted within the spacecraft's avionics bay for protection and connectivity with other onboard systems. To receive GPS signals effectively, antennas are mounted externally on the spacecraft structure, typically on surfaces that maintain clear line-of-sight visibility to GPS satellites. The positioning of GPS antennas is carefully planned to avoid obstructions from rocket components or payload instruments. For CubeSats and smaller spacecraft, compact and lightweight GPS receivers are often integrated with other navigation systems for efficient use of space and power.

Optical navigation systems, including cameras and LiDAR sensors, are mounted externally on the spacecraft to provide direct line-of-sight to celestial bodies, planetary surfaces, or other spacecraft during proximity operations. Cameras used for optical navigation are placed on rigid structures with minimal thermal distortion to ensure image clarity. For LiDAR systems, the positioning depends on their application. For example, LiDAR sensors used for altimetry during planetary landings are mounted on the bottom of the spacecraft, facing the surface. In relative navigation tasks like rendezvous and docking, LiDAR units are positioned on the forward-facing section of the spacecraft to measure distances and velocities to target objects.

For deep space missions, navigation relies heavily on radio transponders, such as the Small Deep Space Transponder (SDST) or IRIS V2. These transponders are integrated within the spacecraft's avionics section, where they are protected and connected to onboard processors. Large antennas, like high-gain antennas (HGAs), are mounted externally to enable communication with the Deep Space Network (DSN). These antennas are often placed on gimbaled platforms, allowing them to reorient and maintain a direct line of communication with Earth as the spacecraft changes orientation. To ensure accuracy, the antennas are positioned far from structures that may cause radio signal interference.

Atomic clocks, which enable precise onboard timing for deep space navigation, are housed in the avionics bay, where they are shielded from temperature fluctuations and radiation. These clocks are integrated with radio transponders to synchronize signal timings for accurate positioning. For small spacecraft, atomic clocks are miniaturized to fit within the constrained volume and power budgets.

Navigation systems are carefully fitted to ensure proper thermal management, as excessive heat or cold can degrade sensor performance. Thermal shielding, insulation, and radiators are incorporated to maintain optimal operating temperatures. Additionally, redundancy is built into navigation systems by incorporating multiple sensors (e.g., star trackers, sun sensors, gyros) to ensure reliability. If one component fails, backups or alternative systems can continue providing navigation and control data.

The placement of rocket navigation systems is determined by their specific functions, environmental requirements, and integration within the spacecraft's overall architecture. Internal components like gyroscopes, accelerometers, and atomic clocks are fitted in the avionics bay for protection and stability. External systems such as star trackers, sun sensors, GPS antennas, optical navigation cameras, and LiDAR sensors are carefully mounted to achieve unobstructed visibility and precision. By strategically fitting these components, navigation systems work cohesively to ensure mission success across launch, orbit, deep-space travel, and planetary landings.

Onboard Computers and Telemetry

Rocket onboard computers and telemetry systems are critical components for the operation, control, and monitoring of rockets and spacecraft. These systems enable real-time data processing, execution of commands, communication with ground stations, and monitoring of vital parameters to ensure mission success. Together, onboard computers and telemetry play a pivotal role in controlling the vehicle's systems, executing flight plans, and transmitting performance data back to Earth.

Onboard Computers

Onboard computers serve as the central processing units (CPUs) of a rocket or spacecraft, executing mission-critical tasks such as vehicle control, trajectory calculations, sensor management, and fault detection [280-282]. These onboard computers must meet stringent requirements for performance, reliability, and resistance to harsh space environments, including radiation, extreme temperatures, and vibrations [283-285].

The onboard computer processes input from various subsystems, including the Guidance, Navigation, and Control (GNC) system, propulsion systems, and environmental sensors [285-287]. It uses this input to make real-time decisions and execute commands. For instance, during the launch phase, the onboard computer continuously calculates the rocket's position, orientation, and velocity while adjusting thrust vectoring and stabilizing the vehicle [288-290].

Modern onboard computers typically use redundant architectures to enhance reliability [108, 291, 292]. Redundancy ensures that if one system fails, a backup system can take over seamlessly, preventing mission failure. Systems like triple modular redundancy (TMR) use three identical processors to compare outputs and identify potential errors caused by radiation or hardware failures [293-295]. Radiation-hardened processors, such as those based on the RAD750 or LEON architecture, are often used to withstand the space environment [296-298].

Onboard computers are also responsible for autonomous operations in missions where real-time communication with ground control is limited or delayed [299-301]. In deep-space missions, for example, the onboard computer must autonomously handle navigation, trajectory corrections, and system adjustments without direct input from Earth due to long communication delays. This autonomy is essential for critical tasks like entry, descent, and landing (EDL) on planetary surfaces or interplanetary manoeuvres [302].

In small satellites and CubeSats, compact and low-power onboard computers are used to meet the size, weight, and power (SWaP) constraints of the platform. These systems integrate advanced processors, memory modules, and interfaces in miniaturized designs, enabling high-performance computing in small spacecraft [299, 301].

Triple Modular Redundancy

Triple Modular Redundancy (TMR) is a fault-tolerant system architecture commonly used in rockets, spacecraft, and other critical aerospace applications to ensure reliability, robustness, and operational continuity in the face of hardware failures or environmental challenges. TMR is specifically designed to mitigate the risks of single-point failures, such as those caused by radiation, mechanical stress, or other anomalies in the harsh environment of space.

At its core, TMR works by triplicating critical components, such as processors, sensors, or other subsystems, and comparing their outputs to detect and isolate errors. By employing majority voting logic, TMR ensures that the system continues functioning even if one component fails. This method enhances mission reliability and reduces the likelihood of catastrophic failures.

In a TMR system, three identical modules (often referred to as "replicas") perform the same computation or task independently. These modules are typically processors, memory units, or other essential electronics. The outputs of the three modules are compared using a voter circuit or voting logic system.

- If all three outputs match, the system proceeds with the agreed-upon output.
- If one module's output differs from the other two, the majority output (chosen by a 2-out-of-3 vote) is accepted as correct.
- The faulty module is identified and isolated, preventing it from influencing subsequent operations.

For example, in the case of a TMR processor, three identical CPUs execute the same calculations in parallel. The voter logic monitors the results and selects the majority output as valid, ensuring system correctness despite any single fault.

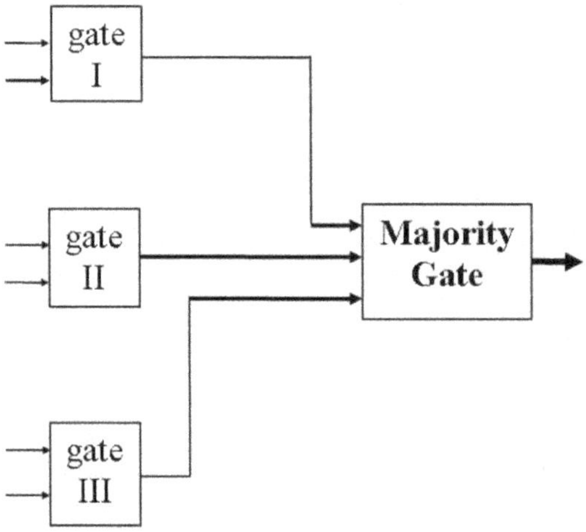

Figure 61: Triple Modular Redundancy Block Diagram. IjonTichyIjonTichy, CC0, via Wikimedia Commons.

Triple Modular Redundancy (TMR) provides significant benefits to rockets by offering robust fault tolerance against hardware failures. In systems where repairs are impossible, such as space missions, TMR ensures that even if one subsystem fails, the remaining two generate the correct output. This fault tolerance allows the mission to continue uninterrupted, which is critical for preventing catastrophic failures. Another major benefit of TMR is its ability to mitigate the effects of radiation. In space, radiation from cosmic rays and solar activity can cause single-event upsets

(SEUs) or transient faults in electronic components. TMR addresses these challenges by detecting incorrect outputs and relying on the unaffected majority outputs to maintain accurate operations.

The high reliability and redundancy offered by TMR further enhance the performance of critical rocket systems. By ensuring that single faults cannot compromise the operation, TMR becomes particularly important for flight control systems, guidance, navigation, and onboard computers, where stability and accuracy are paramount. The real-time error detection and isolation capabilities of TMR provide continuous monitoring and comparison of outputs from the three subsystems. Faulty components are identified without significant delays, allowing smooth and uninterrupted operations. TMR also increases mission safety by adding a layer of assurance that essential systems, such as thrust vector control, trajectory adjustments, and communication subsystems, will remain functional under adverse conditions. This reliability is crucial during high-risk phases of a rocket launch or deep-space mission.

The applications of TMR in rockets span various mission-critical systems. Onboard flight computers rely on TMR to process real-time data for tasks like attitude control, thrust vectoring, and stage separation. If a processor fails, TMR ensures the system remains operational without any loss of functionality. Guidance, Navigation, and Control (GNC) systems, which determine the rocket's trajectory, also utilize TMR. Sensors such as gyroscopes, accelerometers, and star trackers often integrate TMR to detect and correct erroneous readings caused by faults or external disturbances. Communication systems onboard rockets require TMR to maintain reliable communication with ground control. TMR in transponders and processors safeguards data transmission against errors and corruption. TMR also plays a vital role in actuation systems, which control thrust vectoring by adjusting the rocket's nozzle direction to maintain stability. Ensuring actuators operate correctly is critical for following the planned trajectory. Additionally, power management and distribution systems incorporate TMR to guarantee an uninterrupted power supply to critical subsystems, even if a component fails.

While TMR offers significant advantages, it does present some challenges. One major drawback is the increased hardware complexity. Triplicating critical components adds to the system's overall mass, power consumption, and physical footprint, all of which are critical considerations for rockets where minimizing weight is essential. Another challenge is the cost, as TMR requires additional hardware, manufacturing, and rigorous testing, leading to higher development and deployment expenses. The voter circuit, which compares outputs and determines the correct response, can also become a single point of failure. To mitigate this risk, additional redundancy such as triplicated voter circuits or error-correcting techniques may be necessary. In environments with high radiation exposure, TMR may struggle with simultaneous errors across all three modules. While TMR is effective against single faults, it cannot resolve errors caused by a strong radiation burst that affects multiple components. Advanced fault-tolerant methods, such as Error-Correcting Codes (ECC) or memory scrubbing, may be used alongside TMR to address this limitation.

Historically, TMR has been employed in some of the most significant rocket systems. NASA's Saturn V rocket, which launched the Apollo missions to the Moon, relied on triple redundancy in its guidance and control systems to ensure precise navigation. Similarly, the Space Shuttle's General Purpose Computers (GPCs) operated using TMR to manage critical tasks, including flight control and onboard operations. Modern launch vehicles, such as SpaceX's Falcon 9 and ULA's Atlas V, also employ TMR or equivalent redundancy systems to enhance reliability during critical phases like launch, ascent, and orbital insertion. These examples highlight the enduring importance of TMR in ensuring the success and safety of space missions.

Radiation-Hardened Single-Board Computers

Radiation-hardened single-board computers (SBCs) are critical components in spacecraft, satellites, and deep-space missions, where they must endure the harsh conditions of space, including cosmic radiation, extreme temperatures, and physical stresses. These specialized computers are designed to provide high reliability, fault tolerance, and robust computational performance, ensuring mission success in environments where standard electronics would fail.

Radiation-hardened SBCs incorporate advanced design techniques and materials to ensure durability against space radiation. The key technologies and design principles include:

1. **Radiation Hardening by Design (RHBD):** RHBD focuses on building fault-tolerant circuits that can withstand Single Event Effects (SEEs) caused by cosmic rays, such as Single Event Upsets (SEUs), Single Event Transients (SETs), and Latchups (SEL). Techniques like triple modular redundancy (TMR) and error-correcting code (ECC) memory are used to maintain system integrity.
2. **Radiation-Hardened Manufacturing Process:** Radiation-hardened SBCs are fabricated using processes like Silicon on Insulator (SOI) or bulk CMOS with specific doping techniques. These processes enhance the chip's tolerance to radiation by reducing the chance of charge accumulation in transistors.
3. **Thermal and Environmental Resilience:** SBCs are built to function across extreme temperature ranges, typically −55°C to 125°C, ensuring stability in deep space orbits or planetary missions.
4. **Fault Tolerance and Error Detection:** Many SBCs use error correction techniques, such as ECC memory for data integrity, parity checks, and watchdog timers to detect errors and prevent malfunctions caused by radiation-induced faults.
5. **Low Power Consumption:** Radiation-hardened SBCs are optimized for low power operation (5–15 watts) to suit the limited power budgets of spacecraft while delivering high performance.

Examples of Current Radiation-Hardened Single-Board Computers

RAD750 (BAE Systems): The RAD750 is one of the most widely used radiation-hardened SBCs in space applications.

- **CPU Architecture:** PowerPC 750-based processor
- **Performance:** Core clock speeds between 110–200 MHz, delivering up to 266 MIPS (Million Instructions Per Second).
- **Radiation Tolerance:** Can endure 2,000–10,000 grays (200,000–1,000,000 rads) of total ionizing dose (TID).
- **Power Requirements:** Operates at **5–10 watts**.
- **Applications:**
    - The Mars Reconnaissance Orbiter uses the RAD750 for navigation and data processing.
    - NASA's Curiosity Rover and Perseverance Rover on Mars rely on the RAD750 to manage systems, control instruments, and process sensor data.
    - The James Webb Space Telescope (JWST) also employs RAD750 for its onboard computations.

The RAD750 is a highly specialized radiation-hardened single-board computer developed by BAE Systems Electronics, Intelligence & Support for use in spacecraft and satellites that operate in extreme, high-radiation environments [303, 304]. Released in 2001 as the successor to the RAD6000, the RAD750 is designed to withstand the intense radiation, temperature extremes, and other challenging conditions encountered during space missions [303-305].

The RAD750 CPU represents a significant technological leap over the RAD6000, featuring 10.4 million transistors, compared to the RAD6000's 1.1 million [303]. This tenfold increase in transistor count highlights its greater processing capability and improved architecture. The RAD750 is fabricated using 250 nm or 150 nm photolithography and has a die area of 130 mm^2, making it more compact and efficient [303]. It operates with a core clock speed of 110 to 200 MHz and can achieve processing performance of 266 MIPS (Million Instructions Per Second) or more, depending on the configuration [303]. To further enhance its computational efficiency, the RAD750 CPU can include an extended L2 cache, allowing for better memory access speeds and improved overall performance [303].

Figure 62: A radiation hardened version of the PowerPC 750 processor. Manufactured by BAE systems for use in space. Henriok, CC BY-SA 2.0, via Flickr.

One of the defining characteristics of the RAD750 is its exceptional resilience to radiation. The CPU can endure an absorbed radiation dose of 2,000 to 10,000 grays (200,000 to 1,000,000 rads), a level of radiation far exceeding what standard terrestrial electronics can tolerate [303]. This makes the RAD750 ideal for long-term missions in the harsh environments of deep space, where cosmic radiation and solar activity are prevalent [303, 305, 306]. The CPU also boasts a wide operational temperature range, functioning effectively between −55°C and 125°C, ensuring performance in extreme thermal conditions [303, 304]. Despite its durability, the CPU itself consumes only 5 watts of power, a critical consideration for energy-limited space missions [303].

The RAD750 single-board computer system, which includes the CPU and motherboard, maintains similarly robust characteristics. It can endure 1,000 grays (100,000 rads) of radiation exposure and operates reliably within a temperature range of −55°C to 70°C [303]. The complete system requires 10 watts of power, making it relatively energy-efficient given its advanced capabilities [303]. These specifications ensure the RAD750 is well-suited to meet the demanding requirements of space exploration, including deep-space probes, satellites, and planetary rovers [303, 305].

While the RAD750 system offers cutting-edge technology, it comes at a high cost. The price of the RAD750 is comparable to that of its predecessor, the RAD6000, which was listed at $200,000 as of 2002 (approximately $338,797 in 2023 when adjusted for inflation) [303]. However, the final cost of the RAD750 depends heavily on the specific requirements of the customer program and the quantities ordered [303]. Such costs reflect the specialized design, rigorous testing, and reliability assurance necessary for hardware that must perform flawlessly in the unforgiving conditions of space [303, 304].

Technologically, the RAD750 is based on the PowerPC 750 architecture, which is part of the PowerPC 7xx family. Its logic functions and packaging are fully compatible with this processor family, ensuring its integration with existing systems and software [303, 304]. This compatibility also enables the RAD750 to benefit from prior advancements in PowerPC technology while offering enhanced performance and radiation resistance critical for space applications [303, 304].

The RAD750's durability, efficiency, and processing power have made it a standard choice for a wide range of high-profile space missions. It serves as the backbone for the computational needs of spacecraft, ensuring that navigation, data processing, and communication systems remain operational in the harshest environments [303]. The RAD750 is a testament to the importance of radiation-hardened technology in modern space exploration, where even minor system failures can jeopardize entire missions. It remains a key solution for ensuring mission success in Earth orbit, deep space, and planetary exploration [303, 306].

Versal AI Core Radiation-Tolerant (Xilinx/AMD): This FPGA-based radiation-tolerant platform incorporates **AI acceleration** and **adaptive computing** capabilities.

- **Features:** Radiation-hardened programmable logic, integrated AI engines, and adaptable software-defined hardware.
- **Performance:** Offers advanced capabilities like high-speed parallel processing and low-latency data handling.
- **Applications:** Next-generation satellite constellations requiring AI-based onboard image processing or real-time decision-making.

LEON3FT (Cobham Gaisler): The LEON3FT is a fault-tolerant processor core based on the SPARC V8 architecture. It is widely used in space-grade systems:

- **CPU Architecture:** SPARC-based, fault-tolerant design
- **Radiation Tolerance:** Robust to SEEs and optimized for single-event upsets (SEUs) through error detection and correction.
- **Performance:** Supports clock frequencies up to **100 MHz** with integrated fault tolerance.
- **Applications:**
    - The **ESA Sentinel-2 Earth observation satellite** uses LEON3FT for mission-critical operations.
    - It is also employed in cubesats and micro-satellites for cost-effective Earth observation and data processing.

The LEON processor family is a series of radiation-tolerant, 32-bit CPUs based on the SPARC V8 Reduced Instruction Set Computer (RISC) architecture originally developed by Sun Microsystems. The European Space Agency (ESA) initiated the LEON project in 1997, aiming to create an open, portable, and fault-tolerant processor for space missions. Over the years, the processor has undergone significant advancements, with subsequent generations—LEON2, LEON3, LEON3FT, and LEON4—developed by Gaisler Research (now part of Frontgrade, previously Aeroflex and Cobham). The processor cores are described using the VHSIC Hardware Description Language (VHDL), making them highly customizable and suitable for System-on-Chip (SoC) designs.

The LEON processors are designed to operate in harsh space environments, where radiation can cause Single-Event Upsets (SEUs) and transient faults in electronics. To address these challenges, fault-tolerance mechanisms like Triple Modular Redundancy (TMR) and Error Detection and Correction (EDAC) are implemented. The processors are highly configurable using VHDL generics, enabling designers to tailor their performance, fault tolerance, and peripherals to specific mission needs.

All LEON processors implement the SPARC V8 architecture, which supports efficient pipelined execution and reduced instruction sets for high-performance computation. Earlier versions like LEON2 used a five-stage pipeline, whereas later models like LEON3 and LEON4 feature a seven-stage pipeline for enhanced processing capabilities.

The project began with LEON1, a proof-of-concept processor tested on the LEONExpress chip, manufactured using 0.25 µm technology. LEON1 successfully demonstrated fault tolerance in a radiation-rich environment. ESA followed up with LEON2, implemented in devices like the Atmel AT697, which became widely used in satellites.

Subsequent versions introduced significant enhancements. LEON3 and its fault-tolerant variant, LEON3FT, were developed by Gaisler Research. These processors provided improved performance and configurability, introducing support for Symmetric Multi-Processing (SMP) and integrating them with the GRLIB IP Library, which includes advanced peripherals like Ethernet, PCI, and DDR controllers. The latest processor, LEON4, was released in 2010, offering higher performance with features such as static branch prediction, optional level-2 cache, and wider AMBA bus interfaces.

LEON2: The LEON2 processor core features a five-stage pipeline and a set of standard peripherals, including timers, UARTs, memory controllers, and an interrupt controller. The LEON2-FT variant adds fault tolerance through TMR-protected flip-flops and EDAC for internal and external memory. LEON2-FT was used in ESA's Intermediate eXperimental Vehicle (IXV) and China's Chang'e 4 lunar lander, demonstrating its reliability in real-world missions.

LEON3 and LEON3FTL: LEON3 marked a major advancement with its seven-stage pipeline and support for Symmetric Multi-Processing (SMP), allowing multi-core configurations for higher processing capabilities. The LEON3FT variant incorporates SEU error-correction mechanisms for register files and cache memory, making it ideal for space applications where reliability is critical.

LEON3 and LEON3FT are distributed with the GRLIB IP Library, a versatile package that includes peripheral cores like:

- DDR/DDR2 memory controllers
- PCI and Ethernet interfaces

- USB 2.0 and CAN bus controllers
- SPI and I^2C interfaces

This flexibility enables designers to integrate LEON3 into customized SoC solutions for satellites, deep-space probes, and other aerospace applications.

LEON4: The LEON4 processor offers further performance improvements over its predecessors. It features:

- Static branch prediction for optimized instruction flow.
- An optional level-2 cache for faster memory access.
- A 64-bit or 128-bit AMBA AHB interface for high-throughput data transfers.
- Performance improvements of 1.7 DMIPS/MHz, compared to LEON3's 1.4 DMIPS/MHz.

LEON4 provides enhanced capabilities for data-intensive applications like scientific instruments, payload processing, and interplanetary exploration missions.

Radiation-tolerant LEON processors are specifically designed to operate reliably in harsh environments where traditional electronics are prone to failure. These processors employ critical techniques to ensure resilience and reliability under extreme conditions. One of the key methods is Triple Modular Redundancy (TMR), which protects registers and flip-flops by implementing a voting mechanism that ensures output accuracy, even if one component fails. Another essential feature is Error Detection and Correction (EDAC), which is used for both internal caches and external memory to detect and correct single-event upsets (SEUs) caused by radiation. Additionally, the processors incorporate SEU Mitigation mechanisms that autonomously detect and recover from transient faults without requiring external intervention. These combined fault-tolerance techniques allow LEON processors to withstand significant total ionizing doses (TID) of radiation, ensuring consistent performance in applications such as satellites, planetary rovers, and deep-space missions.

LEON processors have been widely adopted in space missions led by ESA, NASA, and other prominent space agencies due to their exceptional reliability and fault tolerance. In missions such as Mars exploration rovers, LEON processors play a critical role in navigation, data processing, and controlling scientific payloads, enabling rovers to traverse and investigate challenging Martian terrain. Similarly, they are used in Earth observation satellites, like Sentinel-2, where the processors facilitate data collection, image processing, and secure communication with ground stations, enabling the delivery of vital environmental and climate information. LEON processors are also central to interplanetary missions, providing fault-tolerant performance for spacecraft like ESA's BepiColombo mission to Mercury and the Solar Orbiter, which operates in extreme radiation conditions near the Sun.

The open-source availability of LEON processors, combined with their dual licensing model, makes them highly attractive for both research and commercial applications. Researchers benefit from the freedom to customize and enhance the design under open-source licenses, while commercial developers can obtain proprietary licenses for integration into advanced products. This flexibility has accelerated the development of innovative and cost-effective space systems, positioning LEON processors as a cornerstone technology for reliable space exploration and scientific advancement.

RAD5545 (BAE Systems): The RAD5545 is an evolution of the RAD750, offering increased performance and higher levels of radiation resilience.

- **CPU Architecture:** Quad-core PowerPC-based processor
- **Performance:** Delivers up to 5.6 GFLOPS for compute-intensive applications.
- **Radiation Tolerance:** Tolerant to up to 1 Mrad TID with SEE protection.
- **Power Consumption:** ~12 watts
- **Applications:**
    - Deep-space probes and high-throughput satellites requiring extensive onboard computation, such as autonomous navigation and AI-based payload processing.

Cobham GR740 (Next-Generation LEON Processor): The GR740 is a quad-core radiation-hardened processor based on the LEON4FT core, designed for space missions requiring enhanced performance.

- **CPU Architecture:** Quad-core SPARC V8 with fault tolerance
- **Performance:** Operates at 250 MHz, providing up to 1,500 MIPS.
- **Radiation Tolerance:** Fault-tolerant design for SEE resilience.
- **Applications:** High-data-rate instruments, advanced payloads, and robotic planetary exploration missions.

The Cobham GR740 is a high-performance radiation-hardened system-on-chip (SoC) developed for space applications [307]. Designed and manufactured by Cobham Gaisler (now part of Frontgrade), the GR740 combines advanced computing power with resilience to operate reliably in the harsh radiation environments of space [307]. It is based on the SPARC V8 architecture, a proven standard in space-grade processors, and is built for use in spacecraft systems, including satellites, planetary exploration missions, and scientific instruments [307].

At the core of the GR740 is a quad-core LEON4FT processor [307]. The LEON4FT (Fault-Tolerant) design offers enhanced performance and fault resilience, ensuring continued operation in environments where single-event upsets (SEUs) are common [307]. Each of the four LEON4FT cores runs at a clock frequency of up to 250 MHz, providing a total processing power of approximately 1.5 DMIPS/MHz per core, making it one of the most powerful radiation-hardened processors available for space applications [307].

The GR740 integrates a Network-on-Chip (NoC) architecture, enabling efficient and scalable communication between its multiple cores and peripherals [307]. This architecture ensures high data throughput and low latency, making the GR740 ideal for systems requiring extensive real-time processing, such as spacecraft command and data handling systems [307].

To further enhance performance, the GR740 features Level 2 cache of up to 2 MB, shared across the four cores [307]. This cache significantly improves processing efficiency by reducing memory access delays, which is particularly important for computationally intensive applications like image processing and scientific simulations [307].

The GR740 is equipped with a comprehensive set of interfaces and peripherals, enabling it to meet the diverse requirements of modern space systems [307]. These include SpaceWire, CAN FD, MIL-STD-1553, Gigabit Ethernet, and a DDR2/DDR3 memory controller [307]. These interfaces provide high-speed, low-latency communication, robust and reliable subsystem integration, and data integrity under radiation exposure [307].

The GR740 is designed to withstand the extreme radiation conditions encountered in space [307]. It incorporates fault-tolerant (FT) mechanisms to mitigate the effects of single-event upsets (SEUs) and other radiation-induced errors [307]. Key fault-tolerant features include Triple Modular Redundancy (TMR), Error Detection and Correction (EDAC), and radiation tolerance up to 100 krad (Si) [307]. These features make the GR740 highly resilient, ensuring it can operate in low Earth orbit (LEO), geostationary orbit (GEO), and deep space missions without compromising reliability [307].

The GR740 is tailored for demanding space applications that require robust processing power, fault tolerance, and efficient communication [307]. It is ideally suited for command and data handling (C&DH) systems, onboard data processing for Earth observation satellites, payload processing in interplanetary missions and scientific instruments, and avionics systems for planetary exploration rovers and landers [307].

Telemetry Systems

Telemetry systems are used to collect, monitor, and transmit data from the rocket or spacecraft back to ground stations. The telemetry system is essential for tracking the vehicle's performance, health, and status throughout all mission phases, including launch, ascent, orbital operations, and re-entry. Telemetry ensures that engineers and mission operators on Earth receive critical information to assess the rocket's behaviour and make adjustments as needed.

Telemetry works by encoding data from onboard sensors, systems, and instruments into signals, which are then transmitted via radio frequencies to ground stations. The received signals are decoded and analysed, providing real-time feedback about the vehicle's operation. Telemetry data typically includes:

- **Flight Dynamics:** Position, velocity, altitude, acceleration, and attitude data.

- **System Health Monitoring:** Temperature, pressure, fuel levels, battery status, and structural stresses.
- **Environmental Parameters:** External conditions such as radiation levels, atmospheric density, and thermal flux.
- **Payload Data:** Information from scientific instruments or mission payloads onboard.

Telemetry systems operate on specific radio frequency bands, such as S-band, X-band, or Ka-band, depending on the mission requirements. These frequencies are chosen for their reliability and ability to transmit data over vast distances with minimal interference. In deep-space missions, large antennas like those in NASA's Deep Space Network (DSN) are used to receive weak telemetry signals from spacecraft millions of kilometres away.

Telemetry systems in rockets and spacecraft operate on specific radio frequency (RF) bands, such as S-band, X-band, or Ka-band, which are chosen based on mission requirements, data transmission needs, and the distance to be covered. These frequency bands define the range of electromagnetic wavelengths used to transmit telemetry signals between the vehicle and ground control stations.

The S-band operates in the frequency range of 2 to 4 GHz and is one of the most widely used bands for telemetry in space missions, especially for near-Earth and low Earth orbit (LEO) applications. The S-band offers a good balance between signal range, reliability, and data rate. It is less susceptible to atmospheric attenuation (signal loss due to Earth's atmosphere) and provides stable communication for transmitting critical telemetry data such as temperature, pressure, and velocity from rockets during launch and early orbital phases. The simplicity and low power requirements of S-band systems make them ideal for many small spacecraft, including CubeSats.

The X-band operates in the 8 to 12 GHz range and is typically used for higher data rate transmissions in missions that require greater bandwidth. It is especially common for deep-space communication and interplanetary missions, as the higher frequency allows more data to be transmitted over long distances with relatively smaller antennas. However, X-band signals are more susceptible to atmospheric attenuation compared to S-band, which is why it is primarily used beyond Earth's atmosphere. For example, NASA's Mars rovers and orbiters use X-band telemetry to send scientific data and images back to Earth via the Deep Space Network (DSN).

The Ka-band operates in the 26 to 40 GHz range and is increasingly used for modern space missions that require very high data rates. The Ka-band offers significantly larger bandwidth compared to S-band and X-band, allowing for faster and more efficient transmission of high-resolution images, video feeds, and large volumes of scientific data. This makes Ka-band ideal for advanced missions, including Earth observation satellites, interplanetary spacecraft, and missions with high-definition imaging requirements. However, Ka-band signals are more prone to atmospheric attenuation, particularly due to rain and water vapor, which limits its use for ground communications in certain weather conditions.

The choice of frequency band depends on the specific mission requirements, including the volume of data to be transmitted, the distance between the spacecraft and ground control, and environmental considerations. For instance, a near-Earth mission may primarily rely on S-band for its cost-effectiveness and reliability, while deep-space missions opt for X-band or Ka-band to handle the long communication distances and higher data demands.

By strategically selecting the appropriate RF band, telemetry systems ensure reliable, high-quality data transmission to ground stations, enabling precise monitoring of the spacecraft's performance and health throughout its mission. Advances in RF technology continue to improve the efficiency and versatility of these bands, supporting increasingly complex space exploration and scientific endeavours.

For rockets during launch and ascent, telemetry is vital for monitoring critical stages such as ignition, stage separation, and engine performance. If anomalies occur, telemetry data allows ground controllers to identify and respond to issues quickly. In cases of catastrophic failure, telemetry systems provide post-flight analysis to determine the root cause.

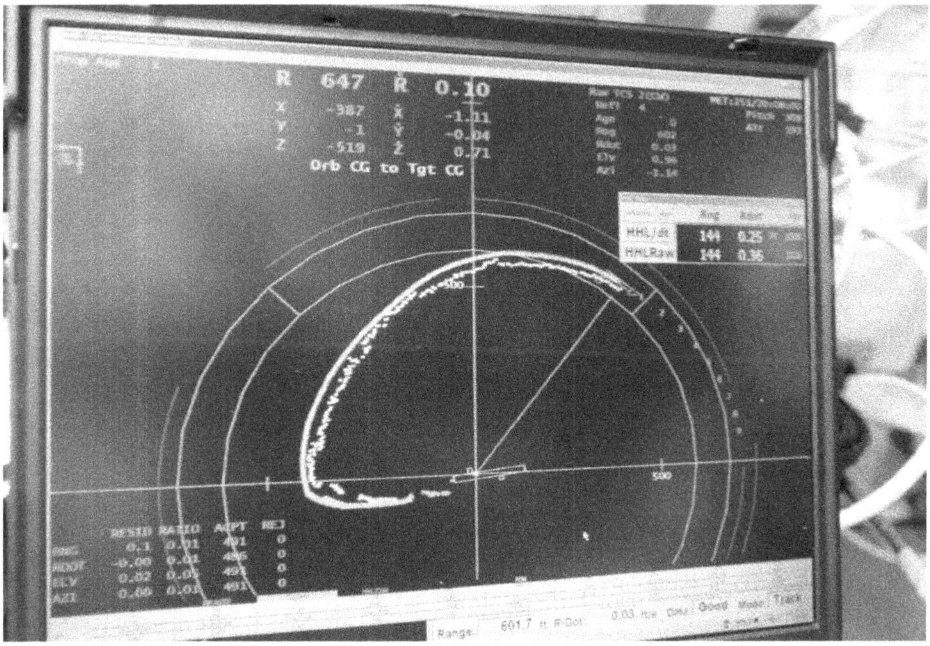

Figure 63: View of a laptop displaying telemetry data during separation from the ISS on STS-128. RPOP. The U.S. National Archives, Public Domain, via Picryl.

Advanced telemetry systems also include bidirectional communication capabilities, enabling the ground station to send commands back to the spacecraft. This two-way communication allows for real-time adjustments, such as trajectory corrections, system reboots, or deployment of payloads and instruments.

Telemetry equipment in rockets is critical for monitoring the vehicle's performance, status, and health throughout its flight. These systems ensure real-time transmission of essential data from the rocket to ground control stations, enabling engineers to evaluate performance, detect anomalies, and make mission-critical decisions. The telemetry data includes vital parameters such as pressure, temperature, velocity, orientation, structural loads, and propulsion system performance. To ensure reliable data communication, telemetry systems integrate sensors, data acquisition systems, transmitters, antennas, and ground-based receivers, all of which work cohesively.

Sensors are the foundational elements of telemetry systems, as they gather physical measurements from the rocket's subsystems. Temperature sensors, including thermocouples and resistance temperature detectors (RTDs), monitor temperature variations in key areas such as combustion chambers, fuel lines, and avionics. Pressure sensors, often strain-gauge-based transducers, track fuel tank pressures, combustion chamber conditions, and hydraulic systems to ensure propulsion performance remains optimal. Accelerometers and gyroscopes, part of the Inertial Measurement Units (IMUs), measure linear acceleration and angular velocity, providing orientation, trajectory, and stability data. Strain gauges measure structural stress and deformation during intense phases like launch and ascent, ensuring the rocket's integrity under high loads. Vibration sensors, such as piezoelectric accelerometers, detect potentially harmful vibrations during ascent and staging. Flow meters monitor the flow rates of fuel and oxidizer to maintain consistent engine performance. These sensors convert physical measurements into electrical signals, which are then digitized and prepared for transmission as telemetry data.

The Data Acquisition System (DAS) plays a central role in telemetry by collecting, processing, and packaging the signals received from the sensors. Analog-to-digital converters (ADCs) transform the analogue signals into digital data for transmission, while multiplexers combine inputs from multiple sensors into a single streamlined data stream. The data is processed by microcontrollers or processors, which format it into telemetry packets. Modern DAS systems utilize fault-tolerant microprocessors, such as the RAD750 and LEON processors, to ensure reliable operation in high-radiation environments. In addition to real-time transmission, onboard data storage systems, like solid-state recorders (SSRs) and flash memory, store telemetry data for post-flight analysis. Data organization often follows standardized protocols like the Consultative Committee for Space Data Systems (CCSDS) to ensure compatibility with ground processing systems.

Transmitters are responsible for sending telemetry data to ground control stations via radio frequency (RF) signals. These devices are lightweight, reliable, and capable of operating in extreme launch conditions. High-frequency (HF) and ultra-high-frequency (UHF) transmitters are typically used for short-range data transmission during launch and ascent. For longer-range communication, S-band and X-band transmitters provide high data rates and are commonly used in modern rockets. Power amplifiers ensure the RF signal has sufficient strength to overcome atmospheric attenuation and reach ground stations, while digital modulation techniques, such as phase-shift keying (PSK) and frequency-shift keying (FSK), encode telemetry data onto carrier signals. To increase reliability, rockets often employ multiple transmitters operating on separate frequencies to provide redundancy in case of signal loss.

Antennas are critical components of telemetry systems, enabling signal transmission and reception. Onboard antennas are specifically designed to withstand extreme launch conditions, including vibrations, shock, and aerodynamic heating, while maintaining reliable communication. Patch antennas are lightweight and low-profile, ideal for transmitting RF signals from smaller rockets and satellites. Omnidirectional antennas provide continuous signal transmission in all directions, ensuring reliable communication regardless of the rocket's orientation. For missions requiring directional communication, such as upper-stage operations or deep-space flights, directional antennas focus signal strength in specific directions. Antennas are strategically placed at multiple locations on the rocket, such as the nose cone, body, and fins, to minimize signal blockage and ensure consistent communication during manoeuvres.

Ground-based telemetry systems play a pivotal role in receiving and processing data transmitted from the rocket. These systems include large parabolic dish antennas or phased-array antennas, which receive telemetry signals over long distances. NASA's Deep Space Network (DSN), for example, uses ground-based antennas to track spacecraft and rockets during interplanetary missions. RF receivers decode the telemetry signals, extracting the data packets for processing. Signal processing systems filter out noise, apply error detection and correction techniques like Forward Error Correction (FEC), and analyse the raw telemetry data. The processed information is displayed on real-time dashboards, providing insights into parameters like engine performance, fuel pressure, altitude, and velocity. Engineers monitor these displays throughout the mission, enabling them to make timely decisions and ensure mission success.

Telemetry systems serve a variety of real-time applications during rocket operations. They enable health monitoring of critical subsystems, such as engines, fuel systems, and avionics, to identify and address anomalies before they escalate. Trajectory tracking provides data on the rocket's altitude, velocity, and position, ensuring it follows the planned flight path. Telemetry also supports flight safety by allowing ground control to detect critical faults and issue commands, such as aborting the mission or activating self-destruct systems to prevent accidents. Additionally, telemetry data stored during the flight is used for post-flight analysis to evaluate system performance, investigate anomalies, and refine designs for future missions.

Modern rockets like SpaceX's Falcon 9, NASA's Space Launch System (SLS), and Blue Origin's New Shepard employ advanced telemetry systems. For example, SpaceX provides real-time telemetry during launches, displaying live data on velocity, altitude, and engine status for both engineering and public monitoring. In deep-space missions, such as NASA's Artemis program, telemetry systems are crucial for transmitting spacecraft health and navigation data over vast distances to Earth, enabling reliable operation far from ground control.

Telemetry systems are essential for the success of rocket missions, providing continuous real-time data on vehicle performance, trajectory, and subsystem health. By integrating sensors, data acquisition systems, transmitters, antennas, and ground-based receivers, telemetry enables engineers to monitor and control rockets throughout their journey. Advances in telemetry technology, such as high-frequency transmitters, enhanced error correction, and compact onboard systems, have significantly improved the accuracy, reliability, and range of telemetry communications, supporting increasingly complex and ambitious space missions.

Integration of Onboard Computers and Telemetry

Onboard computers and telemetry systems work together as part of the spacecraft's avionics suite. The onboard computer manages the vehicle's operations, executes commands, and processes sensor data, while the telemetry system packages and transmits this data to Earth. The computer also receives commands from ground control via the telemetry system and translates these inputs into actionable operations.

For example, during launch, the onboard computer monitors the rocket's flight dynamics, adjusting thrust and control surfaces based on feedback from sensors. Simultaneously, the telemetry system transmits critical flight data, such as engine thrust levels and temperatures, to ensure mission operators can monitor the vehicle's performance. If a deviation is detected, ground control can send corrective commands to the computer via the telemetry link.

In autonomous deep-space missions, onboard computers and telemetry systems work together to achieve precise navigation and communication. The computer handles onboard decision-making and autonomous trajectory adjustments, while the telemetry system transmits progress reports, scientific data, and health information to Earth.

Both onboard computers and telemetry systems face significant challenges, particularly in deep-space environments. Radiation can disrupt electronic components and corrupt data, requiring the use of radiation-hardened hardware. Limited power availability demands that these systems operate efficiently, especially on smaller spacecraft. Additionally, data transmission over long distances, such as interplanetary missions, results in significant signal delays and attenuation.

Recent advancements address these challenges by incorporating more efficient, miniaturized computers with higher processing capabilities, advanced error correction algorithms, and improved redundancy. Modern telemetry systems leverage higher frequency bands (e.g., Ka-band) for increased data rates and use laser communication systems for faster and more reliable data transmission in deep space.

In summary, onboard computers and telemetry systems form the backbone of a rocket's operational capability. Onboard computers handle autonomous control, execute mission-critical tasks, and ensure vehicle stability, while telemetry systems provide real-time monitoring and communication with ground control. Together, these systems enable precise navigation, enhance reliability, and ensure the success of missions across low-Earth orbit, interplanetary travel, and beyond.

Redundancy and Fail-Safes in Avionics Design

Redundancy and fail-safes in avionics design are critical principles aimed at ensuring the safety, reliability, and robustness of aerospace systems, particularly in aircraft, rockets, and spacecraft. These systems operate in high-risk environments where even minor failures can lead to catastrophic consequences. By incorporating redundancy and fail-safes, engineers design avionics systems to maintain functionality, tolerate faults, and recover from failures during critical phases of operation.

Redundancy in Avionics Design

Redundancy is a critical design principle in avionics systems to ensure mission safety and reliability. Several types of redundancy are implemented in avionics design:

Hardware Redundancy: Hardware redundancy involves duplicating or triplicating critical components such as processors, sensors, or power supplies. In a dual-redundant system, two identical components operate simultaneously, with one serving as a backup if the other fails. In a triple-redundant system (often using Triple Modular Redundancy, or TMR), three identical components perform the same function, and a majority-voting mechanism selects the consistent output. This ensures the system can tolerate single-point failures, as demonstrated in rocket guidance and navigation systems using redundant inertial measurement units (IMUs) [308, 309].

Functional Redundancy: Functional redundancy refers to designing different components or subsystems that can perform similar functions if a failure occurs. For example, if a primary navigation system fails, a backup navigation method such as GPS, star trackers, or radar altimeters can take over. This diversity in system architecture ensures alternative solutions are available to complete the mission even if a specific technology fails [310, 311].

Functional redundancy in modern rockets involves designing different components or subsystems capable of performing similar functions to ensure mission success even if one system fails. By using alternative technologies or overlapping capabilities, functional redundancy allows a rocket to continue operating seamlessly, mitigating the risks of single-point failures. This is critical for space missions where reliability is paramount, and repairs are impossible once the vehicle is in flight or deep space.

Functional redundancy works by incorporating independent systems or components that achieve the same objective using different technologies or methods. Unlike hardware redundancy, where identical components back each other up (such as three identical processors in Triple Modular Redundancy), functional redundancy uses diverse systems to ensure continued operation in case of failures. This diversity makes the system more robust, as different technologies are less likely to fail simultaneously due to common faults like radiation or environmental stressors.

For instance, if a primary navigation system using GPS fails due to signal loss or external interference, a secondary navigation system using star trackers, inertial navigation units (IMUs), or radar altimeters can provide the necessary positional and orientation data. Each subsystem works independently and provides similar functionality, ensuring no critical loss of capability.

Examples of Functional Redundancy in Modern Rockets:

1. SpaceX Falcon 9: The Falcon 9 rocket incorporates multiple navigation and control subsystems that provide functional redundancy. The primary navigation system uses Global Positioning System

(GPS) receivers to provide position, velocity, and altitude data. However, if GPS signals are unavailable or disrupted, Falcon 9 can switch to Inertial Measurement Units (IMUs) and onboard flight computers for dead-reckoning navigation. IMUs combine data from accelerometers and gyroscopes to estimate the rocket's movement and orientation without external signals.

Falcon 9 also employs star trackers and horizon sensors during the second stage for precision orientation in orbit. If GPS or IMU data becomes unreliable, these optical systems ensure the vehicle remains correctly oriented for payload deployment.

2. NASA Space Launch System (SLS): The Space Launch System (SLS), NASA's heavy-lift rocket for deep-space missions, incorporates functional redundancy across its avionics, navigation, and guidance systems. The primary system relies on high-accuracy GPS receivers and IMUs for trajectory tracking and control. If these systems experience faults or external disruptions, backup sensors, including star trackers and horizon sensors, can take over for guidance.

The SLS avionics architecture also includes redundant communication links for real-time telemetry and control. Should the primary data bus fail, secondary systems ensure critical flight commands and monitoring data remain accessible.

3. ESA's Ariane 6: The Ariane 6, a next-generation European rocket, uses multiple guidance and navigation systems with functional redundancy. During launch and ascent, Ariane 6 relies on an Inertial Navigation System (INS), which uses accelerometers and gyroscopes to calculate the rocket's velocity and position. If there is an anomaly in the INS, the system can fall back on GPS-based navigation, offering an external positional reference for correction.

For orientation in orbit, Ariane 6 incorporates star trackers that provide accurate attitude determination. Star trackers are independent of GPS or IMU systems, ensuring reliable navigation even in the event of signal loss or internal sensor failures.

4. Blue Origin's New Shepard: Blue Origin's New Shepard suborbital rocket employs multiple redundant systems for flight control. The primary navigation system uses IMUs and GPS receivers to guide the rocket's ascent and descent. However, if GPS becomes unreliable during re-entry, the rocket switches to altimeters and radar sensors to measure altitude and velocity as it prepares for landing.

Additionally, New Shepard incorporates engine gimbaling and reaction control systems (RCS) for stabilization. If the primary thrust vector control (TVC) system malfunctions, RCS thrusters can provide corrective forces to maintain orientation during flight and descent.

5. SpaceX Starship: SpaceX's Starship rocket features functional redundancy across multiple flight systems. For instance, the primary navigation relies on a combination of IMUs, GPS receivers, and star trackers. Should one fail, the other systems compensate to maintain accurate navigation and attitude control.

During re-entry and landing, Starship employs both altimeters and LiDAR systems to measure altitude and terrain features for precise landing. If one sensor type provides unreliable data, the alternative system ensures Starship can still land autonomously and accurately.

Analytical Redundancy: Analytical redundancy uses mathematical models and software algorithms to detect and correct faults within a system. By comparing real-time sensor data to expected behaviour derived from a model, discrepancies can identify faults. In avionics, flight control systems employ analytical redundancy to monitor sensor outputs and cross-check the data with flight dynamics equations to detect anomalies [312, 313].

Analytical redundancy is a sophisticated fault detection and correction method that relies on mathematical models and software algorithms to monitor the behaviour of systems and identify faults. Instead of relying solely on hardware backups (like physical sensors or components), analytical redundancy uses software-based techniques to ensure systems can detect, isolate, and correct faults using real-time data. This approach enhances reliability without adding significant weight, cost, or complexity, which is critical in modern rockets and spacecraft.

Analytical redundancy operates by comparing real-time sensor data with values predicted by mathematical models of the system's expected behaviour. These models are derived using system dynamics equations, which describe how the rocket or spacecraft should behave under specific conditions (e.g., thrust, orientation, altitude, velocity). If the data from a sensor deviates significantly from the modelled prediction, the discrepancy is flagged as a potential fault.

For example, in a flight control system, if an altitude sensor provides a reading of 10 km but the model, based on thrust, acceleration, and previous altitude, predicts 8 km, the system detects a mismatch. Algorithms then isolate the fault, cross-check it with other sensors or redundant data sources, and take corrective actions, such as relying on unaffected sensors or recalibrating the faulty sensor.

Steps Involved in Analytical Redundancy

1. **Mathematical Modelling**: A detailed model of the system's behaviour is developed using equations based on the rocket's physical characteristics, dynamics, and environmental interactions. For instance, models include Newtonian mechanics to predict motion, energy conservation for power systems, and heat transfer models for thermal behaviour.

2. **Real-Time Data Comparison**: During operation, real-time sensor outputs are continuously compared to values predicted by the mathematical model. This real-time analysis is performed using onboard processors capable of handling complex computations.

3. **Fault Detection**: If the difference (or “residual”) between the observed and modelled data exceeds a predetermined threshold, the system identifies a fault. For example, a discrepancy in acceleration data from an IMU can indicate a sensor drift or failure.

4. **Fault Isolation and Correction**: Once a fault is detected, the system determines the faulty component by comparing multiple data sources. Alternative measurements or mathematical estimates can replace the faulty data to ensure the system remains operational.

5. **Recalibration or System Adaptation**: Analytical redundancy systems can recalibrate affected sensors or adjust control outputs based on model estimates to minimize the impact of the fault.

Examples of Analytical Redundancy in Modern Rockets:

1. SpaceX Falcon 9: The Falcon 9 uses analytical redundancy within its flight control system to monitor sensor performance and detect faults. During launch, the rocket relies on Inertial Measurement Units (IMUs) to provide acceleration and angular velocity data, as well as GPS receivers for positional updates. If one IMU sensor shows abnormal readings inconsistent with expected flight dynamics, analytical redundancy algorithms compare its outputs with data from other IMUs, GPS, or thrust vectoring behaviour. This allows the flight control system to isolate the faulty IMU and continue using accurate data to adjust the rocket's trajectory.

For example, if an accelerometer in one IMU begins to drift, analytical redundancy algorithms detect the discrepancy between the predicted motion and observed data, flag the fault, and use data from unaffected IMUs or model estimates to maintain control.

2. NASA Space Launch System (SLS): The SLS employs analytical redundancy to enhance the reliability of its Guidance, Navigation, and Control (GNC) systems. The rocket integrates data from multiple navigation sensors, including IMUs, GPS, and star trackers. Analytical redundancy algorithms compare readings across these sensors and check them against flight dynamics models to identify anomalies.

For instance, if a star tracker loses accuracy due to stray light or thermal drift, the algorithms use real-time IMU and GPS data to cross-check and correct the star tracker's output. This ensures continuous and accurate navigation, even when individual sensors encounter faults.

3. ESA's Ariane 5 and Ariane 6: Ariane rockets use analytical redundancy in their avionics systems to monitor critical components during flight. The Flight Control System (FCS) uses a combination of gyroscopes, accelerometers, and control algorithms to manage rocket stability and trajectory. If a gyroscope provides faulty angular velocity data, the onboard analytical algorithms compare its outputs with acceleration data and dynamic models to detect the fault. The system then relies on unaffected data sources to maintain accurate trajectory control.

4. Boeing CST-100 Starliner: The Starliner spacecraft incorporates analytical redundancy to monitor critical navigation sensors, such as Radar Altimeters and IMUs, during descent and landing. If the radar altimeter data deviates from the expected descent profile, the onboard algorithms use IMU-derived altitude estimates and modelled flight dynamics to validate or replace the faulty altimeter data, ensuring a safe landing.

5. SpaceX Starship: The Starship system incorporates analytical redundancy for its navigation and control systems. The rocket uses multiple navigation sensors, including GPS, IMUs, and star trackers, to guide its ascent and descent. Analytical algorithms compare these data streams with the predicted dynamics of the vehicle, derived from thrust profiles and aerodynamic models. If discrepancies are detected, the algorithms identify the faulty sensor, isolate it, and rely on the remaining data or model estimates to maintain stability and accuracy during the mission.

Cross-Strapping: Cross-strapping ensures that each redundant component in an avionics system can access multiple backup resources. For example, multiple flight control computers can communicate with multiple redundant data buses and actuators, ensuring continuous operation even if a subsystem or connection fails [314].

Cross-strapping is a critical design technique in avionics systems, particularly for modern rockets and spacecraft, to ensure fault tolerance and continuous operation. In this approach, multiple redundant components, such as computers, sensors, data buses, and actuators, are interconnected in a way that allows each redundant element to access multiple backup resources. This design ensures that a single point of failure—be it a component, connection, or subsystem—does not compromise the overall operation of the vehicle. Cross-strapping provides flexibility, resilience, and enhanced system reliability, which are essential for mission-critical systems in rockets.

In a cross-strapped avionics system, redundancy is built not only at the component level but also in the interconnections between systems. Multiple components, such as Flight Control Computers (FCCs), data buses, and actuators, are interconnected in a way that if one component or data path fails, an alternate path or redundant resource can take over seamlessly.

For example, in a typical cross-strapped flight control system:

1. Multiple Flight Control Computers (e.g., FCC A, FCC B, FCC C) are employed to provide redundancy.
2. These computers are interconnected with multiple redundant data buses that transmit commands and data between subsystems.
3. Actuators (responsible for controlling rocket nozzles, fins, or other moving parts) are connected to all the data buses and can accept inputs from any of the flight computers.
4. If one FCC fails or one data bus becomes inoperable, the system automatically reroutes data through the remaining functional buses and computers to maintain control.

This ensures that each critical subsystem has access to multiple control paths, eliminating single points of failure. The design is often cross-verified using error detection techniques to ensure that only reliable data is processed.

Examples of Cross-Strapping in Modern Rockets:

1. SpaceX Falcon 9: SpaceX's Falcon 9 rocket employs cross-strapping to ensure fault tolerance in its flight control and avionics systems. Multiple Flight Control Computers (FCCs) work in tandem, each capable of communicating with redundant sensors, data buses, and actuators. For instance:

- Falcon 9's flight control computers cross-strap their outputs to multiple thrust vector control (TVC) actuators responsible for adjusting the engine nozzle direction.
- If one computer or data path fails, the cross-strapped architecture allows an alternate FCC and data bus to take over, ensuring continuous thrust vectoring and trajectory adjustments.

This level of fault tolerance was critical in ensuring the rocket's reliability during its numerous successful launches, including crewed missions to the International Space Station (ISS).

2. NASA Space Launch System (SLS): NASA's Space Launch System (SLS), designed for deep-space missions, incorporates cross-strapping in its avionics and guidance systems. The SLS features redundant flight computers and data buses that are cross-strapped to ensure continuous operation in the event of a hardware fault.

- Each SLS flight control computer is connected to multiple redundant communication buses that interface with key components like engine controllers, IMUs (Inertial Measurement Units), and hydraulic actuators.
- If a single computer or a communication link fails, the system can reroute control signals through the remaining functional paths to maintain precise control of the rocket during critical phases like launch and ascent.

This architecture ensures the vehicle can autonomously detect, isolate, and mitigate faults without human intervention.

3. Space Shuttle Avionics: The now-retired Space Shuttle utilized a highly robust cross-strapping architecture in its General Purpose Computers (GPCs) and data buses.

- The shuttle employed five redundant flight computers, all cross-strapped to multiple redundant data buses.
- The GPCs were capable of sending commands to multiple actuators, such as those controlling the shuttle's aerodynamic surfaces and engines.
- If a GPC failed, the cross-strapped system allowed the remaining computers to maintain communication with the actuators and continue operations seamlessly.

This approach allowed the Space Shuttle to achieve high levels of reliability over its operational life, even in the event of subsystem failures.

4. ULA's Atlas V and Delta IV: The Atlas V and Delta IV rockets, operated by the United Launch Alliance (ULA), employ cross-strapping in their flight control systems to achieve fault tolerance.

- Redundant Inertial Navigation Systems (INS) and Flight Control Computers are cross-strapped with multiple data buses to ensure constant communication.
- Each FCC can interface with backup buses and provide control inputs to the rocket's actuators, such as thrust vector control systems or stage separation mechanisms.
- If one path or system encounters a fault, cross-strapping ensures the redundant paths and components take over, allowing the mission to proceed uninterrupted.

5. Blue Origin New Shepard: Blue Origin's New Shepard suborbital rocket incorporates a cross-strapped architecture to ensure the safety and reliability of its autonomous flight system.

- Redundant computers and sensors are cross-connected with actuators and control systems, ensuring that even in the event of a fault, the rocket can autonomously maintain its flight path and execute critical manoeuvres like landing the booster.
- This design was integral to New Shepard's repeated success in landing reusable boosters after suborbital missions.

The implementation of these redundancy techniques in avionics design is crucial to achieving the high levels of fault tolerance and reliability required for mission-critical applications [315-317]. Redundancy management strategies and the integration of redundancy into the overall system architecture are active areas of research and development in the avionics domain [318, 319].

Fail-Safes in Avionics Design

Fail-safes are critical design principles in avionics systems to ensure safety and reliability under failure conditions. The key fail-safe design principles in avionics include:

Safe Mode Operations: Avionics systems, particularly spacecraft, are often programmed to enter a safe mode when faults are detected [320]. In safe mode, non-essential systems are shut down, and only core functions like communication and power generation remain active. This prevents mission loss and ensures the system can await recovery commands from ground control.

Safe Mode Operations is a critical fail-safe mechanism in avionics and spacecraft systems, designed to safeguard the vehicle and its payload in the event of detected faults, anomalies, or unexpected conditions. When a fault is identified, the spacecraft autonomously transitions into a safe mode where all non-essential systems are shut down, and only the core subsystems—such as power generation, thermal control, and communication—remain active. This mode prioritizes survival, communication with ground control, and system stability, ensuring the vehicle can be recovered and restored to full operation.

Safe mode is triggered by the Fault Detection, Isolation, and Recovery (FDIR) system, which continuously monitors the health and status of various spacecraft subsystems. Sensors, diagnostic tools, and software algorithms detect anomalies in parameters like power levels, temperature, communication signals, propulsion, or attitude control.

- **Fault Detection**: The avionics system uses redundant sensors and analytical redundancy (e.g., comparing real-time data with expected system behaviour) to identify deviations, such as voltage drops, excessive heat, or communication loss. For example, if a star tracker fails to provide valid orientation data, the system identifies a fault in the guidance subsystem.
- **Isolation of the Fault**: Once a fault is detected, the system isolates the problematic subsystem or component. Redundant components or pathways are activated if available. For example, a faulty reaction wheel may be disabled, and a backup wheel or alternative control method like magnetic torquers is brought online.

- **Transition to Safe Mode**: The spacecraft automatically shuts down all **non-essential functions**, such as science instruments, payloads, or secondary communication channels. The focus shifts to maintaining:
    - **Power Generation**: Ensuring solar panels or power systems remain operational to sustain the vehicle.
    - **Attitude Control**: Maintaining spacecraft orientation to ensure proper alignment of solar arrays or communication antennas.
    - **Communication**: Establishing a stable link with ground control to relay fault diagnostics and await recovery instructions.
    - **Thermal Management**: Keeping critical components within operational temperature ranges.
- **Ground Command Recovery**: Once in safe mode, the spacecraft sends diagnostic data to mission control, where engineers analyse the fault and devise corrective actions. Commands are then sent to restore full functionality.

Examples of Safe Mode Operations in Modern Rockets and Spacecraft:

1. NASA's Mars Reconnaissance Orbiter (MRO): NASA's Mars Reconnaissance Orbiter, launched in 2005, has entered safe mode multiple times during its mission. For instance, in 2009, the spacecraft autonomously detected an anomaly in its onboard computer system and transitioned into safe mode.

- In this state, MRO shut down its scientific instruments, maintained orientation to keep its solar panels pointed at the Sun for power, and prioritized communication with Earth.
- Engineers on the ground diagnosed the fault, reloaded the necessary software, and restored normal operations, allowing the mission to continue.

2. SpaceX Crew Dragon: SpaceX's Crew Dragon spacecraft features an advanced safe mode to protect crew members and critical systems.

- If a fault is detected in the life support, propulsion, or navigation systems, the spacecraft enters safe mode to maintain power, attitude control, and communication with ground teams.
- For example, during the In-Flight Abort Test, the Crew Dragon spacecraft demonstrated its ability to autonomously stabilize itself and communicate with ground control after a high-stress event, showcasing the robustness of its safe mode capabilities.

3. ESA's Solar Orbiter: ESA's Solar Orbiter, launched in 2020, is designed to study the Sun's atmosphere and solar wind. If an anomaly occurs—such as extreme thermal fluctuations or communication issues—it enters safe mode.

- In safe mode, the spacecraft maintains its thermal control systems to prevent damage, keeps its solar arrays aligned for power, and reorients its high-gain antenna toward Earth for communication.
- This capability ensures that mission-critical data can be sent to Earth for analysis, and appropriate recovery measures can be implemented.

4. Hubble Space Telescope: The Hubble Space Telescope has entered safe mode several times over its decades-long mission. Safe mode is triggered when anomalies occur in systems like gyroscopes, power supply, or onboard software.

- When Hubble enters safe mode, its scientific instruments are turned off, and the spacecraft positions its solar arrays to maintain power while keeping its antenna oriented toward Earth.
- For example, in 2021, a power control unit issue caused Hubble to enter safe mode. Ground control was able to analyse the issue, implement a workaround, and successfully restore the telescope's operations [321].

5. New Horizons Probe: NASA's New Horizons spacecraft, which flew by Pluto in 2015, entered safe mode just days before its closest approach.

- The spacecraft autonomously detected an anomaly in its onboard computer system and transitioned into safe mode, shutting down science operations but maintaining power and communication with Earth [322, 323].
- Mission engineers quickly resolved the issue and restored normal operations, allowing New Horizons to continue its historic flyby and collect critical scientific data.

Safe mode operations are essential for ensuring mission success, particularly for long-duration or deep-space missions where human intervention is delayed or impossible. The benefits include:

- **Fault Containment**: Preventing faults in one subsystem from cascading into other systems.
- **Mission Preservation**: Maintaining the spacecraft's health and stability while awaiting recovery commands.
- **Data Integrity**: Ensuring communication systems remain active to transmit fault diagnostics and operational status.
- **System Survival**: Protecting critical components like power generation and thermal control systems, which are vital for spacecraft longevity.

Graceful Degradation: Avionics systems are designed for graceful degradation, where functionality is gradually reduced while maintaining critical operations during a failure [324]. For example, if one engine fails during a rocket launch, the avionics system can redistribute thrust to the remaining engines, enabling the vehicle to continue its trajectory with reduced performance.

Graceful Degradation in avionics systems refers to the design principle where, during a failure, non-essential functionalities are gradually reduced or disabled while maintaining critical operations to ensure mission success. Instead of a total system shutdown, the system adapts by reallocating resources or redistributing workloads to sustain core operations. This approach ensures that the vehicle can continue functioning under degraded conditions, reducing the risk of catastrophic mission failure. Graceful degradation is particularly vital in rockets and spacecraft, where redundancy and adaptability are critical for handling unforeseen faults or failures in-flight.

Graceful Degradation includes:

1. **Failure Detection and Isolation**: The avionics system, equipped with Fault Detection, Isolation, and Recovery (FDIR) mechanisms, continuously monitors critical subsystems for anomalies. Sensors, software algorithms, and analytical redundancy models are employed to detect faults in components such as engines, sensors, or communication systems.

2. **Resource Redistribution**: Once a failure is detected, the avionics system prioritizes critical operations by redistributing available resources. For example, in the case of propulsion system faults, thrust from healthy engines can be adjusted to compensate for the loss of performance.

3. **Gradual Reduction of Functionality**: Non-essential systems, such as secondary payloads, scientific instruments, or auxiliary functions, are disabled or scaled down to preserve power and computational resources for critical operations. For instance, a spacecraft may shut down non-mission-critical experiments to prioritize power for communication and attitude control.

4. **Mission Continuation with Reduced Capability**: Despite the degradation, the system continues operating at a reduced level of performance. For rockets, this could mean achieving a slightly altered trajectory or orbit rather than mission abort. For spacecraft, essential functions like power generation, communication, and stabilization are prioritized to extend operational life.

Examples of Graceful Degradation in Current and Modern Rockets:

1. SpaceX Falcon 9 – Engine-Out Capability: The Falcon 9 rocket, developed by SpaceX, is designed with engine-out capability, a prime example of graceful degradation. The rocket's first stage is equipped with nine Merlin engines. If one engine fails during ascent, the avionics system detects the fault and redistributes thrust among the remaining healthy engines to compensate.

- During a Falcon 9 flight in 2012, one of the nine Merlin engines shut down mid-flight due to an anomaly. The onboard avionics system automatically adjusted the thrust of the remaining engines to keep the rocket on its intended trajectory. While the secondary payload was affected, the primary mission—delivering the Dragon spacecraft to the International Space Station (ISS)—was successfully completed.

- This redundancy and adaptive response mechanism ensure mission continuity, even under partial failure conditions.

2. NASA's Space Shuttle – Engine and Thruster Adaptation: The retired Space Shuttle had systems designed for graceful degradation in case of engine or thruster failure. If one of the three RS-25 main engines experienced reduced performance or failure during launch, the shuttle's avionics system would detect the issue and adjust the throttle levels of the remaining engines to maintain ascent stability and trajectory.

- Additionally, during orbit manoeuvres, if one of the Orbital Manoeuvring System (OMS) thrusters failed, the avionics system redistributed workload to the remaining thrusters, ensuring mission objectives could still be achieved with slightly reduced precision.

3. ULA Atlas V – Redundant Propulsion Control: The Atlas V, developed by United Launch Alliance (ULA), incorporates mechanisms to handle partial engine or subsystem failures. If anomalies occur in the rocket's RD-180 main engine during the ascent phase, the avionics system adapts by slightly altering the engine's throttle and burn time to compensate for reduced performance.

- This design ensures the rocket can still deliver its payload into the desired orbit, albeit with modified parameters, without an immediate mission abort.

4. Ariane 5 – Power Management for Satellite Payloads: The Ariane 5 rocket, used primarily for delivering satellites into geostationary transfer orbit, demonstrates graceful degradation in its power systems. If a fault occurs in the power distribution unit, the avionics system prioritizes essential components—such as engine control, telemetry, and attitude stabilization—while reducing power to secondary systems.

- This approach prevents total power loss and ensures the payload remains protected until the vehicle achieves orbit.

5. Orion Spacecraft – Attitude Control Redundancy: NASA's Orion spacecraft, designed for deep space exploration, incorporates graceful degradation into its Attitude Control System (ACS). If one or more reaction control thrusters fail, the avionics system redistributes the workload to remaining thrusters while maintaining spacecraft stability and alignment.

- For instance, during re-entry or trajectory corrections, Orion can continue operating with degraded thruster performance, ensuring the vehicle stays on course.

Fail-Operational and Fail-Passive Systems: Fail-operational systems are designed to continue functioning fully even after a failure, such as redundant control computers in flight control systems. Fail-passive systems, on the other hand, degrade functionality safely without causing major disruptions, like an autopilot system disengaging and handing control back to the pilot.

Fail-Operational and Fail-Passive Systems are key concepts in avionics design, ensuring safety, reliability, and mission success during potential failures. Both approaches involve building redundancy and intelligent response mechanisms into systems to handle faults, but they differ in their degree of continued functionality after a failure.

Fail-operational systems are designed to remain fully functional after a single failure. These systems include built-in redundancy where backup components or subsystems seamlessly take over if a primary system fails. Fail-operational systems are particularly critical in scenarios where even brief interruptions can lead to catastrophic consequences, such as during launch, ascent, or orbital insertion.

In modern rockets, redundant control computers and parallel subsystems enable fail-operational behaviour. These systems detect faults using Fault Detection, Isolation, and Recovery (FDIR) mechanisms, allowing the faulty component to be isolated while the backup takes over with minimal disruption.

Watchdog Timers and Monitoring: Avionics systems often include watchdog timers and monitoring circuits that detect unresponsive components or software [325]. If a failure is detected, the system triggers a reset or switches to a redundant backup, ensuring quick recovery from faults and minimizing operational downtime.

Watchdog Timers and Monitoring are essential components in avionics systems that ensure the reliability and fault tolerance of modern rockets. These mechanisms serve as automatic fault detection and recovery tools by monitoring system health, identifying unresponsive components or software, and triggering corrective actions like resets or redundancy switches. The implementation of watchdog timers minimizes operational downtime and ensures that critical systems remain functional during flight.

A watchdog timer (WDT) is a hardware or software-based timer that operates continuously, expecting periodic "heartbeats" or "refresh signals" from monitored components or software systems. If the watchdog timer does not receive this expected signal within a predefined time limit, it assumes the system has failed or become unresponsive. At this point, the WDT triggers a corrective action such as:

1. **System Reset**: The affected component or software subsystem is restarted to restore functionality.
2. **Redundancy Switch**: The watchdog can switch operations to a redundant backup system to ensure mission continuity.
3. **Fault Isolation**: The unresponsive component is isolated to prevent further disruption to the system.

In modern rockets, watchdog timers are tightly integrated into the avionics, flight control, and guidance systems to ensure any temporary faults or system hang-ups do not lead to mission failure. Monitoring circuits are often paired with WDTs to cross-check system outputs, sensor data, and communication lines for anomalies.

Fault Detection, Isolation, and Recovery (FDIR) Systems

Fault Detection, Isolation, and Recovery (FDIR) systems are critical to the safety, reliability, and autonomy of rockets and spacecraft. These systems are designed to detect faults during a mission, identify or isolate the faulty component, and implement recovery measures to ensure continued operations. By integrating FDIR into various rocket subsystems, the vehicle can autonomously handle unexpected failures, minimize disruptions, and maintain mission objectives without immediate ground intervention.

FDIR operates in three key stages: Fault Detection, Isolation, and Recovery. These processes are closely integrated into a rocket's avionics architecture, which includes flight computers, sensors, actuators, and communication systems.

1. Fault Detection

Fault detection involves continuously monitoring critical subsystems for anomalies or deviations from expected behaviour. This is achieved using:

- **Sensor Data**: Inputs from temperature, pressure, vibration, and gyroscopic sensors are monitored to detect abnormal readings.
- **Redundancy Checks**: Data from multiple redundant components (e.g., flight control computers or navigation systems) are compared to ensure consistency.
- **Analytical Models**: Real-time data is compared to predicted behaviour derived from software models, and discrepancies signal a potential fault.
- **Watchdog Timers**: These detect unresponsive systems or delays in software execution and trigger fault alerts.

For example, if a **pressure sensor** in a propulsion system reports values outside an acceptable range, the FDIR system flags it as a fault.

2. Fault Isolation

Once a fault is detected, the system identifies (isolates) the faulty component or subsystem to prevent it from impacting other systems. Fault isolation is performed using:

- **Cross-Checking**: Outputs from redundant components are compared to pinpoint which one has failed.
- **Logical Analysis**: Fault isolation algorithms use decision trees or voting logic (e.g., Triple Modular Redundancy) to identify the malfunctioning system.
- **Data Filtering**: Abnormal data is isolated to prevent it from propagating and affecting mission performance.

For example, in a rocket's **navigation system**, if one of three star trackers provides inconsistent data, FDIR isolates the faulty tracker and relies on the remaining healthy sensors for navigation.

3. Fault Recovery

Recovery is the process of restoring functionality, often by implementing predefined corrective actions. Recovery mechanisms include:

- **System Resets**: Restarting a software module or component to clear transient faults.
- **Redundancy Switching**: Transferring operations to backup components or redundant systems.
- **Graceful Degradation**: Allowing the system to operate at reduced functionality while critical operations are maintained.
- **Safe Mode Operations**: Shutting down non-essential systems and activating minimal operations to prevent further damage.

For example, if a primary flight control computer fails during a launch, the FDIR system will automatically switch to a redundant computer to ensure continuous operation.

Fault Detection, Isolation, and Recovery (FDIR) systems are deeply integrated into the core architecture of modern rockets and spacecraft to ensure mission reliability. FDIR systems continuously monitor, diagnose, and address anomalies across critical subsystems, maintaining operational stability during all phases of flight.

The avionics system is central to FDIR functionality, comprising flight computers and communication networks that manage data processing and fault detection. The software embedded in these systems continuously monitors inputs from sensors and subsystems, including propulsion, navigation, and power systems. For example, SpaceX's *Falcon 9* employs integrated avionics computers that detect and recover from faults in propulsion systems, thrust vector control, and telemetry networks, ensuring smooth operation even during high-stress phases like launch and ascent.

Redundant architectures are a cornerstone of FDIR, ensuring that critical functions remain unaffected in the event of a component failure. Redundancy includes Triple Modular Redundancy (TMR) for flight control and guidance computers, which compare outputs from three parallel systems to isolate and correct errors. Additionally, cross-strapped systems allow multiple redundant components—such as sensors, actuators, and data buses—to share resources, so backups seamlessly take over when primary systems fail. NASA's *Space Launch System (SLS)* exemplifies this approach, using redundant flight computers and communication buses to ensure fault isolation and recovery during critical launch and orbital insertion phases.

Extensive sensor networks play a vital role in FDIR integration, providing real-time monitoring of critical parameters such as engine performance, temperature, pressure, thrust, and vehicle attitude. Accelerometers, gyroscopes, GPS, and star trackers feed continuous data to the avionics system, which compares the inputs against predefined thresholds or predicted values. Any deviations from normal behavior automatically trigger FDIR protocols, isolating the affected subsystem and initiating recovery measures.

Health management software embedded in avionics systems enhances FDIR capabilities by analyzing sensor data, detecting anomalies, and implementing corrective actions. This software often uses machine learning algorithms or rule-based logic to predict faults before they occur, thereby enabling preventive recovery. By diagnosing faults and recommending corrective actions, the software ensures the rocket remains operational under adverse conditions.

Communication links between the rocket and ground control provide another layer of FDIR functionality. Onboard systems transmit real-time telemetry data to ground control stations, where engineers monitor subsystem health and performance. When faults cannot be resolved autonomously, emergency communication protocols trigger manual intervention from ground control, allowing engineers to reconfigure systems or adjust recovery strategies.

The design and testing phases of rocket development incorporate system-wide fault trees to anticipate failure scenarios. Fault trees map out all potential faults, their cascading effects, and the corresponding recovery actions. These predefined recovery procedures are programmed into the FDIR software, automating the detection, isolation, and correction of failures during flight. By planning for worst-case scenarios, engineers ensure that the FDIR system can manage a wide range of anomalies with minimal disruption.

PART 3

Construction and Testing

Chapter 8

Building Rockets: From Design to Assembly

Translating Designs into Production

Translating rocket designs into production is a complex process that bridges engineering concepts with tangible, operational hardware. It requires meticulous planning, extensive collaboration, and advanced manufacturing capabilities to ensure the final rocket meets performance, safety, and reliability standards. The process can be broken into several key stages: design finalization, prototyping, testing, manufacturing, quality assurance, and integration. These stages collectively ensure that a design evolves from concept to a fully operational rocket.

Design Finalization

The process of translating rocket designs into production begins with design finalization, where the conceptual design is refined into detailed, actionable plans. Engineers complete the detailed design phase by developing technical drawings, precise 3D models, and comprehensive specifications for each subsystem of the rocket, such as propulsion, avionics, structures, and thermal control systems.

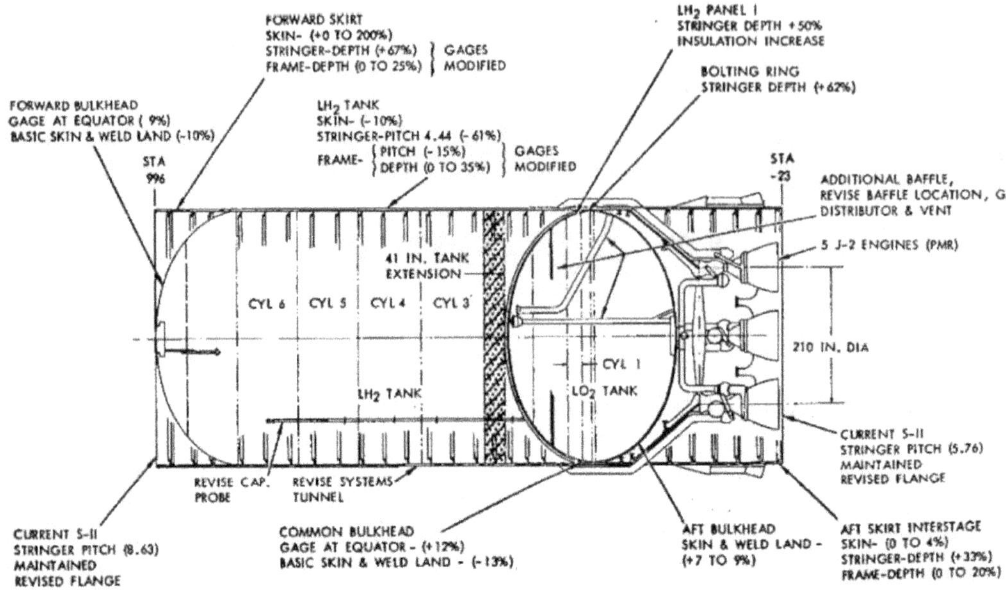

Figure 64: A drawing of the MS-II rocket stage from a NASA study. The MS-II rocket stage was intended to be used in some Saturn MLV rocket variants. The Saturn MLV is a rocket designed as an upgrade of the Saturn V. NASA, Public Domain, via Picryl.

In this stage, detailed CAD modelling plays a critical role. Advanced software tools like *CATIA*, *Siemens NX*, and *SolidWorks* are used to create high-fidelity 3D models of the rocket and its components. These models serve as the foundation for both production and quality verification. Every subsystem is modelled to ensure seamless integration, accuracy, and alignment with performance requirements.

To validate and optimize the design, engineers perform simulation and analysis using computational tools. Finite Element Analysis (FEA) is utilized to simulate structural stress and mechanical loads, ensuring the rocket can withstand forces during launch, ascent, and re-entry. Simultaneously, Computational Fluid Dynamics (CFD) helps analyse aerodynamic performance, identifying how the rocket will interact with atmospheric forces, while thermal analysis evaluates temperature effects on critical components. These simulations are essential for identifying potential design weaknesses and fine-tuning performance before physical production begins.

The Design for Manufacturing (DFM) process follows to ensure that the final design can be manufactured efficiently. Engineers evaluate materials, assembly techniques, and production methods to identify ways to simplify manufacturing, minimize material waste, and reduce production costs. This step often involves iterative adjustments to the design to strike a balance between performance, quality, and production feasibility.

Throughout this phase, collaboration between design engineers and manufacturing teams is essential. Close coordination ensures that the technical specifications align with real-world production capabilities. Manufacturing constraints, such as tooling availability, machining precision, and material sourcing, are reviewed alongside the finalized design to prevent issues during production. By working together, these teams ensure that the rocket design is not only optimized for performance but also practical for scalable and cost-effective production.

Prototyping and Model Building

Once the design is finalized, the next step involves prototyping and model building to verify the design's manufacturability and functionality. Prototypes or scale models play a critical role in testing various aspects of the rocket's design before moving to full-scale production.

Subscale models are often built as small-scale prototypes to conduct aerodynamic testing in wind tunnels. These models allow engineers to analyse airflow, pressure distribution, and drag forces under controlled conditions, simulating how the rocket will perform during flight. Subscale models can also be used for static engine testing to evaluate combustion performance and thrust generation without requiring full-size hardware.

Figure 65: Prototype for a fusion rocket developed by MSNW (2000 to 2012) using NASA funding. Alan Boyle, GeekWire and MSNW Inc (now defunct)., CC BY-SA 4.0, via Wikimedia Commons.

Component prototyping focuses on manufacturing specific parts of the rocket to test their individual performance and compatibility. Advanced techniques like additive manufacturing, or 3D printing, are frequently employed to produce complex components, such as engine nozzles, fuel tanks, and avionics enclosures. 3D printing accelerates the prototyping process by reducing lead times, enabling engineers to refine intricate parts efficiently and cost-effectively.

In addition to subscale and component-level prototypes, test articles are built as full-scale models or mock-ups for structural testing and system integration. Full-scale test articles are subjected to rigorous stress, vibration, and thermal tests to validate their structural integrity under launch conditions. These large prototypes also allow engineers to assess how different subsystems—such as propulsion, avionics, and thermal control—interact when integrated into the overall vehicle design.

Prototyping provides a crucial opportunity for teams to identify design flaws early in the process. By testing critical components and validating manufacturing techniques, engineers can address issues before committing to full-scale production. This iterative approach ensures that the final rocket design meets performance requirements and adheres to production feasibility, paving the way for successful manufacturing and operation.

Manufacturing Process

Once prototypes are tested and validated, the manufacturing process transitions into full-scale production. This phase involves producing individual rocket components, assembling subsystems, and ensuring all parts conform to strict design specifications. Modern rocket manufacturing leverages advanced technologies and highly specialized processes to achieve precision, reliability, and efficiency.

The process begins with material procurement, where specialized materials capable of withstanding the extreme environments of rocket launches are sourced. High-strength alloys such as titanium and Inconel are commonly used for critical components like engine parts, while advanced composites and ceramics provide lightweight strength for structural sections and thermal protection.

Component manufacturing encompasses various fabrication techniques tailored to the specific requirements of rocket parts. Precision machining is employed for manufacturing components like turbopumps, engine parts, and valves. These parts are produced using computer numerical control (CNC) machines that ensure exact tolerances and superior accuracy. Additive manufacturing, or 3D printing, is increasingly utilized to create complex, lightweight structures such as combustion chambers, nozzles, and intricate cooling channels. This method not only reduces production time but also minimizes material waste. For structural elements such as rocket bodies, fuel tanks, and fairings, composite fabrication techniques using carbon fibre-reinforced composites provide high strength with reduced weight. Welding and joining processes, including precision methods like friction stir welding, are used to bond critical sections, such as fuel tanks and engine components. For example, NASA's Space Launch System (SLS) employs friction stir welding to ensure the integrity of its massive cryogenic fuel tanks.

Figure 66: Rocket component assembly. SpaceX, CC0, via Pexels.

During subsystem assembly, individual components are integrated into larger, functional systems. The propulsion system involves assembling engines, turbopumps, valves, and fuel lines, which are then rigorously tested to ensure performance and reliability under operating conditions. Avionics systems, including flight computers, navigation systems, and telemetry equipment, are carefully assembled and calibrated to guarantee precise control, communication, and data collection during flight. Structural assembly involves constructing the rocket's airframe, interstage sections, and payload fairings, ensuring these elements can endure the mechanical stresses of launch and flight while protecting critical subsystems and payloads.

Figure 67: Assembly of rocket core at aerospace factory. SpaceX, CC0, via Pexels.

The manufacturing process is characterized by tight quality control and testing protocols to ensure every component meets stringent safety and performance standards. This precision and attention to detail are vital for delivering rockets capable of withstanding extreme forces, high temperatures, and the demanding conditions of space travel. By integrating cutting-edge technologies such as additive manufacturing and advanced composites, modern rockets achieve greater efficiency, reduced weight, and improved reliability, enabling complex missions like those undertaken by SpaceX, Blue Origin, and NASA.

Testing and Validation

Before full-scale production is finalized, rigorous testing and validation processes are conducted to ensure that each component, subsystem, and the rocket as a whole meets stringent performance, reliability, and safety requirements. This phase is critical for ensuring that the rocket can endure the extreme conditions of launch, flight, and space operations.

Component testing is the first step, where individual subsystems are evaluated in isolation to verify their functionality and performance. Rocket engines, for example, undergo static fire tests to validate thrust levels, combustion stability, and fuel efficiency. Avionics systems are individually tested for their ability to manage navigation, flight control, and telemetry functions accurately. Each sensor, valve, and structural part is checked to ensure it performs as designed under expected conditions.

In structural testing, critical components such as fuel tanks, fairings, and airframes are subjected to mechanical stress and load simulations to verify their structural integrity. These tests simulate the extreme forces, vibrations, and aerodynamic pressures experienced during launch and ascent. Components are pushed to their design limits to ensure they can endure both expected and unexpected conditions, reducing the risk of in-flight failure.

Integrated systems testing brings together multiple subsystems to evaluate how they interact and function as a complete unit. These tests include vibration tests, which simulate the intense shaking during rocket ignition and ascent, and thermal vacuum tests, which replicate the temperature extremes and vacuum conditions of space. Electrical tests ensure that all avionics and control systems work seamlessly together and can withstand electromagnetic interference. Integrated systems testing is vital for identifying potential issues that may not appear during isolated component tests.

Flight simulations play a crucial role in modelling the rocket's behaviour across all phases of the mission. Advanced software tools simulate everything from launch conditions and atmospheric dynamics to stage separation and payload deployment. These simulations allow engineers to refine the design, test control algorithms, and evaluate the rocket's performance under different scenarios. Virtual models help predict how the rocket will react to real-world challenges, such as crosswinds or variations in engine performance.

For instance, SpaceX conducts extensive static fire tests on its Merlin engines before integrating them into the Falcon 9 rocket to validate their performance and reliability. Each Falcon 9 booster also undergoes a full static fire test on the launch pad before an actual mission. Similarly, NASA's Space Launch System (SLS) core stage undergoes comprehensive Green Run tests, where the entire propulsion system is fired and evaluated to ensure it operates correctly over the full duration of simulated flight conditions.

Figure 68: SpaceX's Crew Dragon is at NASA's Plum Brook Station in Ohio, ready to undergo testing in the Reverberant Acoustic Test Facility. SpaceX, CC0, via Wikimedia Commons.

The culmination of testing and validation ensures that rockets are ready for launch, capable of performing reliably under immense stresses and delivering mission success. This meticulous process of testing every level—from individual components to integrated systems—helps identify and resolve potential issues early, minimizing risks and increasing confidence in the rocket's operational performance.

Quality Assurance and Inspection

Throughout the production process, rigorous quality control measures are implemented to ensure that every component meets the highest standards of safety, performance, and reliability. These measures are essential for identifying flaws early and ensuring that the rocket can perform as designed under extreme conditions.

Non-destructive testing (NDT) plays a vital role in quality assurance. Techniques such as X-ray imaging, ultrasonic testing, and laser scanning are employed to detect internal defects or irregularities in components without causing any damage. These methods allow engineers to inspect critical parts like engine nozzles, fuel tanks, and structural elements for cracks, voids, or imperfections that could compromise performance.

Dimensional inspections are conducted to ensure that components meet precise design tolerances. Advanced tools such as coordinate measuring machines (CMMs), laser trackers, and optical scanners are used to verify the exact dimensions of fabricated parts. Even minor deviations from specified measurements can impact the overall assembly and functionality of the rocket, making precision essential at every step.

Maintaining traceability is another critical aspect of quality control. Detailed records are meticulously kept to track the origins of materials, manufacturing processes, and inspection results for each component. This ensures that every stage of production complies with stringent safety and quality regulations. Should an issue arise, traceability allows engineers to quickly pinpoint the cause, assess its impact, and implement corrective measures.

For instance, Blue Origin's New Shepard rocket undergoes rigorous inspections at each stage of manufacturing to meet strict flight readiness criteria. Components are subjected to multiple layers of quality checks, from material verification to final assembly validation, ensuring that each system operates flawlessly before the rocket is cleared for launch.

These comprehensive quality control processes are essential for guaranteeing that every component functions reliably under the extreme stresses of launch, flight, and re-entry. By combining advanced inspection techniques with thorough record-keeping, manufacturers ensure the safety, consistency, and performance of modern rockets, reducing the risk of failure and enhancing mission success.

Final Integration and Assembly

In the final integration and assembly stage, the fully tested subsystems are brought together to create the complete rocket. This phase takes place in clean, climate-controlled environments to prevent contamination and ensure optimal working conditions. Each major component, including the first stage, second stage, engines, payload fairing, and other critical systems, is carefully assembled to ensure precise alignment and compatibility.

During this process, all electrical, fluid, and mechanical interfaces are connected and rigorously verified to guarantee seamless integration. Engineers perform comprehensive checks to confirm that every connection, sensor, and system operates as intended when combined. Functional tests are conducted on the fully assembled rocket to simulate real-world operations. These tests include fuelling simulations to evaluate propellant loading systems and system activations to confirm the functionality of avionics, engines, and control systems. The end-to-end tests ensure that the rocket performs reliably as a single, cohesive unit.

Figure 69: Assembly and construction of a module. SpaceX, CC0, via Pexels.

Modern rockets, such as SpaceX's Falcon 9, are assembled horizontally, a method that allows for more efficient handling and inspection during integration. Once assembled, the rocket is transported to the launch site on specialized transporters. At the launch site, it is carefully raised into a vertical position for final pre-launch checks. These final checks include comprehensive system verifications, environmental tests, and rehearsals to ensure the rocket is ready for a successful launch. By meticulously integrating and testing each subsystem, engineers ensure that the rocket meets all performance and safety requirements before it embarks on its mission.

Launch Site Preparation and Final Checks

Once the rocket is fully assembled, it is transported to the launch site for final preparations, marking the critical last phase before launch. The rocket is carefully moved using specialized transporters to ensure its safety and stability during transit. At the launch pad, teams begin a series of rigorous checks and procedures to prepare the vehicle for flight.

The fuelling and pressurization process is one of the first steps, where propellant systems are loaded with liquid fuel or cryogenic oxidizers and pressurized to meet flight-ready conditions. This step ensures the propulsion system is fully operational and able to deliver the required thrust during launch. Engineers closely monitor tank pressures, temperatures, and flow rates to prevent anomalies during the fuelling process.

Next, final systems checks are conducted to validate the functionality of critical subsystems. Flight computers, telemetry equipment, propulsion systems, and avionics are thoroughly tested to confirm they are performing as expected. These checks are essential for ensuring that all onboard systems are synchronized and capable of responding to commands during the mission.

To further ensure readiness, countdown simulations or dry runs are performed. These rehearsals simulate the complete launch sequence, enabling engineers and ground crews to verify the timing and functionality of all systems in real-time conditions. This includes activating systems, conducting communication checks, and ensuring all components work seamlessly together as the countdown progresses. Any issues identified during the simulations are addressed immediately to minimize risks.

An example of this process is NASA's Space Launch System (SLS) rocket, which undergoes a Wet Dress Rehearsal. During this procedure, the full countdown sequence is simulated while the rocket is loaded with actual propellants. The rehearsal allows teams to validate fuelling operations, countdown processes, and system responses under realistic conditions, ensuring the rocket is fully prepared for launch.

Through these meticulous preparations and final checks, engineers ensure the rocket is ready to withstand the extreme conditions of launch and successfully complete its mission. The combination of careful fuelling, systems verification, and simulation ensures that no detail is overlooked before the vehicle lifts off the pad.

Figure 70: Falcon 9 and Dragon lift off from Launch Pad 39A for CRS-10. SpaceX, CC0, via Wikimedia Commons.

Assembly Processes for Rockets and Spacecraft

Subsystem Assembly

The subsystem assembly process begins with building, testing, and integrating individual components into their respective sections. Each subsystem plays a critical role in ensuring the rocket or spacecraft functions reliably throughout its mission. Subsystems are developed with high precision to meet stringent performance and safety requirements.

The propulsion systems form the core of a rocket's power. Critical components such as engines, turbopumps, fuel tanks, valves, and plumbing systems are assembled and rigorously tested. For example, SpaceX's Merlin engines used in the Falcon 9 rocket undergo individual tests to validate thrust performance, combustion stability, and efficiency before they are integrated into the rocket's first stage. These tests ensure the propulsion system can operate flawlessly under extreme conditions.

Avionics systems, which include flight computers, sensors, navigation systems, and communication equipment, are built and calibrated to deliver precise control over the rocket's operations. These systems are responsible for managing flight control, processing vast amounts of data, and transmitting telemetry to ground stations. Every component, such as inertial measurement units

(IMUs), GPS receivers, and star trackers, is tested for accuracy and calibrated to interact seamlessly with the overall flight software.

The structural components are manufactured to provide the rocket's framework and house critical systems. Fuselage sections, fuel tanks, interstage segments, and payload fairings are produced using advanced manufacturing techniques. For instance, NASA's Space Launch System (SLS) employs friction stir welding to join large fuel tank segments with exceptional precision and strength. Composite fabrication techniques and additive manufacturing are also used to create lightweight yet robust structures capable of withstanding high mechanical loads.

The thermal control systems are integrated to safeguard the rocket and spacecraft from extreme temperature variations during flight. Components such as heat shields, thermal blankets, and radiators are added to protect against the intense heat generated during launch, re-entry, or deep-space operations. For missions like NASA's Orion spacecraft, thermal protection systems are critical to ensuring the safety of the crew and payload.

Figure 71: Subsystem assembly. SpaceX, CC0, via Pexels.

Each subsystem undergoes rigorous testing to validate its functionality, durability, and compatibility with other components. This process involves both standalone testing of individual components and integrated testing to ensure seamless performance within the broader system. By thoroughly verifying each subsystem, engineers ensure the rocket or spacecraft can withstand the demanding conditions of spaceflight and meet mission requirements.

Major Section Integration

The process of major section integration involves assembling the rocket or spacecraft into its primary sections in clean, climate-controlled environments to prevent contamination and ensure precision. This phase is critical for combining previously assembled subsystems into larger units, preparing the vehicle for final integration and testing.

For multistage rockets, each stage is assembled as a distinct unit, complete with engines, fuel tanks, and avionics systems. In the case of SpaceX's Falcon 9, the reusable first stage is integrated with its cluster of nine Merlin engines, which provide the thrust necessary for liftoff and ascent. The second stage, or upper stage, is assembled separately and prepared to carry the payload to its designated orbit. Each stage undergoes additional checks to ensure proper alignment, fuel system integrity, and avionics readiness.

Payload integration is another crucial step, where the mission's payload, such as satellites, crew capsules, or scientific instruments, is prepared independently before being carefully integrated into the rocket's payload fairing. The payload fairing, often referred to as the rocket's "nose cone," protects the payload from aerodynamic forces and environmental conditions during launch. For example, NASA's Artemis missions integrate the Orion spacecraft into the upper stage of the Space Launch System (SLS) rocket, ensuring a secure connection and proper alignment for orbital insertion.

Interstage sections connect the various rocket stages and play a critical role during flight. These sections house mechanisms for stage separation, as well as electrical and fluid connections that enable communication and fuel transfer between stages. Precision alignment is essential to ensure smooth and reliable separation of stages during ascent. Faults in the interstage design or alignment can lead to mission failure, making this part of the integration process particularly rigorous.

During major section integration, engineers perform extensive verification to confirm structural integrity, alignment, and compatibility across all components. By assembling the rocket into these key sections—stages, payload, and interstages—teams create a cohesive and functional launch vehicle capable of withstanding the forces and complexities of spaceflight. This phase brings the rocket closer to final integration, where all sections are combined to form the complete flight-ready vehicle.

Electrical and Mechanical Interfaces

During the integration process, all mechanical, electrical, and fluid interfaces between subsystems are meticulously connected and verified to ensure seamless operation. Mechanical alignment is a critical step where structural sections are joined and precisely bolted together. Engineers work with millimetre-level accuracy to guarantee aerodynamic stability and structural integrity, which are essential for withstanding the extreme forces during launch, ascent, and flight.

Electrical connections are established to integrate avionics systems, sensors, and actuators across the vehicle. Wiring harnesses, power cables, and data buses are connected to facilitate seamless communication between various subsystems. These connections ensure that systems like flight computers, navigation sensors, and telemetry equipment can operate in unison. For instance, SpaceX rockets use cross-strapped systems that allow redundant pathways for electrical signals, enhancing reliability in case of a single point of failure.

Figure 72: Subsystem connection. SpaceX, CC0, via Pexels.

The integration of fluid and propellant systems is another critical phase. Fuel and oxidizer lines, valves, and pressure vessels are carefully interconnected to enable proper delivery of propellants to the engines. Leak checks are rigorously conducted to verify the system's integrity and ensure there are no compromises in the pressurized fuel and oxidizer systems. This step is particularly vital for the performance and safety of the propulsion system, as even minor leaks can disrupt engine operation.

Through meticulous verification and testing, engineers confirm that all interfaces—mechanical, electrical, and fluid—are fully functional and meet design specifications. This thorough process ensures that every system is interconnected, stable, and ready to perform flawlessly during the rocket's mission.

System Integration and Validation

Once the major sections and interfaces are assembled, the rocket or spacecraft undergoes rigorous integration testing to ensure that all subsystems and components work seamlessly together as intended. This phase is critical for identifying and addressing any discrepancies or malfunctions before launch. Engineers conduct a series of comprehensive tests to validate the vehicle's performance, safety, and reliability under real-world conditions.

Functional testing is the first step, where key systems such as propulsion, avionics, communication, and navigation are activated and tested for end-to-end functionality. Each subsystem is monitored to confirm it operates correctly and integrates smoothly with other components. Propulsion systems are ignited in controlled settings to ensure proper thrust and stability, while avionics systems are tested to verify communication between sensors, flight computers, and control mechanisms.

The vehicle's structural integrity is validated through stress tests, where the integrated structures are subjected to mechanical loads similar to those experienced during launch and flight. Engineers assess how well the rocket or spacecraft can endure the intense forces and pressures it will face during its mission, ensuring there are no weak points in the design.

To simulate the extreme launch environment, vibration and acoustic testing is performed on the fully assembled vehicle. The rocket is subjected to intense vibrations and noise levels that replicate lift-off and ascent conditions. This test ensures that sensitive components, like avionics systems and payloads, remain operational and undamaged under these harsh circumstances.

The spacecraft or payload is also tested in a thermal vacuum chamber to simulate the conditions of space, including the vacuum environment and extreme temperature variations. This test verifies that thermal control systems function correctly to protect the vehicle and its components during spaceflight, ensuring consistent performance in both high-heat and cryogenic conditions.

For example, NASA's Space Launch System (SLS) undergoes Green Run testing for its core stage. This process involves firing the stage's systems in an integrated environment, including running the engines and evaluating all critical systems, to confirm they are flight-ready. Such comprehensive validation processes are vital for ensuring mission success, as they allow engineers to address any potential issues before the vehicle is cleared for launch.

Final Integration and Rollout

In the final phase of rocket preparation, the fully integrated vehicle undergoes a series of steps to ensure it is launch-ready. This stage begins with end-to-end verification, where all systems—including avionics, propulsion, and communication—are subjected to final functional tests. Engineers power up critical systems, simulate operations, and conduct fuelling rehearsals to confirm the rocket's readiness. These comprehensive checks ensure that all components interact seamlessly and that no anomalies remain before the vehicle is transported to the launch site.

Once verified, the rocket or spacecraft is moved to the launch site using specialized transport systems. The method of transport depends on the vehicle's design and launch site location. Rockets such as SpaceX's Falcon 9 are assembled horizontally at the integration facility before being transported to the pad. Once at the launch site, the rocket is carefully raised into an upright position for final preparations. Vertical transportation, as seen with NASA's Space Launch System (SLS), involves the vehicle being moved upright using massive crawler-transporters that can handle its immense weight and size.

Figure 73: Transportation of assembled rocket. SpaceX, CC0, via Pexels.

At the launch pad, the rocket undergoes a final round of launch pad preparation. Fuelling and pressurization systems are tested to ensure the proper flow of propellants, and countdown rehearsals—like wet dress rehearsals—are conducted to simulate the full launch sequence. Engineers verify that the rocket, ground systems, and infrastructure are fully synchronized and operational. Once all systems are confirmed for launch readiness, the vehicle is secured on the pad, where it awaits the countdown to ignition.

Figure 74: Ready for launch on launch pad. SpaceX, CC0, via Pexels.

This final integration and rollout phase is a culmination of extensive testing, validation, and precision engineering, ensuring that the rocket is ready to safely and successfully execute its mission.

The assembly process for rockets and spacecraft combines cutting-edge engineering, rigorous testing, and precision manufacturing. Each subsystem is carefully integrated, tested, and validated to ensure the vehicle's reliability during critical mission phases. Through advanced assembly techniques, such as modular integration, additive manufacturing, and real-time testing, modern rockets like SpaceX's Falcon 9 and NASA's SLS demonstrate the success of meticulously planned assembly processes, enabling safe and efficient space missions.

Saturn V Manufacturing and Transportation Process

The Saturn V manufacturing and transportation process was a massive logistical and engineering achievement, reflecting the complexity and scale of the Apollo Program. The Saturn V rocket, a 3-stage launch vehicle, was meticulously designed, built, transported, and assembled by numerous contractors across the United States. Each stage of the rocket, as well as the command and service module, lunar module, and the instrumentation section, played an essential role in the vehicle's ascent, lunar operations, and safe return of the crew [326].

The Saturn V's first stage, the S-IC, was the largest and heaviest section of the rocket. Standing 42 meters (138 feet) tall with a 10-meter (33-foot) diameter, it was powered by five powerful F-1 Rocketdyne engines. This stage lifted the vehicle to an altitude of 67 kilometres (42 miles) before separation. Due to its massive size, Boeing manufactured the first stage at the Michoud Assembly Facility in New Orleans, Louisiana. Transportation posed a significant challenge, as the stage was too large for standard road or rail transport. The solution was to use barges. The S-IC was transported via the Mississippi River to the Gulf of Mexico and then along the Intra-Coastal Waterway to reach NASA's Kennedy Space Center in Florida [326].

The second stage, the S-II, measured 24.87 meters (81.6 feet) in height and maintained the same diameter of 10 meters (33 feet) as the first stage. Unlike the first stage, it used five J-2 Rocketdyne engines powered by liquid hydrogen (LH2) and liquid oxygen (LOX). The use of hydrogen as a propellant required significant design innovations due to its extremely low operating temperature of −252.8°C (-423.04°F). North American Aviation, awarded the contract for the second stage, manufactured it in Seal Beach, California. Transportation was again a challenge due to its size, so the stage was moved by sea from California to the Stennis Space Center in Mississippi for testing. From there, it continued its journey to Kennedy Space Center along the Intra-Coastal Waterway [326].

The third stage, the S-IVB, was smaller than the first two, measuring 17.86 meters (58.6 feet) in height with a 6.6-meter (21.7-foot) diameter. It was powered by a single J-2 engine that fired twice—once to place the vehicle into Earth orbit and again to initiate translunar injection. Douglas Aircraft Company manufactured the S-IVB at Huntington Beach, California. Due to its smaller size, the third stage could be transported by air using NASA's Super Guppy, a specially-designed aircraft capable of carrying oversized payloads. This innovation streamlined transportation and allowed for efficient delivery of the stage to Kennedy Space Center [326].

The Command and Service Module (CSM) was a critical element of the Apollo Program, housing astronauts and facilitating operations in lunar orbit. The combined CSM was 11 meters (36.2 feet) in length and 3.9 meters (12.8 feet) in diameter. The command module, which contained the crew cabin, was 3.48 meters (11.4 feet) long, while the service module, containing the propulsion and support systems, measured 7.49 meters (24.6 feet). North American Aviation (later North American Rockwell) built the CSM at its Downey, California facility. Transporting the CSM was simplified by using the Super Guppy, allowing it to be flown directly to Kennedy Space Center, where it underwent final integration with the rest of the rocket [326].

The Lunar Module (LM) was the two-stage spacecraft that carried astronauts from lunar orbit to the Moon's surface and back to the command module. The LM was 7.04 meters (23.1 feet) tall with a 4.22-meter (13.8-foot) diameter. Its descent stage acted as the landing platform, while the ascent stage carried the astronauts back to the CSM. The LM was built by the Grumman Aircraft Company in Bethpage, New York. Like the CSM, the Lunar Module was transported by air using NASA's Super Guppy. It was stored between the Saturn V's third stage and the service module during launch, within the Spacecraft-to-LM adapter [326].

The Instrumentation Section was the guidance and control hub of the Saturn V rocket. Without this system, the vehicle could not orient itself or follow a precise trajectory. This circular unit was 1 meter

(3.2 feet) tall with a 6.7-meter (22-foot) diameter and housed the guidance, navigation, and telemetry systems. The International Business Machines (IBM) company built the instrumentation section at its Space Systems Center in Huntsville, Alabama. Once complete, it was transported by air using the Super Guppy, ensuring a secure and timely delivery to Kennedy Space Center [326].

Once all stages and components arrived at Kennedy Space Center, they underwent final inspections and were prepared for vertical assembly. NASA's Vehicle Assembly Building (VAB), specially constructed for the Apollo Program, played a crucial role in this process. The massive structure featured four high bays, allowing up to four Saturn V rockets to be assembled simultaneously. Each stage was stacked vertically on the Mobile Launcher, with the first stage at the bottom, followed by the second stage, third stage, instrumentation section, and spacecraft components. All mechanical, electrical, and fluid systems were connected and tested to ensure seamless integration [326].

Once fully assembled, the Saturn V was transported to the launch pad using the Crawler Transporter, a massive tracked vehicle designed to carry the rocket and its mobile platform. The crawler moved at a slow, precise pace to ensure stability during the journey to Launch Complex 39 [326].

The manufacturing and transportation process of the Saturn V rocket was a monumental effort involving collaboration among multiple contractors across the United States. Each stage and component was meticulously designed, produced, tested, and transported using innovative methods such as barge transport, air delivery via the Super Guppy, and the purpose-built Vehicle Assembly Building. This complex yet efficient process set the foundation for modern rocket manufacturing and assembly, demonstrating the engineering prowess required to accomplish the historic Apollo missions [326].

SpaceX Manufacturing and Assembly Process

The SpaceX manufacturing and assembly process reflects a blend of advanced engineering, streamlined production methods, and operational efficiency. Currently, the company focuses on two major launch vehicles: the Falcon 9, a proven workhorse for orbital missions, and the Starship Super Heavy, a next-generation rocket designed for deep-space exploration and full reusability. By examining the production and assembly methods for both rockets, it becomes evident how SpaceX has embraced modern manufacturing techniques while adhering to trends established during the era of the Saturn V [326].

The Falcon 9 rocket is one of the most successful and widely used launch vehicles of the modern era. Designed to carry payloads of 22,800 kilograms (50,300 pounds) to Low Earth Orbit (LEO) or 8,300 kilograms (18,300 pounds) to Geostationary Transfer Orbit (GTO), it is renowned for its partial reusability. The Falcon 9's first stage can autonomously return to Earth and land, making refurbishment and reuse possible, significantly reducing launch costs [326].

The primary manufacturing for the Falcon 9 occurs at SpaceX's main facility in Hawthorne, California. This sprawling factory serves as the production hub where the rocket's key components are fabricated, including the Merlin engines, fuselage, interstages, and avionics. Advanced manufacturing techniques such as precision machining, welding, and additive manufacturing (3D printing) are employed to produce components with tight tolerances. The integration of the engines, fuel tanks, and flight systems ensures that the stages are nearly complete before transportation [326].

Unlike the Saturn V, which was assembled vertically, the Falcon 9 follows a horizontal integration process. Once the rocket's components are built, they are transported by road using specially designed trucks. These trucks deliver the stages to integration facilities located at launch sites, such as Kennedy Space Center, Cape Canaveral Space Force Station, and Vandenberg Space Force Base. At these locations, the Falcon 9's first and second stages are mated, and the rocket is prepared for payload integration. After final checks, the horizontally assembled rocket is hoisted to a vertical position on the launch pad, ready for flight [326].

This horizontal integration method allows for rapid assembly and reduces the infrastructure costs associated with vertical construction. The Falcon 9's modular design also simplifies transportation, making it highly adaptable for frequent launches and refurbishment. By centralizing manufacturing at Hawthorne while performing final assembly at launch sites, SpaceX optimizes production timelines and minimizes logistical challenges [326].

SpaceX's Starship Super Heavy is poised to become the largest and most powerful orbital launch vehicle ever constructed. It consists of two main sections: the Super Heavy Booster, which forms the rocket's lower stage, and the Starship spacecraft, which serves as the upper stage and can carry crew and cargo. Combined, the vehicle stands 120 meters (393 feet) tall. Both stages are designed to be fully reusable, aligning with SpaceX's mission to reduce the cost of space exploration dramatically [326].

The majority of Starship's manufacturing and assembly occurs at Starbase, SpaceX's dedicated production and testing facility located in Boca Chica, Texas. Starbase serves as both a factory and a test site, allowing SpaceX to produce, integrate, and launch Starship prototypes efficiently. The rocket's stainless-steel fuselage is fabricated on-site using advanced welding and manufacturing techniques, ensuring durability and reusability in extreme conditions. Unlike traditional aluminium-based rockets, stainless steel provides strength while withstanding the high temperatures experienced during re-entry [326].

Although most of Starship's components are produced at Starbase, the Raptor engines—critical to the rocket's propulsion system—are manufactured separately. These engines are built at SpaceX's engine production facility in Hawthorne, California, and undergo rigorous testing at the McGregor Test Facility in Texas before being shipped to Boca Chica for integration. The Raptor engines, powered by liquid methane (CH4) and liquid oxygen (LOX), represent a departure from traditional rocket fuels and are designed for reuse, efficiency, and deep-space missions [326].

The Starship Super Heavy is assembled vertically on massive launch mounts at Starbase. Unlike Falcon 9's horizontal integration, the sheer size of Starship necessitates vertical construction. Components such as the Super Heavy Booster and Starship spacecraft are stacked using cranes, and integration includes connecting fuel lines, avionics, and engine mounts. Prototypes undergo frequent static fire tests to validate engine performance and system readiness [326].

Starship is currently being launched from Starbase's orbital test pad, where SpaceX conducts development flights to refine the rocket's design. However, plans are underway to construct additional launch infrastructure. SpaceX is actively building a dedicated Starship launchpad at Kennedy Space Center's Launch Complex 39A, a historic site used for the Apollo missions and Falcon Heavy launches. Another potential launchpad at Launch Complex 49 in Florida is under consideration to support Starship's ambitious launch cadence [326].

The manufacturing and assembly processes for the Falcon 9 and Starship echo trends established during the construction of rockets like the Saturn V. Both programs involve geographically dispersed production, with specialized facilities for manufacturing engines, stages, and avionics. However, SpaceX has revolutionized the process by embracing reusability, modular designs, and modern manufacturing techniques such as 3D printing and vertical stacking for Starship [326].

While the Saturn V relied heavily on barge and air transport for its massive components, SpaceX uses road transportation for the Falcon 9 and localized production at Starbase for Starship. Additionally, the integration of digital tools, automated testing, and real-time feedback has accelerated development timelines, allowing SpaceX to iterate rapidly through prototype designs [326].

SpaceX's manufacturing and assembly processes reflect a seamless blend of innovation and efficiency. The Falcon 9's success demonstrates the power of reusability and streamlined integration, while the Starship Super Heavy project pushes the boundaries of modern engineering. By centralizing production where feasible and employing advanced techniques, SpaceX continues to redefine how rockets are built, tested, and launched [326].

Integration of Subsystems and Payloads

Subsystem integration begins with assembling the major systems of the rocket, including propulsion, avionics, structural elements, and thermal control systems. Each subsystem is constructed, tested, and verified individually before being incorporated into the larger assembly.

The propulsion system is one of the most complex integrations, involving engines, turbopumps, fuel tanks, valves, and plumbing systems. These components must be aligned and interconnected to enable the proper flow of fuel and oxidizer. Leak tests and pressure tests are conducted to ensure the integrity of the fluid systems. For instance, SpaceX integrates its Merlin engines into the first stage of the Falcon 9 rocket with precision, testing each engine for thrust performance and combustion stability before installation.

The avionics system, which includes flight computers, sensors, navigation equipment, and communication systems, is another critical subsystem. During integration, wiring harnesses and data buses are connected to allow seamless communication between components. Cross-strapping techniques are often used to provide redundancy, ensuring that a failure in one pathway does not compromise the entire system. Avionics integration is tested through functional simulations and hardware-in-the-loop testing to validate the system's ability to manage flight operations and telemetry.

The rocket's structural components, such as fuselage sections, interstages, and payload fairings, are integrated to form the rocket's body. These sections are joined using advanced welding techniques like friction stir welding for tanks and other load-bearing structures. Composite materials, often used in modern rockets, are fabricated and aligned to ensure lightweight strength and durability. For example, NASA's Space Launch System (SLS) uses precision welding to integrate its core stage.

Payload integration occurs after the primary subsystems have been assembled. The payload is often a satellite, crew capsule, or scientific instrument, housed within the rocket's payload fairing. This process begins with the payload's preparation, which includes final testing and encapsulation to protect it from environmental factors like contamination and mechanical vibrations.

Figure 75: In the mobile service tower on Pad 17-B at Cape Canaveral Air Force Station, the second half of the fairing, at right, moves toward the waiting THEMIS spacecraft. The first half has already been put in place. NASA, Public Domain, via Picryl.

A rocket's payload is the part of the rocket that carries the mission's objectives into space [327]. Payloads can include astronauts, spacecraft, supplies, scientific instruments, and communication equipment [327]. The type of payload a rocket carries depends on the mission, which could be for weather monitoring, planetary exploration, or other purposes [327].

The term "payload" comes from a seafaring term for cargo on a ship that produced revenue [327]. Payloads are an essential part of space exploration, allowing us to learn about other planets, galaxies, and the conditions of space [327].

Payload ratios can be significantly increased by performing air launches, where the rocket is launched into the air from an airplane or balloon [164]. Airplane-type air launches have been used throughout the history of rocket launches [164].

Payloads can include instruments for various scientific objectives, such as observing the Sun [328, 329], studying the Moon's subsurface [330], monitoring atmospheric conditions [331], and exploring Mars [332]. The payload design and development must consider the mission objectives and requirements [173, 333, 334].

Payload systems also involve technologies such as inertial measurement units [335], robotics [336], and computer systems [337-339] to support the mission operations and data collection.

Once prepared, the payload is mated to the rocket's upper stage. This involves mechanical attachment and the connection of electrical and data interfaces. Engineers perform thorough alignment checks to ensure that the payload is correctly positioned and secure. The payload's interfaces are tested for proper communication with the rocket's systems, verifying that commands, telemetry, and power supply function as intended.

After integrating the subsystems and payload, the entire assembly undergoes rigorous testing. Functional tests ensure that all systems work together seamlessly, including the activation of propulsion, avionics, and telemetry. Structural tests verify the rocket's ability to withstand the mechanical loads experienced during launch and flight. Vibration and acoustic tests simulate the intense forces of liftoff to ensure that both the rocket and payload remain stable and undamaged.

Once subsystem and payload integration is complete, the rocket is prepared for transport to the launch pad. At the pad, final checks are performed to confirm that all systems are operational. Functional tests are repeated to ensure the rocket is launch-ready, and the payload fairing is closed, encapsulating the payload for flight.

Chapter 9
Testing and Validation

Static Fire Tests and Engine Trials

Launch vehicle system tests are critical evaluations conducted to ensure that rockets and their ground support equipment (GSE) are ready for a safe and successful journey to orbit. These tests, which typically occur before launch, simulate various aspects of the mission, including fuelling, engine performance, and communication between ground systems and the vehicle. They help identify and address potential issues, thereby reducing the risk of failure during the actual launch. Two key types of tests, wet dress rehearsals (WDR) and static fire tests, play a central role in these assessments.

Wet Dress Rehearsal (WDR)

The wet dress rehearsal is a comprehensive test that involves loading liquid propellants, such as liquid oxygen and liquid hydrogen, into the rocket's fuel tanks. The term "wet" refers to the use of these cryogenic propellants, which mimic actual launch conditions. During a WDR, engineers validate the rocket's fuelling processes, including the flow of propellant, pressurization of tanks, and the functionality of valves, pumps, and ground connections. However, the engines are not ignited during this test.

WDRs are often performed on both prototype rockets under development and operational vehicles preparing for launch. They allow engineers to verify the readiness of the fuelling system, the integration of ground and flight systems, and the ability to handle the rocket under cryogenic conditions. For instance, SpaceX routinely conducts WDRs on its Falcon 9 and Falcon Heavy rockets to ensure that all components work seamlessly. During these rehearsals, the launch countdown sequence may also be simulated to evaluate the coordination between ground control and onboard systems.

Figure 76: Falcon 9 with CRS-3 Dragon wet dress rehearsal. SpaceX, CC0, via Wikimedia Commons.

Step-by-Step Process for Wet Dress Rehearsal (WDR):

1. Preparation and Setup

- Ensure the rocket is fully assembled, integrated, and transported to the launch pad.
- Perform preliminary checks on the rocket and ground support equipment (GSE), including electrical, mechanical, and communication systems.
- Verify that propellant storage tanks, fuelling lines, and pressurization systems at the launch pad are operational.
- Confirm that the countdown clock, telemetry systems, and communication links with mission control are functioning.

2. Rocket Positioning and Securing

- Secure the rocket on the launch mount to prevent movement during the test.
- Attach ground umbilicals to the rocket for fuelling, pressurization, and communication.

3. Countdown Simulation Initialization

- Begin the simulated countdown sequence to mimic the actual launch procedure.
- Activate the rocket's avionics and onboard systems to test their communication with ground control.

4. Propellant Loading

- Begin loading the rocket's propellant tanks with cryogenic fuel (e.g., liquid oxygen and liquid hydrogen) and oxidizers.
- Monitor the flow rates, pressures, and temperatures to ensure the fuelling process aligns with operational requirements.
- Test valves, pumps, and sensors to verify proper operation during fuelling.

5. Tank Pressurization

- Pressurize the propellant tanks to flight-ready conditions.
- Validate the performance of pressurization systems and their ability to maintain stable pressure levels.

6. Simulated Launch Countdown

- Proceed through the countdown sequence, including all steps leading up to engine ignition.
- Activate critical systems, such as telemetry, navigation, and propulsion controls.
- Verify the synchronization of launch control and onboard systems.

7. Abort or Hold Commands

- Pause the countdown at the predetermined point just before engine ignition.
- Validate that the rocket can respond to a hold or abort command from mission control.

8. Propellant Draining and Safing

- Begin draining the propellant from the tanks once the test is complete.
- Safely depressurize the tanks and disconnect the fuelling lines.
- Inspect ground and rocket systems for any anomalies or damage during the fuelling and pressurization process.

9. Post-Test Inspections

- Conduct a thorough inspection of the rocket and ground support equipment.
- Review telemetry data and analyse sensor readings to identify any discrepancies or areas of concern.
- Document findings and address any issues before proceeding to further testing or the actual launch.

10. System Reset

- Reset all rocket and ground systems to their original state, ensuring readiness for subsequent tests or launch preparations.
- Confirm that the rocket is stable and secure after the test.

Example Applications

SpaceX regularly uses the WDR process for its Falcon 9 and Falcon Heavy rockets. During these rehearsals, the company tests propellant loading, tank pressurization, and countdown sequences to ensure readiness for upcoming launches. Similarly, NASA's Space Launch System (SLS) conducts WDRs to simulate full launch conditions, including liquid hydrogen and oxygen fuelling.

Static Fire Tests

Static fire tests build upon wet dress rehearsals by including engine ignition while the rocket remains securely anchored to the launch mount. This process involves firing the engines at full thrust for a short duration, typically ranging from 3 to 7 seconds, though longer durations have been conducted. For example, SpaceX's Falcon Heavy performed a 12-second static fire test before its maiden launch in February 2018. These tests generate critical data on engine performance, including thrust levels, propellant flow rates, pressure, and temperature gradients.

Static fire tests serve multiple purposes. They validate the startup sequence of the engines, ensure that the propulsion system is functioning correctly, and provide insights into the specific criteria that will determine whether to proceed with the actual launch. Rockets can undergo static fire tests with or without the payload attached, depending on mission requirements and safety considerations.

Figure 77: Gilmour Space large hybrid engine test-fire. GilmourSpace, CC BY-SA 4.0, via Wikimedia Commons.

Step-by-Step Process for Static Fire Tests:

1. Preparation and Setup

- **Rocket Integration**: Ensure the rocket is fully assembled with all relevant stages, engines, and subsystems integrated.
- **Launch Pad Setup**: Secure the rocket to the test stand or launch mount. Attach ground support equipment (GSE), such as fuelling lines, pressurization systems, and umbilicals.
- **Preliminary Inspections**: Conduct checks on structural, mechanical, electrical, and telemetry systems to confirm readiness for the test.
- **Safety Measures**: Establish safety zones around the test site and ensure all personnel and equipment are positioned at a safe distance.

2. Propellant Loading

- **Fuelling**: Load liquid propellant (e.g., liquid oxygen, kerosene, liquid hydrogen) into the rocket's tanks. Follow precise protocols to monitor flow rates, pressure, and temperatures.
- **Tank Pressurization**: Pressurize the propellant tanks to operational levels. Validate that the pressurization system is functioning correctly and can maintain stable conditions.
- **Cryogenic Testing**: If cryogenic propellants are used, confirm that the rocket's thermal insulation and venting systems operate effectively.

3. Ignition System Check

- Perform checks on the rocket's ignition systems, including pyrotechnic or electrical igniters, to ensure they are operational and ready for engine start.

4. Simulated Countdown Sequence

- Initiate a countdown sequence to mimic launch conditions.
- Activate avionics, telemetry, and flight control systems to simulate real-time operational scenarios.
- Synchronize ground control systems with the rocket's onboard systems.

5. Engine Startup and Ignition

- Ignite the rocket engines while the rocket remains securely fastened to the test stand or launch mount.
- Allow the engines to fire at full thrust for a predetermined duration, typically 3–12 seconds (or longer for specific tests).
- Monitor engine parameters such as thrust levels, combustion stability, pressure, temperature, and propellant flow rates.

6. Real-Time Data Collection

- Use onboard and ground-based sensors to collect data on engine performance, structural loads, and thermal behaviour.
- Validate that all critical systems are functioning as expected and within predefined parameters.

7. Shutdown Sequence

- Shut down the engines and begin the cool-down process for the propulsion system.
- Safely depressurize the propellant tanks and disconnect fuelling and pressurization systems.

8. Post-Test Inspections

- Conduct a thorough inspection of the rocket, engines, and ground support equipment.
- Analyse telemetry and sensor data to identify any anomalies or performance issues.
- Document test results and determine if the rocket is ready for the next phase (e.g., additional testing or launch preparation).

9. System Reset

- Reset all systems to their original state, ensuring the rocket and ground equipment are prepared for subsequent tests or the actual launch.
- Replenish any used propellant or consumables, if necessary.

Example Applications

- **SpaceX**: SpaceX performs static fire tests for every Falcon 9 booster before launch. These tests verify engine performance and validate readiness for missions, including payload deployment and crewed flights.
- **NASA SLS**: The Space Launch System's core stage underwent a full-duration static fire during its Green Run test, which evaluated all four RS-25 engines firing for over eight minutes, simulating launch conditions.
- **Blue Origin**: For the New Shepard and New Glenn rockets, Blue Origin conducts static fire tests to validate engine performance and safety protocols before progressing to flight.

Use of WDR and Static Fire Tests

Not all launch providers regularly perform WDRs or static fire tests for their vehicles. However, companies like SpaceX have made these tests a standard part of their launch preparation. SpaceX performs a static fire test for every new booster and for each reused booster to ensure reliability. These tests are often combined with WDRs, creating a comprehensive pre-launch evaluation. For instance, in January 2018, SpaceX conducted multiple WDRs and a static fire test for its Falcon Heavy, marking a critical step toward its successful maiden flight.

Such rigorous testing protocols are essential for identifying and mitigating risks. The information gained from these tests helps refine the "go/no-go" criteria used during the launch countdown, increasing confidence in the vehicle's readiness.

Potential Anomalies and Risks

Failures during WDRs and static fire tests can have significant consequences [106]. For example, during a Falcon 9 static fire test on September 1, 2016, a catastrophic failure occurred due to a breach in the cryogenic helium system, leading to the explosion of the rocket and its payload, the AMOS-6 satellite, as well as extensive damage to the launch pad [106]. Similarly, in 2024, a structural failure between the rocket and its test stand during the static fire test of Space Pioneer's Tianlong-3 first stage resulted in an unintended launch, ending with the rocket crashing into nearby mountains [106].

Unintentional launches during static fire tests have also been observed historically, such as the Viking 8 rocket in 1952, which accidentally lifted off and flew for 55 seconds before being terminated [106]. These types of failures can be attributed to various factors, including structural issues, propulsion system malfunctions, and control system failures [282, 340].

The risks associated with WDRs and static fire tests are not limited to launch failures. Façade fire tests have shown that combustible exterior wall systems, such as exterior thermal insulation, sandwich panels, and photovoltaic sheets, can lead to the propagation of fire along the façade, posing a significant threat to nearby structures [341]. Additionally, fire exposure can degrade the mechanical properties of materials used in rocket components, such as GFRP (Glass Fiber-

Reinforced Polymer) reinforcement and concrete elements, reducing their load-bearing capacity and structural integrity [342-346].

To mitigate these risks, researchers have developed various approaches, including numerical simulations to assess the blast resistance of reinforced concrete slabs under fire conditions [347], and the design and implementation of thrust vector control (TVC) test systems to improve the reliability of rocket propulsion systems [348]. Additionally, the development of algorithms for early fire detection, such as the GOES Early Fire Detection (GOES-EFD) algorithm, can help in the early identification and response to potential fire hazards [349].

Prototyping and Development Testing

Static fire tests and propellant loading exercises are also conducted on prototype stages rather than fully assembled vehicles. For example, SpaceX routinely tests individual Starship and Super Heavy booster stages to validate their performance before integrating them into a full stack. These tests provide valuable insights into engine and structural performance, enabling iterative improvements in rocket design.

Test Facilities and Standardization in Ion Engine Development

The development and qualification of ion engines are heavily reliant on specialized vacuum test facilities designed to replicate the operating environment of space. These facilities aim to provide conditions as close as possible to those experienced in orbit, enabling the precise evaluation of thruster performance. For smaller ion engines, commercial off-the-shelf vacuum test systems can often meet requirements. However, engines in the kilowatt power range necessitate custom-designed facilities. The costs associated with constructing and maintaining such test setups are significant, often representing a financial challenge, particularly for small and medium-sized enterprises (SMEs). User fees for these facilities can reach thousands of dollars per test day due to their high operational expenses [228].

The size and capability of test facilities scale with the power output of the engines. Medium-sized facilities, such as the JUMBO facility at JLU, are ideal for engines up to about 5 kW. Facilities like this, with a diameter of 2.6 meters and a length of 5.5 meters, achieve a pumping speed of 150,000 litres per second for xenon propellant. However, high propellant consumption by certain engine types, such as Hall Effect Thrusters (HETs), can push the limits of these facilities. Larger engines require significantly more robust setups, which are fewer in number due to their high acquisition and maintenance costs [228].

Three critical aspects of test facility design include pumping speed, energy dissipation, and vacuum quality. For engines in the 5 kW range, maintaining a high vacuum with pressures better than 1×10^{-5} mbar necessitates advanced pumping systems, typically cryogenic pumps. These pumps ensure minimal contamination by avoiding oil mist, a common issue with oil diffusion pumps. Cryogenic pumps operate at temperatures below 50 K to effectively condense xenon propellant. The design of cryogenic surfaces must account for heat dissipation from the thruster and the vacuum chamber walls, which can be mitigated by pre-cooling techniques or reflective shielding. Proper geometric arrangement of pumps is also critical, as it influences the spatial pressure distribution and affects the thruster's performance by limiting backflow of neutral propellant and minimizing collision-induced plume scattering [228].

Thrusters in the high power range release their energy as a directed beam of fast-moving charged particles. To accommodate this energy, beam dumps are utilized to convert kinetic energy into heat safely. These beam dumps, typically composed of water-cooled graphite panels, minimize material sputtering due to graphite's low sputter yield. Chevron-style arrangements of these panels help direct sputtered material toward pumping systems, ensuring efficient removal and maintaining a low-pressure environment [228].

Diagnostic tools integrated into test facilities include Faraday probes for beam current measurement, retarding potential analysers for ion acceleration data, and thermal imaging equipment for temperature analysis. These tools facilitate in-depth evaluation of thruster performance. Advanced setups, like the Advanced Electric Propulsion Diagnostics (AEPD) platform developed by the European Space Agency (ESA), incorporate such tools to assess and compare the performance of different thruster technologies [228].

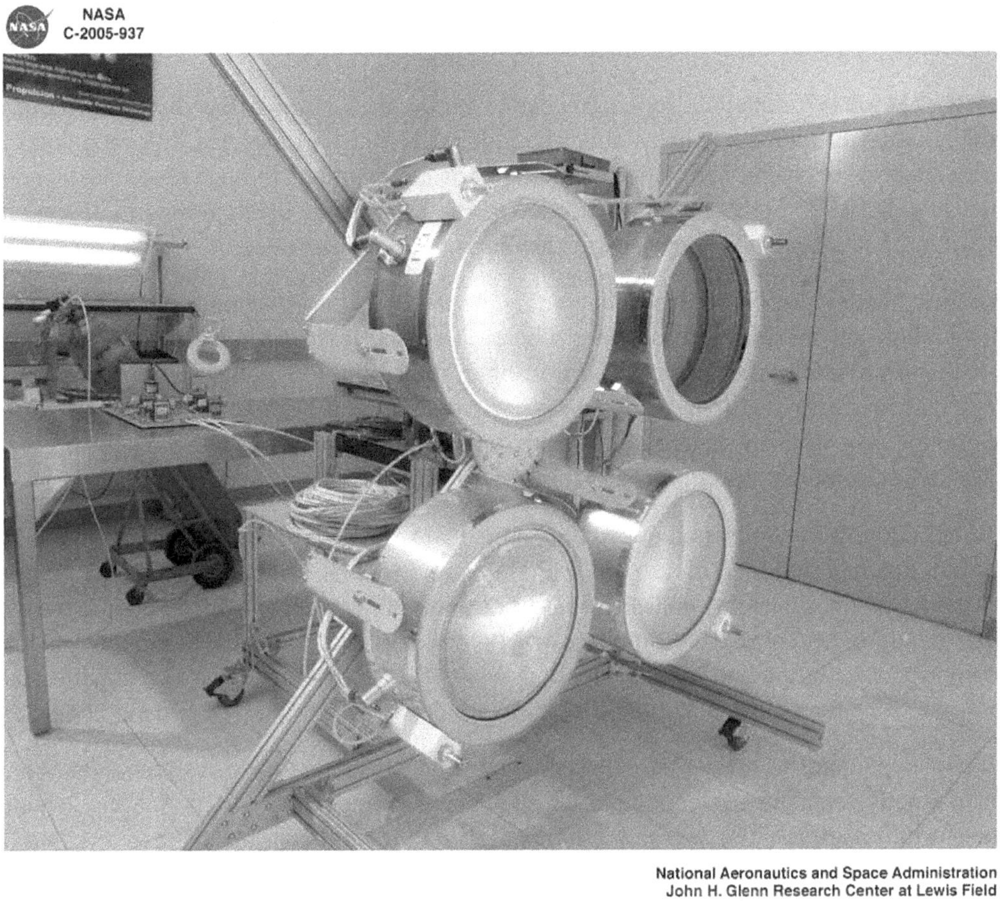

Figure 78: NEXT (NASA Evolutionary Xenon Thruster) ion thruster array being assembled for testing. Defense Visual Information Distribution Service, Public Domain, via Picryl.

Standardization of test conditions remains a crucial yet challenging topic within the electric propulsion community. Differences in facility size, pumping configurations, and diagnostics complicate comparisons across tests. While certain practices, such as thrust and pressure measurement protocols, are well-documented, others, like the interaction between thruster plumes and chamber walls, resist standardization. Numerical simulations using Direct Simulation Monte Carlo (DSMC) or Particle-in-Cell (PIC) models are instrumental in bridging this gap by providing validated frameworks for extrapolating test data to actual space conditions [228].

Numerical simulations employing Direct Simulation Monte Carlo (DSMC) and Particle-in-Cell (PIC) methods are advanced computational techniques crucial for modelling particle and plasma behaviours in environments where conventional analytical approaches are inadequate. These methods bridge the gap between experimental test conditions in vacuum facilities and the actual conditions of space, allowing researchers to predict and validate the performance of propulsion systems, such as ion and Hall-effect thrusters, in the near-perfect vacuum of space where particle and electromagnetic interactions dominate.

The DSMC method is a statistical approach used to simulate gas flows at low densities, particularly in scenarios where traditional fluid dynamics equations like Navier-Stokes fail. This technique treats gas as a collection of discrete particles and models their interactions through probabilistic collision mechanisms. DSMC is particularly effective for rarefied gas flows, such as those encountered in spacecraft propulsion plumes or the exhaust environments of ion engines.

On the other hand, PIC is a computational technique for simulating plasma dynamics, focusing on the interactions between charged particles, such as ions and electrons, and electromagnetic fields. By combining particle motion with grid-based field equations, PIC enables detailed analysis of complex plasma phenomena, including ion acceleration, plume behaviour, and charge-exchange collisions. PIC is especially valuable in regions where electromagnetic fields significantly affect particle dynamics, making it a primary tool for analysing thruster behaviour.

Both DSMC and PIC methods are applied to simulate vacuum environments that mimic the sparse particle interactions and long mean free paths characteristic of space. DSMC primarily handles neutral gas flows and collisions, while PIC addresses the dynamics of charged particles in the plasma exhaust. In plume dynamics, DSMC models how neutral propellant molecules expand and interact with chamber walls or the ambient background, while PIC simulations delve into ion and electron interactions within the plume, accounting for their interactions with magnetic and electric fields.

One critical application of these methods is mitigating facility effects. Since vacuum test chambers cannot perfectly replicate the conditions of space, DSMC and PIC models quantify issues like backflow from neutral particles and interactions with chamber walls. These insights allow researchers to adjust test results and predict true in-space performance. Both methods are also essential for predicting thrust and efficiency, as they model ion acceleration, charge-exchange collisions, and energy distribution, providing critical data for validating experimental results and refining thruster designs.

Validation and extrapolation form another cornerstone of these methods. Data collected in vacuum facilities are compared to simulation outputs to ensure accuracy. Once validated, DSMC and PIC models can extrapolate results to predict performance in actual space conditions, where experimental testing is impossible.

The advantages of DSMC and PIC are numerous. They provide highly accurate insights by capturing particle-level interactions neglected in traditional fluid dynamics models. Their flexibility allows them to simulate a wide range of conditions, from dense plasma regions near the thruster to rarefied gas flows in the far-field plume. By isolating specific physical processes, these methods help identify inefficiencies or potential failure modes, leading to optimized propulsion system designs.

For instance, PIC simulations are extensively used to model ion thruster grid erosion caused by charge-exchange ions, aiding in grid design optimization for durability and efficiency. DSMC models help predict the distribution of neutral xenon in the exhaust plume and its effects on spacecraft surfaces. Similarly, PIC is employed to study electric and magnetic field interactions in Hall-effect thrusters, while DSMC evaluates the impact of backscattering and neutral gas flows on performance. Both methods are integral to understanding how spacecraft surfaces interact with ion and neutral particle deposition, guiding the development of protective coatings and improved satellite configurations.

Despite their benefits, DSMC and PIC simulations face challenges. Their computational intensity requires high-performance computing resources to manage the large number of particles and grid points involved. Validation remains a critical step, as discrepancies between facility conditions and actual space environments can complicate comparisons. Additionally, there is an ongoing need for standardization in simulation protocols and benchmarking tools to ensure consistency across research facilities.

By integrating DSMC and PIC simulations into the design and testing process, researchers can significantly reduce development times, refine propulsion systems, and enhance the reliability of space missions. These methods are indispensable for advancing the capabilities of modern space exploration technologies.

Efforts to streamline test processes and establish standardized practices are ongoing, with organizations like NIST, PTB, and NPL positioned to contribute significantly. Publications addressing thrust measurement, pumping speed calculations, and flow diagnostics already provide valuable guidelines for achieving consistency. As space agencies and metrology institutes refine these practices, the development cycle for ion engines is expected to shorten, reducing the time from conceptualization to market deployment. This advancement is essential for enabling more efficient, reliable propulsion systems for future space exploration missions [228].

Wind Tunnel Testing for Aerodynamics

Wind tunnel testing is a critical phase in the development of rockets, providing valuable insights into their aerodynamic performance under simulated atmospheric conditions. This testing helps engineers refine the rocket's design to ensure stability, reduce drag, optimize trajectory, and prevent potential aerodynamic instabilities during ascent. The process involves using scaled models or full-size rocket components in controlled airflow environments to gather data on aerodynamic forces and moments.

The purpose of wind tunnel testing in rocket development is to understand and optimize the aerodynamic performance of the vehicle. One of its primary objectives is to measure aerodynamic forces such as lift, drag, and side forces acting on the rocket as it moves through the atmosphere. Engineers analyse pressure distributions and identify areas of high stress or instability, which are critical for ensuring the rocket's structural integrity during flight.

Stability and control are also essential aspects of wind tunnel testing. The tests verify the rocket's ability to maintain stability across a range of speeds, angles of attack, and flight phases. Control surfaces, such as fins and flaps, are evaluated to ensure they function effectively, allowing the rocket to adjust its trajectory as needed during ascent or descent.

Wind tunnel testing is crucial for simulating high-speed aerodynamic heating and identifying areas subject to extreme temperatures. This is particularly important in supersonic and hypersonic flight regimes, where shockwave formation can create significant aerodynamic challenges. By studying these effects, engineers can develop solutions to mitigate adverse impacts on the rocket's performance and structural integrity.

Another vital goal is the optimization of the rocket's design. Wind tunnel testing helps refine the vehicle's shape, including nose cone geometry, payload fairings, and interstage sections, to minimize aerodynamic drag and improve efficiency. The experimental data gathered during these tests also validate computational models, such as Computational Fluid Dynamics (CFD), ensuring the accuracy of simulations and guiding further design improvements.

Wind tunnel testing encompasses various types of assessments tailored to specific phases of a rocket's flight and the aerodynamic challenges it encounters. These tests are critical for ensuring the rocket's design can withstand and perform optimally under diverse conditions during ascent and beyond.

Subsonic testing focuses on simulating low-speed conditions, typically representing the initial phase of the rocket's ascent [350, 351]. This phase involves analysing the aerodynamic behaviour of components like fins, fairings, and payload adapters, which play essential roles in stability and drag reduction [43, 350, 351].

Transonic testing addresses the complex aerodynamic behaviour as the rocket approaches and passes through the speed of sound [352, 353]. This regime is challenging due to rapid changes in pressure and airflow, leading to increased drag and potential instability [352, 353]. This testing is crucial for ensuring a smooth transition through this critical speed range [352, 353].

Supersonic and hypersonic testing examines the rocket's behaviour at speeds exceeding Mach 1 and Mach 5, respectively [354, 355]. At these velocities, shockwave interactions and aerodynamic heating become significant challenges [354, 355]. These tests evaluate stability, aerodynamic efficiency, and thermal impact on components like the nose cone and leading edges [354, 355].

Pressure and force distribution testing employs advanced techniques like pressure-sensitive paints or embedded sensors to map surface pressure distributions [352, 356]. This allows identifying the aerodynamic loads on critical components, providing data to optimize the structural design [352, 356].

Each wind tunnel test plays a pivotal role in validating and refining the rocket's aerodynamic performance, bridging the gap between computational models and real-world conditions [350-352, 357, 358]. These tests serve to ensure the rocket can achieve its mission objectives with precision and reliability [350-352, 357, 358].

The wind tunnel testing process is a meticulously planned and executed series of steps designed to evaluate and refine the aerodynamic performance of rockets. This process ensures that the vehicle can meet the challenges of its mission and withstand the dynamic forces encountered during flight.

The first step in the process is model preparation. Scaled-down models of the rocket are manufactured with exceptional precision, often including intricate details like control surfaces, fairings, interstage sections, and payload adapters. These models are designed to replicate the aerodynamic properties of the actual rocket as closely as possible. Materials used for the models must be selected to ensure accurate simulation of aerodynamic behaviours, such as surface roughness and rigidity, to provide reliable test results.

Once the model is ready, it is integrated into the test setup within the wind tunnel. This typically involves mounting the model on a sting or balance system, which provides structural support and measures aerodynamic forces and moments. Advanced sensors and pressure taps are strategically placed on the model's surface to gather detailed data on pressure distributions and airflow interactions. The mounting system ensures that the model is held securely while allowing for precise adjustments to simulate different flight conditions.

The next phase involves flow simulation, where airflow at various velocities is generated within the wind tunnel to mimic the rocket's real-world flight conditions. This phase is critical for evaluating the vehicle's performance across multiple flight scenarios. Parameters such as speed, angle of attack, and yaw angle are systematically varied to simulate different phases of flight, from subsonic lift-off to supersonic and hypersonic velocities during ascent. The wind tunnel's capability to control airflow conditions with precision allows engineers to replicate the complex aerodynamic environment the rocket will experience.

During the testing, data collection is a key focus. Sensors capture measurements of lift, drag, side forces, and moments acting on the rocket, providing quantitative data on its aerodynamic performance. Visualization tools like smoke trails, dye injection, or advanced schlieren photography are employed to reveal airflow patterns, turbulence, and shockwave structures. These visual techniques offer valuable insights into how air interacts with the rocket's surface, highlighting areas of concern such as high drag regions or potential instability.

The final step is analysis and iteration. Engineers analyse the collected data to evaluate the rocket's aerodynamic efficiency, stability, and overall performance. Computational tools are often used to compare test results with theoretical models, ensuring consistency and identifying discrepancies. Based on the findings, design modifications are proposed to address identified issues or improve performance. The iterative nature of this process means that additional wind tunnel tests are often conducted to validate the effectiveness of these design changes.

The wind tunnel testing process is a cornerstone of rocket development, providing engineers with critical insights into the aerodynamic characteristics of their designs. By combining precision modelling, advanced measurement techniques, and rigorous analysis, this process ensures that rockets can achieve optimal performance and reliability during their missions.

Figure 79: A Highly Maneuverable Aircraft Technology (HiMAT) inlet model installed in the test section of the 8-by 6-Foot Supersonic Wind Tunnel at the National Aeronautics and Space Administration (NASA) Lewis Research Center. NASA, CC0, via GetArchive.

Modern rocket programs extensively utilize wind tunnel testing to refine and validate their designs, ensuring optimal performance and safety during various phases of flight.

SpaceX, for example, employs comprehensive wind tunnel testing for its Falcon 9 and Starship vehicles. These tests focus on enhancing aerodynamic efficiency during both ascent and re-entry phases. For the Starship, particular attention is given to its heat shield tiles and body flaps. Hypersonic testing of these components ensures stability and thermal resilience as the vehicle re-enters Earth's atmosphere at extreme velocities, critical for achieving SpaceX's goal of full reusability.

NASA's Space Launch System (SLS) has undergone rigorous wind tunnel testing at transonic and supersonic speeds. These tests are essential to evaluate the aerodynamic effects of its large solid rocket boosters and to assess the integration of the Orion spacecraft. The data collected helps refine the vehicle's design to ensure smooth transitions through critical flight regimes, where aerodynamic forces are at their peak.

Blue Origin also incorporates wind tunnel testing into the development of its New Shepard and New Glenn rockets. These tests optimize the rockets' shapes to minimize drag and enhance aerodynamic performance during ascent and recovery. For New Shepard, testing is crucial to ensure smooth and controlled landings of both the booster and crew capsule.

Similarly, the European Space Agency (ESA) employs wind tunnel testing to perfect the design of the Ariane 6 rocket. Focus areas include the nose cone and payload fairings, which are vital for protecting payloads and reducing aerodynamic loads during ascent. These refinements contribute to the Ariane 6's reliability as a commercial launch vehicle.

Wind tunnel testing provides several key advantages in rocket development. It offers reliable experimental data that complements computational simulations, enabling engineers to cross-verify results. By identifying potential design flaws early in the development process, it reduces costly rework during later stages. Furthermore, wind tunnel testing allows engineers to replicate specific conditions that might be challenging to model with computational tools, such as complex turbulence or shockwave interactions. Ultimately, these tests improve safety and reliability by validating aerodynamic performance across all flight regimes, ensuring that rockets meet the stringent demands of space exploration.

Launch Simulations and Failure Analysis

Launch simulations and failure analysis are integral parts of rocket development and mission preparation. These processes ensure the reliability, safety, and success of a launch by allowing engineers to predict system behaviour, identify potential issues, and implement corrective measures before a rocket ever leaves the ground. Together, these techniques reduce risks and enhance the overall performance of launch vehicles.

Launch simulations are virtual or physical representations of the conditions a rocket will encounter during its mission [359]. They model various phases of flight, including ignition, ascent, staging, and payload deployment, and evaluate the performance of systems under realistic operational conditions [360]. Advanced computational tools and high-fidelity models simulate forces such as gravity, aerodynamics, and thrust, as well as environmental factors like wind shear and atmospheric density [360]. Engineers use launch simulations to validate designs, refine guidance and control algorithms, and test the integration of subsystems [359]. These simulations help predict the rocket's trajectory, stability, and the effectiveness of thrust vectoring systems [360]. Hardware-in-the-loop (HIL) simulations further integrate physical components, such as flight computers or actuators, into the simulation to ensure hardware and software work seamlessly together under simulated flight conditions [359].

Failure analysis involves studying potential or actual failures to identify their causes and develop mitigation strategies [361]. This process begins with identifying failure modes through techniques like fault tree analysis (FTA) and failure mode and effects analysis (FMEA) [362-364]. Engineers assess how individual component failures could propagate and impact the entire system [362-364]. Redundancy is then introduced in critical systems, such as flight control computers or propulsion subsystems, to ensure continued operation even if one component fails [365]. Analytical tools, including finite element analysis (FEA), are used to simulate structural loads and identify weak points in the design [360]. If a failure occurs during testing or an actual mission, failure analysis shifts to diagnosing the root cause [366].

Launch simulations and failure analysis are closely linked, as simulations often include failure scenarios to test a rocket's ability to respond to anomalies [359]. Engineers can simulate events like engine shutdowns or control system failures to evaluate redundancy and fault tolerance [359]. These tests inform design improvements and operational procedures [359]. Both processes also benefit from iterative refinement, as data gathered from simulations and real-world tests feed back into updated models, enabling more accurate predictions and robust designs [359].

Modern launch vehicles, including SpaceX's Starship, Blue Origin's New Glenn, and NASA's Artemis missions, rely heavily on launch simulations and failure analysis to optimize designs and minimize risks [359, 367]. These processes ensure rockets can operate under extreme conditions, providing confidence in their ability to safely transport payloads and crews to space [359, 367]. The integration of advanced tools like artificial intelligence (AI) and machine learning further enhances these processes by analysing vast datasets from simulations and past failures to identify patterns and predict potential issues with unprecedented accuracy [366, 367].

PART 4

Applications of Rockets

Chapter 10

Space Exploration

Rockets for Planetary Exploration (Rovers, Orbiters, Landers)

Rockets designed for planetary exploration play a critical role in advancing our understanding of the solar system. These vehicles are developed to transport and deploy specialized instruments, such as rovers, orbiters, and landers, to distant planetary bodies. Each type of payload serves a unique purpose, enabling humanity to study planetary surfaces, atmospheres, and other celestial phenomena in unprecedented detail. These missions demand robust engineering, cutting-edge technology, and meticulous planning to address the challenges of launching, navigating, and operating in diverse planetary environments.

Orbiters are spacecraft designed to enter a stable orbit around a planet or moon, providing a broad and sustained view of the target body. They are equipped with advanced imaging systems, spectrometers, and communication instruments to map surfaces, study atmospheric compositions, and relay data back to Earth.

Figure 80: A view of the Solar Orbiter STM. This is part of a set of photos showing the Solar Orbiter Structural Thermal Model (a full scale model of the spacecraft, which will be used for testing), taken at the Airbus Defence & Space facility in Stevenage in March 2015. UCL Mathematical and Physical Sciences from London, UK, CC BY 2.0, via Wikimedia Commons.

Examples include NASA's Mars Reconnaissance Orbiter, which captures high-resolution images of the Martian surface, and ESA's Juno mission, which studies Jupiter's magnetic fields and atmospheric dynamics. The deployment of orbiters requires precise rocket launches and complex trajectory planning, including gravitational assist manoeuvres, to ensure successful insertion into the target orbit.

Figure 81: At the Astrotech facility in Titusville, Fla., a crane moves NASA's Lunar Reconnaissance Orbiter, or LRO, toward a stand in the foreground. NASA, CC0, via Picryl.

Rovers are mobile exploration platforms that traverse planetary surfaces, conducting detailed analyses of geology, mineralogy, and potential biosignatures. These vehicles are equipped with robotic arms, cameras, spectrometers, and drills to collect and analyse samples. Notable examples include NASA's Perseverance rover, which is exploring the Martian surface for signs of ancient life, and China's Yutu-2 rover, which operates on the Moon's far side. Rockets carrying rovers must ensure a soft and accurate landing on the planetary surface, often using descent stages, parachutes, or retropropulsion systems to minimize impact forces.

Figure 82: NASA's Perseverance Mars rover took these 2 selfie versions over a rock nicknamed "Rochette," on September 10, 2021, the 198th Martian day, or sol of the mission. Two holes are visible where the rover used its robotic arm to drill rock core samples. . NASA/JPL-Caltech, Public domain, via Wikimedia Commons.

Figure 83: Yutu-2 rover on lunar surface. CSNA/Siyu Zhang/Kevin M. Gill, CC BY 2.0, via Wikimedia Commons.

Landers are stationary spacecraft designed to touch down on a planet or moon and conduct localized studies. They are often equipped with instruments to analyse soil composition, seismic activity, or atmospheric properties. Examples include NASA's InSight lander, which studies Mars' interior, and ESA's Huygens probe, which landed on Saturn's moon Titan to study its atmosphere and surface. Landers rely on precision entry, descent, and landing (EDL) systems to ensure they arrive intact and in the correct orientation for operation.

Figure 84: This is NASA InSight's first full selfie on Mars. It displays the lander's solar panels and deck. On top of the deck are its science instruments, weather sensor booms and UHF antenna. NASA/JPL-Caltech, Public domain, via Wikimedia Commons.

Rocket systems for planetary exploration are engineered to overcome significant challenges. These include achieving escape velocity to break free from Earth's gravity, navigating vast interplanetary distances, and surviving the harsh environments of space. Propulsion systems are optimized for high efficiency, often using liquid fuels or hybrid engines capable of long-duration burns for course corrections. Navigation and guidance systems incorporate autonomous technologies to manage in-flight adjustments, as the time delay in communication with Earth makes real-time control impractical.

For missions targeting planetary bodies with atmospheres, rockets must also account for aerodynamic forces during entry. Heat shields and ablative materials protect payloads from the intense heat generated during atmospheric entry, while parachutes or retro-thrusters control the descent speed. For airless bodies, precision landing is achieved using advanced radar and optical navigation systems.

Rocket Design and Construction Fundamentals

Planetary exploration missions often involve international collaboration and are built on decades of scientific and engineering advancements. Agencies like NASA, ESA, Roscosmos, and CNSA play pivotal roles in developing rockets and spacecraft for these missions. For example, the Mars Sample Return mission, a collaboration between NASA and ESA, plans to retrieve samples collected by Perseverance and return them to Earth using a combination of orbiters, landers, and ascent vehicles.

Rockets for planetary exploration embody humanity's quest to explore the unknown. They not only expand our scientific knowledge but also inspire innovations in technology, engineering, and international cooperation. Through these missions, we continue to uncover the mysteries of our solar system and prepare for future endeavours, including crewed missions to other planets.

Rocket systems used to transport and deploy rovers, orbiters, and landers are meticulously engineered to ensure that these payloads can achieve their mission objectives in the harsh and diverse environments of space. These systems encompass launch vehicles, propulsion systems, guidance and navigation technologies, and entry, descent, and landing mechanisms tailored to the unique requirements of each type of payload.

The first critical component of any planetary exploration mission is the launch vehicle. Rockets like NASA's Atlas V, SpaceX's Falcon Heavy, or ESA's Ariane 5 are commonly used to carry spacecraft to their intended destinations. These rockets are equipped with powerful first and second stages to achieve the necessary escape velocity, enabling payloads to break free from Earth's gravity. Launch vehicles for such missions are chosen based on the mass of the payload, the distance to the target body, and the trajectory required.

For instance, the Atlas V rocket launched NASA's Mars rovers like Perseverance and Curiosity, while the Ariane 5 was instrumental in deploying ESA's Rosetta orbiter to study Comet 67P. These rockets often feature upper stages with cryogenic engines, like the Centaur upper stage, which provide precise burns to place the payload into interplanetary trajectories.

Once in space, propulsion systems onboard the spacecraft handle the journey to the target planet or moon. Chemical propulsion systems, such as bipropellant engines, are often used for major trajectory corrections, while electric propulsion systems, like ion thrusters, are utilized for fine adjustments over long durations.

For example, the Dawn spacecraft, which explored the asteroid belt, used ion propulsion to travel between Vesta and Ceres. These systems offer high efficiency, consuming less fuel while generating continuous thrust to gradually adjust the spacecraft's trajectory. Propulsion systems are crucial for orbiters to enter stable orbits around their target bodies and for landers to execute precision landings.

Guidance and navigation systems ensure that the rocket and spacecraft remain on their planned trajectories. During the launch phase, onboard computers and sensors monitor the rocket's velocity, altitude, and position, making real-time adjustments to maintain the correct path. Advanced inertial measurement units (IMUs) and star trackers are used for mid-course corrections during interplanetary travel, while deep-space navigation relies on signals from Earth-based Deep Space Network (DSN) antennas to triangulate the spacecraft's position.

For missions involving rovers or landers, these systems play a vital role in ensuring precise entry, descent, and landing. NASA's Perseverance rover, for instance, employed a sophisticated Terrain-Relative Navigation (TRN) system, using onboard cameras to compare the landing site with preloaded maps, enabling the spacecraft to autonomously steer toward a safe landing zone.

Rovers and landers face the critical challenge of surviving the transition from space to the surface of a planet or moon. Entry, descent, and landing (EDL) systems are specifically designed to manage this process, which varies depending on the target environment. For planets with atmospheres, such as Mars, EDL systems often include heat shields, parachutes, and retropropulsion. Heat shields protect the payload from the intense heat of atmospheric entry, while parachutes slow the descent. In some cases, rockets or sky cranes are used for final touchdown, as demonstrated by the Curiosity and Perseverance rovers.

On airless bodies like the Moon, descent systems rely entirely on retropropulsion and precision guidance. NASA's Apollo Lunar Module used throttleable descent engines to land astronauts safely on the lunar surface, while recent missions like China's Chang'e landers employed advanced radar and optical systems for autonomous landing.

Retropropulsion is a widely used technique in space missions to control the descent and landing of spacecraft or rockets [368]. This method involves firing engines or thrusters in the opposite direction of motion to counteract the vehicle's velocity, enabling safe landing or controlled manoeuvring [368].

The mechanics of retropropulsion involve generating thrust in the opposite direction of travel, which creates a deceleration force [368]. This is achieved through the controlled ignition of rocket engines or thrusters that expel propellant to produce thrust [368]. The amount of deceleration depends on factors such as the thrust produced, the mass of the vehicle, and its velocity [368]. Retropropulsion is particularly effective in environments where aerodynamic drag is insufficient or unavailable, such as on airless moons or during the final stages of a descent through a planet's atmosphere [368].

Retropropulsion plays a crucial role in planetary exploration missions, where precision and safety are paramount [369, 370]. For example, NASA's Mars missions, such as the landing of the Perseverance and Curiosity rovers, utilized retropropulsion during their descent [369]. After a parachute reduced initial velocity, retropropulsion was employed to slow the descent further and facilitate the deployment of the sky crane system, which gently lowered the rovers to the surface [369]. For airless bodies like the Moon, where no atmosphere exists to assist with deceleration, retropropulsion is the primary method for controlling descent, as demonstrated by NASA's Apollo Lunar Module [369].

Modern advancements in spaceflight have also popularized the use of retropropulsion for enabling the reuse of rocket components, such as SpaceX's Falcon 9 and Falcon Heavy rockets [299, 368]. These boosters utilize retropropulsion to land back on Earth after launch, reorienting themselves mid-flight and firing their engines to reduce speed, stabilize their descent, and make precision landings [299, 368].

Retropropulsion is also used for orbital adjustments and controlled deorbiting of spacecraft [368]. When a satellite or spacecraft needs to lower its orbit or prepare for atmospheric re-entry, retropropulsion can precisely reduce orbital velocity, which is critical for missions that require targeted reentry or disposal of satellites to minimize space debris [368].

However, retropropulsion presents several engineering challenges, such as the consumption of large amounts of propellant, which impacts the spacecraft's overall design and payload capacity [368]. Heat management is another significant concern, as the engines firing during descent generate intense heat, which, combined with atmospheric friction, can stress the spacecraft's thermal protection systems [368]. Precision control is also vital, particularly for missions that require pinpoint landings, requiring advanced guidance and navigation systems, coupled with real-time sensor feedback, to modulate engine thrust and maintain stability during retropropulsion manoeuvres [368].

Different payloads require unique systems to ensure mission success. Orbiters, for instance, are equipped with thrusters for orbital insertion and station-keeping, maintaining their position relative to the target planet. Landers include legs or platforms designed to absorb the impact of touchdown, and rovers feature deployment systems that allow them to safely disembark from their landing platforms.

For example, the Mars Pathfinder mission used airbags to cushion the landing of its rover, Sojourner, while ESA's Philae lander on the Rosetta mission deployed harpoons and screws to anchor itself on the surface of a comet.

Planetary exploration missions are high-stakes endeavours with little room for error. Rocket systems are built with extensive redundancy to ensure reliability. This includes duplicate propulsion systems, redundant guidance computers, and backup power supplies to mitigate the risks of failure during critical mission phases. For instance, NASA's Viking landers on Mars employed redundant fuel tanks and engines to ensure a safe descent and landing.

Recent advancements in rocket technology have introduced innovations like reusable boosters and modular spacecraft. SpaceX's Falcon 9 has demonstrated the feasibility of reusing first-stage boosters, significantly reducing launch costs for planetary missions. Similarly, modular spacecraft architectures, such as ESA's Solar Orbiter, allow components to be integrated and tested independently, enhancing mission flexibility.

Human Spaceflight: ISS, Moon, and Mars Missions

Human spaceflight requires highly advanced and reliable rocket systems to safely transport astronauts and their payloads to destinations like the International Space Station (ISS), the Moon, or Mars. These systems include launch vehicles, spacecraft, propulsion, life support, and entry-descent-landing (EDL) technologies, each tailored to the specific challenges of the mission.

Human spaceflight to the International Space Station (ISS) relies on a range of advanced launch systems designed to safely transport astronauts and supplies. The Russian Soyuz Rocket has been a cornerstone of ISS missions for decades, offering reliability through its three-stage design powered by kerosene and liquid oxygen. The Soyuz spacecraft itself consists of an orbital module, a descent module, and a service module, ensuring secure transport and serving as an escape module while docked. Another significant player is SpaceX's Falcon 9, which operates under NASA's Commercial Crew Program. The partially reusable Falcon 9, equipped with advanced Merlin engines, has become a cost-effective and efficient solution for ISS resupply and crew transportation. Similarly, the Atlas V, developed by United Launch Alliance, supports ISS missions by launching Boeing's CST-100 Starliner spacecraft, which is designed to ferry astronauts and cargo with advanced safety and avionics systems.

Figure 85: The International Space Station as seen from the departing Space Shuttle Discovery during STS-119. National Aeronautics and Space Administration (Q23548), Public domain, via Wikimedia Commons.

For lunar missions, the need for more powerful rockets is paramount to overcome Earth's gravity and reach the Moon. NASA's Space Launch System (SLS) is the primary heavy-lift vehicle for Artemis missions, capable of transporting astronauts aboard the Orion spacecraft. This rocket leverages immense thrust from solid rocket boosters and RS-25 engines fuelled by liquid hydrogen and liquid oxygen. SpaceX's Starship, currently under development, represents another revolutionary approach to lunar missions. Designed as a fully reusable spacecraft, Starship offers unprecedented payload capacity and in-orbit refuelling capabilities, making it a pivotal part of NASA's Human Landing System (HLS) program for Moon exploration.

Missions to Mars present even greater challenges, demanding advanced systems for long-duration flights and substantial payload capacities. SpaceX's Starship is central to these efforts, designed with habitats, cargo bays, and in-situ resource utilization (ISRU) capabilities to produce fuel on Mars for return journeys. Additionally, space agencies are investigating modular heavy-lift rockets and emerging technologies like Nuclear Thermal Propulsion (NTP) to increase efficiency and reduce travel time to the Red Planet.

Nuclear Thermal Propulsion (NTP) is a technology that holds significant potential for advancing space exploration [61, 371, 372]. At the core of an NTP system is a nuclear reactor that generates heat through controlled nuclear fission [371-373]. This heat is then used to rapidly expand and expel a propellant, typically liquid hydrogen, through a rocket nozzle to produce thrust [371-373].

One of the primary advantages of NTP is its exceptional efficiency, with a specific impulse two to three times greater than that of chemical rockets [61, 371]. This increased efficiency translates to reduced propellant requirements, enabling longer-duration missions and shorter travel times [371]. For instance, a Mars mission using NTP could cut the travel time from the typical 9 months to approximately 4-6 months [61, 372].

NTP systems also offer flexibility, with a high thrust-to-weight ratio that allows for rapid acceleration and deceleration, enabling manoeuvres such as orbital insertion, mid-course corrections, and planetary landing assist [61, 371-373]. This versatility makes NTP a compelling choice for deep-space exploration missions, including sending spacecraft to the outer planets or even interstellar probes [61, 372].

Despite the promising potential of NTP, several challenges must be addressed [371]. Developing robust reactor designs that can withstand the harsh environment of space, ensuring material durability at high temperatures and radiation levels, and addressing safety and regulatory concerns are critical areas of focus [371, 372, 374-376].

The historical context of NTP technology dates back to the 1960s and 1970s, when NASA and the U.S. Atomic Energy Commission conducted extensive research and testing under the NERVA program [62, 374, 377]. While the program was eventually discontinued, the renewed interest in NTP, driven by ambitious goals for human exploration of Mars and beyond, has led to ongoing research and development efforts by NASA, industry partners, and other countries [61, 70, 371, 372].

Nuclear Thermal Propulsion represents a transformative step forward in space exploration, offering unparalleled efficiency, performance, and the potential to enable safer and more efficient deep-space missions [61, 371, 372]. While significant technical and regulatory challenges remain, the ongoing research and development efforts are paving the way for the future use of this technology [61, 70, 371, 372].

Spacecraft systems are tailored to mission objectives and environments, whether for Earth orbit, lunar operations, or deep space exploration. For ISS missions, the Soyuz spacecraft has long been a reliable option, featuring integrated modules for transport, habitation, and service functions. SpaceX's Dragon spacecraft, fully autonomous with a capacity for seven astronauts, incorporates advanced life support systems and is designed for reusability. Boeing's Starliner offers similar capabilities, serving as a safe and efficient transport vehicle to the ISS.

Lunar missions require deep-space-capable spacecraft, such as NASA's Orion spacecraft. Orion combines a robust crew module with the European Service Module (ESM) to provide propulsion, power, and life support for Artemis missions. Meanwhile, SpaceX's Human Landing System (HLS) variant of Starship is designed to carry astronauts between lunar orbit and the Moon's surface, supporting extended operations.

For Mars missions, spacecraft systems must sustain astronauts for months in transit. SpaceX's Starship is again a central player, featuring advanced life support and habitat designs for long-term survival. Additionally, modular transit habitats equipped with radiation shielding and artificial gravity systems are under development to ensure crew safety and mission success.

Propulsion systems play a crucial role in human spaceflight, providing the thrust needed to escape Earth's gravity and enabling efficient transit in space. Chemical propulsion remains the primary method for launch and ascent, with liquid or solid propellants powering engines like the RS-25 on the SLS and the Raptor on SpaceX's Starship. For interplanetary missions, electric propulsion, such as ion thrusters, is gaining attention for its fuel efficiency and capability for long-duration burns, particularly for cargo transport. Nuclear Thermal Propulsion (NTP), which uses a nuclear reactor to heat propellants, is also being explored for Mars missions, offering higher efficiency and reduced travel times compared to chemical systems.

Human spaceflight demands robust life support systems to maintain astronaut health and safety. Atmosphere control systems regulate oxygen levels, remove carbon dioxide, and manage humidity. Water recovery systems recycle water from waste and condensation, ensuring a sustainable supply for long missions. Food systems rely on freeze-dried and packaged provisions to sustain astronauts during extended periods in space. These systems are critical for missions to the ISS, Moon, and Mars, where resupply opportunities are limited or non-existent.

The atmosphere control system is one of the most critical components of a life support system, as it is responsible for creating and maintaining a habitable environment for astronauts in space. This system replicates Earth's breathable air composition while efficiently removing harmful gases and regulating humidity levels.

Atmosphere control systems are designed to maintain oxygen levels at approximately 21% and use nitrogen as an inert buffer gas to closely simulate Earth's atmospheric conditions. These systems must account for the gas exchange processes caused by human respiration, including oxygen consumption and carbon dioxide production. Excess carbon dioxide is removed using chemical scrubbers, such as lithium hydroxide, or through regenerative systems like the Sabatier process, which converts CO_2 into water and methane, ensuring a continuous supply of clean air.

The construction of atmosphere control systems incorporates a modular design to enhance flexibility and functionality. They include storage tanks for oxygen and nitrogen, CO_2 scrubbers, and humidity control units. Advanced spacecraft, such as NASA's Orion and SpaceX's Dragon, employ high-efficiency sensors to continuously monitor gas concentrations, providing real-time feedback for necessary adjustments. For long-duration missions, oxygen generation through water electrolysis, which splits water into oxygen and hydrogen, is a vital feature to ensure sustainability.

Extensive testing is performed to validate the performance and reliability of atmosphere control systems. Vacuum chamber simulations replicate microgravity and the spacecraft's cabin conditions, allowing engineers to assess the system's capability to maintain a stable atmosphere. Fault tolerance and redundancy are rigorously evaluated to ensure that the system operates continuously, even in the event of an emergency, safeguarding astronaut health and mission success.

Water is an essential resource in space, and recovery systems are meticulously designed to recycle every available source, including condensation, urine, and sweat. This recycling ensures a sustainable supply, which is particularly critical for long-duration missions where resupply opportunities are limited or non-existent.

Water recovery systems are developed with advanced multi-stage purification processes. These typically include filtration to remove particulate matter, distillation to separate impurities through heat, and microbial treatment to eliminate harmful microorganisms. NASA's Water Recovery System (WRS) aboard the International Space Station exemplifies this technology, achieving remarkable efficiency by reclaiming over 90% of the water used aboard the station.

The construction of these systems relies on materials that are lightweight and corrosion-resistant, such as titanium and specialized polymers. These materials are ideal for handling the high-purity water required in space environments and for preventing contamination. The systems themselves comprise various components, including pumps to circulate water, filters to trap impurities, distillers for purification, and ultraviolet (UV) sterilization units to maintain water safety. Advanced systems may also incorporate forward osmosis or electrochemical processes to enhance efficiency and improve overall water recovery rates.

Once processed, the reclaimed water is stored in dedicated tanks, where real-time sensors continuously monitor its purity. This water is then distributed throughout the spacecraft for various essential uses, such as drinking, food rehydration, and even oxygen generation through electrolysis. By integrating these water recovery systems seamlessly into spacecraft operations, missions can maintain a reliable and sustainable water supply, supporting both the crew and critical systems during extended stays in space.

Food systems for space missions are designed to meet the unique challenges of providing adequate nutrition in environments with limited storage, long shelf-life requirements, and the complexities of meal preparation in microgravity or reduced gravity. These systems must balance the need for compact, lightweight provisions with the necessity of maintaining taste, nutritional value, and variety over extended missions.

Astronaut diets are meticulously planned to include a mix of freeze-dried, vacuum-sealed, and thermostabilized food items. These options are lightweight and compact, allowing for efficient storage while retaining essential nutrients and flavours over long durations. For future missions, particularly to the Moon or Mars, the cultivation of fresh produce using hydroponic or aeroponic systems aboard spacecraft or habitats is being explored as a way to supplement stored provisions with fresh vegetables and herbs, adding both nutritional and psychological benefits.

The construction of food storage systems involves the use of advanced thermal insulation and vacuum-sealing technologies to preserve the contents for the duration of the mission. Packaging materials are engineered to minimize waste and weight while being user-friendly in microgravity. For instance, rehydratable meal kits are equipped with specialized nozzles that allow astronauts to inject water into the packaging to prepare their meals conveniently and safely. This design ensures that meal preparation is both efficient and compatible with the constraints of space.

Food systems undergo extensive testing to ensure they meet the stringent demands of space travel. Nutritional analysis ensures that dietary requirements are met, while taste testing helps maintain variety and satisfaction for the crew. Shelf-life studies determine the longevity of food items under storage conditions, and vibration and vacuum tests verify their durability during launch and in the space environment. Additionally, packaging materials are tested for their resistance to radiation exposure and extreme temperature fluctuations, ensuring that food remains safe and consumable throughout the mission. These comprehensive measures ensure that food systems support astronaut health and performance, even on the longest and most challenging spaceflights.

For missions to the ISS, Moon, or Mars, life support systems are designed to operate autonomously, with minimal maintenance. These systems are integrated into the spacecraft or habitat's architecture, ensuring seamless operation under harsh conditions. Advanced computer models simulate mission scenarios to optimize system layouts, redundancy, and interdependence among subsystems. For example, water recovered from the humidity control system can be directed to oxygen generation units, closing resource loops.

As missions extend beyond low Earth orbit to the Moon and Mars, life support systems are evolving to incorporate in-situ resource utilization (ISRU). For instance, Martian missions may leverage local water ice for oxygen and water production. Additionally, bioregenerative systems, including algae or microbial reactors, are being explored to create self-sustaining ecosystems capable of producing oxygen, recycling water, and growing food.

The entry, descent, and landing phase is one of the most challenging aspects of space missions, particularly for lunar and Mars exploration. Parachute systems are commonly used to decelerate spacecraft during Earth or Mars re-entry, while retropropulsion engines enable soft landings, as demonstrated by SpaceX's Falcon 9 and Mars landers. Heat shields protect spacecraft from the intense frictional heat during atmospheric entry, dissipating energy and ensuring structural integrity. For precision landings, NASA has employed the sky crane system, used successfully for the Curiosity and Perseverance rovers, to lower payloads gently onto the Martian surface.

Human spaceflight relies on the integration of advanced rocket systems, spacecraft technologies, and support infrastructure to achieve mission objectives. Whether servicing the ISS, landing on the Moon, or venturing to Mars, each component is carefully designed to ensure safety, reliability, and efficiency in some of the most extreme environments known to humanity.

Deep Space Missions: Interstellar Probes and Telescopes

Rocket systems for deep space missions, such as interstellar probes and telescopes, are specialized to handle the unique challenges of delivering payloads beyond Earth's orbit and into interstellar or deep-space environments. These systems integrate advanced propulsion technologies, precise navigation capabilities, and robust launch vehicles to ensure mission success in the vast and remote regions of space.

Deep space missions rely on powerful launch vehicles to overcome Earth's gravity and deliver payloads with precision to their intended trajectories. These missions typically require heavy-lift rockets capable of providing the significant thrust needed to propel large payloads, such as interstellar probes and space telescopes, into high-energy escape orbits.

Rockets like NASA's Space Launch System (SLS), SpaceX's Falcon Heavy, and the Ariane 5 and 6 are commonly utilized for these demanding tasks. Each of these systems is engineered to handle the complexities of launching and delivering sophisticated payloads for deep space exploration. For example, the James Webb Space Telescope (JWST) was launched aboard an Ariane 5 rocket, a mission that required precise trajectory adjustments to place the telescope into orbit at the second Lagrange point (L2), approximately 1.5 million kilometres from Earth. Similarly, NASA's Voyager and Pioneer missions relied on powerful launch systems to achieve the necessary escape velocities for interstellar exploration, enabling these probes to travel beyond the solar system and provide groundbreaking data about distant regions of space.

Once in space, specialized propulsion systems take over to navigate and propel the spacecraft toward its deep-space destination. Among these systems, chemical propulsion plays a critical role in executing high-thrust manoeuvres essential for precise trajectory corrections and planetary flybys. These systems use chemical reactions to generate rapid and powerful thrust, making them ideal for situations requiring quick and decisive adjustments.

For instance, NASA's Voyager probes relied on small chemical thrusters for course corrections and attitude control during their historic journeys through the outer planets. These thrusters allowed the spacecraft to adjust their trajectories and maintain proper orientation for scientific observations and communication with Earth, ensuring the success of their extended missions across vast distances in the solar system.

Electric propulsion systems, including ion and Hall-effect thrusters, are fundamental in deep-space exploration due to their high efficiency and capability for sustained thrust over long durations. These systems operate by ionizing a propellant, such as xenon, and accelerating the ions using electric or magnetic fields to generate thrust. This technology enables spacecraft to perform continuous, low-thrust manoeuvres, making it ideal for missions requiring significant orbital adjustments or travel across vast distances. A prominent example is NASA's Dawn mission, which used ion propulsion to explore the asteroid belt, visiting the protoplanet Vesta and the dwarf planet Ceres. Similarly, the BepiColombo mission employs electric propulsion for complex orbital adjustments during its journey to Mercury.

Gravity assist manoeuvres are another critical technique used in deep-space missions. By utilizing the gravitational pull of planets, spacecraft gain additional velocity and adjust their trajectories without expending significant onboard propellant. This method has been employed effectively in numerous missions. The Voyager probes used the gravitational fields of Jupiter and Saturn to accelerate toward interstellar space, while the New Horizons mission to Pluto leveraged Jupiter's gravity to dramatically increase its speed, enabling a timely arrival at its distant target.

For missions requiring extended durations or targeting extreme distances, nuclear propulsion systems, such as nuclear thermal propulsion (NTP) or nuclear electric propulsion (NEP), are under development. These systems offer high efficiency and the potential for faster travel times to interstellar destinations. In addition to propulsion, deep-space missions often depend on nuclear-based power systems. Radioisotope thermoelectric generators (RTGs) are commonly used to provide a reliable power source in environments where solar energy is insufficient, ensuring the functionality of spacecraft systems far from the Sun. These technologies collectively enable humanity's reach into the farthest corners of our solar system and beyond.

Payload systems for deep-space missions are designed to carry sophisticated scientific instruments or observatories tailored to specific mission objectives. These systems include space telescopes and interstellar probes, each requiring specialized engineering to achieve their goals and withstand the challenges of deep-space environments.

Space telescopes, such as the James Webb Space Telescope (JWST), are equipped with advanced technologies to operate effectively in the cold, radiation-rich vacuum of space. These include highly precise pointing systems for accurate observation, cutting-edge optical components for capturing distant celestial phenomena, and thermal control systems to maintain functionality in extreme temperatures. Rockets launching such payloads must deliver them to stable orbits, such as Lagrange points, where gravitational forces balance to minimize the energy needed for station-keeping, allowing telescopes to focus on scientific exploration.

Interstellar probes, like Voyager 1 and 2, are designed for missions that extend beyond our solar system. These probes require highly reliable communication systems to transmit data over vast distances and must be equipped with durable instruments capable of operating autonomously for decades. Launch vehicles for these missions must achieve escape velocities sufficient to leave the gravitational pull of the Sun, ensuring that the probes can continue their journeys into interstellar space. These payload systems represent the cutting edge of technology, enabling humanity to study distant worlds and the broader universe.

Precise navigation and communication systems are essential for deep-space missions to ensure that spacecraft reach their intended destinations and maintain reliable communication with Earth. These systems must operate effectively over vast distances, often requiring advanced technologies to handle the challenges of interplanetary and interstellar travel.

Navigation systems rely on a combination of tools and technologies to maintain the spacecraft's orientation and trajectory. Star trackers and gyroscopes are commonly used for attitude control and precise orientation. Star trackers capture images of stars and compare them to onboard star maps to determine the spacecraft's position and orientation, while gyroscopes measure angular velocity to maintain stability and assist with manoeuvring. These systems work in tandem to provide accurate navigation, even in the absence of direct Earth-based control.

Communication with spacecraft in deep space is achieved using the Deep Space Network (DSN), a global system of ground-based antennas that supports long-distance communication. High-gain antennas on the spacecraft are designed to transmit data over billions of kilometres, ensuring that mission-critical information, such as scientific observations and health status, can be received on Earth. The DSN's ability to handle weak signals and its precise tracking capabilities make it indispensable for deep-space exploration.

Autonomous navigation systems are also becoming increasingly important for missions where delays in communication with Earth due to the vast distances involved could hinder real-time decision-making. These systems use celestial bodies, such as planets, moons, and stars, as reference points to calculate the spacecraft's position and adjust its trajectory autonomously. The New Horizons mission, which successfully navigated to Pluto and beyond, exemplifies the effectiveness of such systems. By combining advanced navigation and communication technologies, deep-space missions can achieve unprecedented accuracy and reliability in exploring the outer reaches of our solar system and beyond.

Deep-space missions operate in environments with extreme temperatures and intense radiation. Rocket systems are designed to ensure that payloads, such as space telescopes or probes, are insulated and protected during launch and transit. Thermal shielding and radiation-hardened electronics are employed to safeguard instruments during interplanetary and interstellar travel.

Chapter 11
Satellite Deployment

Types of Satellites (Communication, GPS, Weather, Earth Observation)

Satellites are artificial objects launched into orbit to perform various functions that support modern life and scientific exploration. They can be categorized based on their specific purpose and the type of data or services they provide. Key types include communication satellites, GPS satellites, weather satellites, and Earth observation satellites, each playing a distinct role in technology, navigation, meteorology, and environmental monitoring.

An artificial satellite is a human-made object, typically a spacecraft, placed into orbit around a celestial body. Satellites serve diverse purposes, including relaying communication, forecasting weather, providing navigation through GPS, broadcasting, conducting scientific research, and observing Earth. Military applications include reconnaissance, early warning, signals intelligence, and potentially weapon delivery. Other satellites include rocket stages that position payloads into orbit and defunct satellites that have become non-functional.

Most satellites, except passive ones, have onboard power generation systems such as solar panels or radioisotope thermoelectric generators (RTGs) to operate their equipment. Communication with ground stations is typically facilitated through transponders. Many satellites use standardized designs, or "buses," to reduce costs and streamline production, with CubeSats being a popular small-scale model. Groups of similar satellites often operate in coordinated constellations. Due to the high cost of launching satellites into space, they are engineered to be lightweight yet durable. Communication satellites, in particular, function as radio relay stations, equipped with multiple transponders, each supporting significant bandwidth.

Satellites are launched into orbit by rockets, positioned high enough to avoid atmospheric drag and orbital decay. Once in orbit, they can adjust or maintain their trajectory using propulsion systems such as chemical or ion thrusters. As of 2018, approximately 90% of Earth's satellites operate in low Earth orbit (LEO) or geostationary orbit. Geostationary satellites remain stationary relative to a fixed point on the ground, while Sun-synchronous orbits are preferred by imaging satellites for consistent lighting conditions during global scans. With the growing number of satellites and debris in space, the risk of collisions has become increasingly significant. While most satellites orbit Earth, others are placed around celestial bodies such as the Moon or Mars, or follow complex orbits like halo or Lissajous paths.

Earth observation satellites collect data for reconnaissance, mapping, and monitoring natural phenomena, such as weather, oceans, and forests. Space telescopes use the vacuum of space to observe celestial objects across the electromagnetic spectrum. Communication satellites enable data transmission to remote areas, and their predictable orbits and signal delays are vital for navigation systems like GPS. Space probes are specialized satellites designed for robotic exploration beyond Earth, while space stations function as crewed satellites for extended missions.

The Soviet Union's launch of Sputnik 1 on October 4, 1957, marked the beginning of artificial satellite technology [378]. As of December 31, 2022, there were 6,718 operational satellites orbiting Earth, with 4,529 owned by the United States (3,996 of them commercial), 590 by China, 174 by Russia, and 1,425 by other nations [379].

The launch of Sputnik 1 heralded the era of satellite remote sensing [378]. Since then, hundreds of Earth-observing satellites have been launched, providing both iconic views and unprecedented scientific insights [378]. Subsequent satellites, such as Explorer 7 and the TIROS series, initiated the study of Earth's radiation budget and climate from satellites [380].

The development of satellite technology has had a significant impact on various fields, including search and rescue operations [381], space weather monitoring [382], and ground-based astronomical observations [383]. Satellite constellations, such as Starlink, are being proposed in large numbers to provide communication, global monitoring, and space observation [384]. However, these constellations also pose challenges, such as the need for coordinated efforts to address technological limits in spacecraft operations and space traffic management [384].

The increasing number of satellites in orbit has also led to concerns about their impact on ground-based astronomical observations. Efforts are being made to mitigate the effects of satellite brightness and trails on astronomical surveys, such as the Rubin Observatory's Legacy Survey of Space and Time [385].

Orbit and Altitude Control in Satellites

Orbit and altitude control systems are essential for maintaining a satellite's position and orientation in space. Without these systems, satellites would drift off-course due to gravitational influences, atmospheric drag (for low Earth orbit satellites), and radiation pressure, rendering them ineffective for communication, observation, or scientific research. Satellites rely on propulsion systems for orbit maintenance and reaction mechanisms for attitude control, ensuring precise alignment with their intended targets.

Satellites in low Earth orbit (LEO) are most influenced by Earth's gravitational variations, magnetic field, and solar radiation pressure. To mitigate these effects, satellites use chemical or ion propulsion systems to adjust or maintain their orbits. Satellites in higher orbits, such as geostationary orbit (GEO), experience more pronounced effects from the gravitational pull of other celestial bodies like the Moon and the Sun. Ultra-reflective coatings are often applied to satellites to minimize damage from ultraviolet (UV) radiation, protecting sensitive instruments and ensuring longevity.

Chemical thrusters, commonly used for orbital adjustments, utilize monopropellant or bipropellant systems. Monopropellant systems rely on single-component fuels like hydrazine, which combusts when exposed to a catalyst. Bipropellant systems use two components, such as monomethylhydrazine (MMH) and dinitrogen tetroxide (N_2O_4), which ignite spontaneously upon contact due to their hypergolic properties. These systems provide the high thrust necessary for rapid orbital adjustments but are less efficient in terms of propellant usage.

Ion propulsion systems, such as Hall-effect thrusters, offer an alternative for satellites requiring high propellant efficiency. These systems generate thrust by ionizing a propellant, usually xenon, and accelerating the ions through an electric field. Xenon is ideal for ion propulsion due to its high atomic mass, inert nature, and ability to be stored as a high-pressure liquid. Although ion thrusters produce significantly lower thrust (approximately 0.5 Newtons), they can operate for extended durations, making them suitable for precise, long-term orbital manoeuvres and deep-space missions.

Power Systems in Satellites

Satellites require reliable power sources to operate their onboard systems, including propulsion, communication, and scientific instruments. The majority of satellites generate electricity using solar panels, which convert sunlight into electrical energy. Satellites designed for deep-space missions, where sunlight is limited, often rely on radioisotope thermoelectric generators (RTGs), which produce power through the decay of radioactive isotopes.

Solar panels on satellites are mounted using slip rings, which allow the panels to rotate and align perpendicularly to the Sun, maximizing power generation. This rotational capability ensures optimal energy absorption throughout the satellite's orbit. However, because sunlight is blocked during periods such as launch or when the satellite passes through Earth's shadow, onboard batteries are essential for storing energy. Modern satellites predominantly use lithium-ion batteries due to their high energy density, lightweight nature, and long cycle life. Older designs often employed nickel-hydrogen batteries, which, while robust, have largely been replaced by more efficient alternatives.

These power systems are carefully integrated to ensure continuous energy availability, enabling satellites to maintain communication, propulsion, and operational stability regardless of their position relative to the Sun.

Types of Satellites by Orbit

Satellites are placed in specific orbits based on their intended function, coverage needs, and mission requirements. These orbits determine how the satellite interacts with the Earth, other celestial bodies, and its operational environment. Satellites can orbit Earth or follow interplanetary trajectories toward destinations in the solar system. The most common classifications of satellite orbits are based on their altitude and the characteristics of their motion relative to Earth. The primary types of satellite orbits include Low Earth Orbit (LEO), Medium Earth Orbit (MEO), Geostationary Orbit (GEO), Sun-Synchronous Orbit (SSO), and Geostationary Transfer Orbit (GTO).

Low Earth Orbit (LEO) Satellites: Satellites in Low Earth Orbit operate at altitudes between approximately 160 to 1,500 kilometres above Earth's surface. These satellites complete a full orbit in about 90 to 120 minutes, allowing them to circle the planet up to 16 times per day. LEO is particularly well-suited for applications like remote sensing, high-resolution Earth observation, scientific research, and rapid data transmission.

LEO provides flexibility in satellite paths and enables frequent coverage of specific areas. However, its proximity to Earth limits the area covered by each satellite. To address this, satellite constellations, like SpaceX's Starlink or EOS SAT, are deployed in LEO to provide global coverage by working collaboratively. Such networks are critical for sectors like precision agriculture, environmental monitoring, and global internet connectivity.

Medium Earth Orbit (MEO) Satellites: Medium Earth Orbit lies between LEO and GEO, typically at altitudes ranging from 5,000 to 20,000 kilometres. MEO is often used for navigation and positioning services like GPS, Galileo, and GLONASS. These satellites operate with orbital periods of two to 12 hours, offering a balance between the fast orbits of LEO and the extensive coverage of GEO.

MEO satellites require fewer units to achieve global coverage compared to LEO, but they face longer signal delays and slightly reduced data transmission rates. Recent advancements in high-throughput MEO constellations enable faster and more reliable communication for commercial and government users.

Geostationary Orbit (GEO) Satellites: Geostationary Orbit satellites are positioned 35,786 kilometres above the equator and appear stationary relative to the Earth's surface. This unique characteristic results from their orbital period matching Earth's rotational period (23 hours, 56 minutes, and 4 seconds).

GEO satellites are ideal for continuous communication services, such as television broadcasting, telephony, and meteorology. For instance, GEO weather satellites monitor cloud patterns and track regional weather phenomena, offering real-time updates. The downside of GEO satellites is the significant signal delay due to their distance from Earth, which can impact real-time applications.

Sun-Synchronous Orbit (SSO) Satellites: Sun-Synchronous Orbit satellites travel from pole to pole at altitudes between 600 and 800 kilometres. These satellites are calibrated to cross specific points on Earth's surface at the same local solar time during each orbit. This consistency in lighting conditions makes SSO satellites invaluable for Earth observation, environmental monitoring, and disaster management.

Applications of SSO satellites include studying weather systems, predicting cyclones, monitoring wildfires, and assessing deforestation. Although SSO satellites offer high imaging accuracy, their smaller coverage area necessitates deploying multiple units for continuous monitoring.

Geostationary Transfer Orbit (GTO) Satellites: Geostationary Transfer Orbit is an intermediary path used to transition satellites into geostationary orbit. After launch, rockets often place payloads in GTO, from which satellites use their propulsion systems to reach GEO. This method minimizes the fuel and resources required to achieve geostationary positioning. Rockets like SpaceX's Falcon 9 frequently use this approach for cost-efficient satellite deployments.

Other Orbit Types: Other orbits, such as Highly Elliptical Orbits (HEO), Polar Orbits, and Lagrange Points (L-points), are selected based on specific mission objectives. HEOs are used for communication in high-latitude regions, while polar orbits are ideal for global mapping. Lagrange points serve as stable locations for space telescopes or scientific instruments to observe celestial phenomena.

In all cases, the choice of orbit is dictated by the satellite's mission, coverage requirements, and the data it needs to collect or transmit, ensuring optimal performance and cost-effectiveness.

Communication Satellites

Communication satellites enable global communication by transmitting signals over vast distances, connecting remote areas, and supporting television, internet, and telephony [386-388]. Positioned in geostationary or medium Earth orbit (MEO), these satellites remain stationary relative to the Earth's surface or move in predictable patterns, making them ideal for consistent signal coverage [388, 389]. They use transponders to receive, amplify, and retransmit signals across designated frequency bands [390, 391].

Examples of communication satellites include Intelsat and SpaceX's Starlink satellites, which aim to provide global broadband coverage [386, 387, 392]. Intelsat satellites have been used extensively for communication services in various countries, including developing nations [393]. Starlink, a satellite internet constellation operated by SpaceX, is designed to provide high-speed, low-latency broadband internet access to users worldwide, particularly in remote and underserved areas [387, 392].

The use of satellite communication systems is particularly important for the Internet of Remote Things (IoRT), where sensors and actuators are distributed over a wide area, often in remote locations without access to terrestrial networks [388]. Satellite communication can provide the necessary connectivity for these applications, enabling data collection and control [388].

Satellite communication systems face various challenges, such as rain-induced attenuation, which can disrupt signal transmission [391, 394]. Factors like location, satellite-specific characteristics, and the choice of frequency bands (e.g., Ku-band, Ka-band) must be carefully considered when designing satellite communication systems to mitigate these effects and ensure reliable service [391, 394].

Advancements in satellite communication technology, such as the use of MIMO (Multiple-Input Multiple-Output) techniques, have the potential to improve channel capacity and meet the growing demand for data transmission [390, 395]. Additionally, the development of inter-satellite links (ISLs) in global navigation satellite systems (GNSS) can enhance the capabilities of these systems, enabling autonomous navigation and communication [396].

The evolution of the international communication regime has also influenced the satellite communication landscape. The transition from state-owned or private monopolies to privatization and liberalization has led to changes in organizations like the International Telecommunication Union (ITU) and Intelsat, which have had to adapt to the new market dynamics [397, 398].

GPS Satellites

GPS satellites form the backbone of navigation systems used worldwide [399, 400]. These satellites are typically in medium Earth orbit (MEO) and provide location, velocity, and timing data to devices on the ground, sea, and air [399, 400]. GPS satellites use atomic clocks to broadcast precise timing signals, which are received and interpreted by GPS receivers to calculate their position [399, 401].

The U.S. GPS constellation, Europe's Galileo, and Russia's GLONASS are examples of global navigation satellite systems (GNSS) that provide essential services for transportation, military operations, and disaster management [399, 400, 402]. These GNSS constellations have been rapidly developing, with over 100 satellites now in orbit [402]. The networking of these four major GNSS systems (GPS, Galileo, GLONASS, and BeiDou) has significantly increased the number of visible satellites, improving positioning availability, continuity, and accuracy [402-404].

GNSS signals can be affected by various factors, such as ionospheric effects, interference, and scintillation, which can degrade the positioning accuracy [405, 406]. To mitigate these issues, techniques like multi-GNSS positioning, GNSS augmentation systems (e.g., EGNOS), and advanced signal processing methods have been developed [407-409].

GNSS technology has found widespread applications in various fields, including transportation, surveying, agriculture, mining, aviation, and disaster management [403, 410, 411]. The high-precision positioning and timing capabilities of GNSS have enabled new applications, such as precise point positioning (PPP) for high-accuracy geodetic and engineering applications [412, 413].

Weather Satellites

Weather satellites play a crucial role in monitoring Earth's atmospheric conditions and providing critical data for weather forecasting and climate studies [414, 415]. These satellites operate in two main orbits: geostationary orbit and polar orbit [414, 416].

Geostationary satellites, such as NOAA's GOES satellites, provide continuous coverage of specific regions, allowing for real-time monitoring of weather patterns and the development of severe weather conditions [414, 416]. These satellites use advanced sensors, such as the Advanced Baseline Imager (ABI), to capture high-resolution images and measure atmospheric parameters like temperature, humidity, and wind patterns [416, 417].

Polar-orbiting satellites, on the other hand, offer global data collection by circling the Earth in a north-south direction [415, 418]. Examples of these satellites include NASA's Aqua and Terra, which carry instruments like the Visible Infrared Imaging Radiometer Suite (VIIRS) to monitor various environmental parameters [415].

The data collected by these weather satellites is essential for improving weather forecasting and understanding long-term climate trends [415]. The availability of this long-term, stable, and global dataset has allowed climate researchers to study the Earth's ecosystem and carbon cycle dynamics over time scales where climate change signals become discernible [415].

In addition to weather monitoring, weather satellites also play a crucial role in space weather observations, providing data on solar activity and its impact on the Earth's atmosphere and ionosphere [419, 420]. This information is crucial for understanding and mitigating the effects of space weather on satellite systems and other technological infrastructure [419, 420].

Earth Observation Satellites

Earth observation satellites gather detailed information about the planet's surface, including land use, vegetation, water bodies, and urban development. These satellites are typically placed in low Earth orbit (LEO) for high-resolution imaging [421]. They employ optical, radar, and multispectral sensors to collect data used in agriculture, disaster management, urban planning, and environmental conservation [422].

Examples of such Earth observation satellites include the European Space Agency's Sentinel satellites under the Copernicus program and NASA's Landsat series, which have been pivotal in monitoring Earth's changing landscapes over decades [423]. These satellites provide a global view and continuous measurements that are unmatched in their ability to capture the dynamics and variability of Earth processes [424].

The collaboration of multiple Earth observation satellites has strengthened the Earth's observation ability, expanded coverage, and increased the number of observation targets, which is beyond what a single satellite could accomplish [425]. The planned growth in the number of Earth observation satellite missions and the increasing use of complex instruments at high spatial resolution and often with high revisit cycles have also increased the pressure to clarify the legal basis of the offer of Earth observation data [426].

Earth observation satellites have made considerable progress in advancing our understanding of the Earth system, contributing to numerous disciplines such as hydrology, climatology, meteorology, oceanography, and biology [427]. The combination of data from Landsat-8 and Sentinel-2 Earth observation missions, for example, produces dense time series of multispectral images that are essential for monitoring the dynamics of land-cover and land-use classes across the Earth's surface with high temporal resolution [428, 429].

In recent years, the capacity to survey the Earth's surface has also been enhanced by the use of constellations of nano-satellites, which can provide high-resolution data at a lower cost and with greater flexibility [430]. Additionally, the development of new data transmission technologies, such as Ka-band 5-Gbps data transmission, is enabling Earth observation satellites to transmit large amounts of data to the ground more efficiently [431].

Each type of satellite serves a specialized purpose, contributing to advancements in science, technology, and everyday life. Together, they form a global network that supports communication, navigation, weather prediction, and a deeper understanding of our planet.

Rocket Systems for Satellite Launches (LEO, MEO, GEO Orbits)

The design and operation of rocket systems for satellite launches are tailored to the specific orbital requirements of the payload, which vary significantly between Low Earth Orbit (LEO), Medium Earth Orbit (MEO), and Geostationary Orbit (GEO). These differences affect the rocket's propulsion, trajectory planning, and staging configurations to ensure successful deployment.

Launching to Low Earth Orbit (LEO) involves placing satellites at altitudes ranging from 160 to 1,500 kilometres above Earth's surface. Achieving this requires reaching a velocity of approximately 7.8 km/s, allowing the satellite to maintain a stable orbit. Rockets designed for LEO missions are engineered to deliver high thrust during the initial ascent phase to overcome Earth's gravity and achieve the necessary velocity.

LEO missions typically employ two or three-stage rockets. The first stage provides the majority of thrust during the atmospheric ascent, while the second or third stages complete the orbital insertion and fine-tune the payload's deployment. Propulsion systems in these rockets often include liquid-fuelled engines, such as the Merlin engines used in SpaceX's Falcon 9, which offer precision and restart capabilities for orbital adjustments. Solid-fuelled boosters are also commonly used, providing reliable and powerful thrust during launch, as seen in vehicles like the Ariane 5 or the Indian PSLV.

The payload fairing plays a critical role in LEO launches, protecting satellites from aerodynamic forces and atmospheric heating during ascent. Once the rocket reaches the upper atmosphere, the fairing is jettisoned to reduce weight and allow the payload to operate unencumbered.

Several rockets are frequently used for LEO missions. SpaceX's Falcon 9 is a leading example, launching constellations like Starlink. Rocket Lab's Electron is designed for small satellite deployments, while India's PSLV is renowned for its reliability in Earth observation missions.

The trajectory for LEO launches begins with a near-vertical ascent to escape the dense lower atmosphere. As the vehicle gains altitude, the trajectory transitions to a shallower angle, allowing the rocket to build horizontal velocity and achieve orbital speed. For specific orbits, such as polar or sun-synchronous paths, launch sites near the equator or at polar locations are often utilized to maximize efficiency and coverage.

Launching to Medium Earth Orbit (MEO) involves placing satellites at altitudes ranging from 5,000 to 20,000 kilometres above Earth's surface. Achieving these orbits requires velocities between 3.9 and 5.5 km/s, depending on the specific altitude. Rockets designed for MEO missions typically employ three stages to efficiently achieve the higher altitudes and velocities required. The third stage or a dedicated kick stage is crucial, providing the precise energy needed to insert the payload into its intended orbit.

Propulsion systems for MEO missions often use liquid-fuelled engines, which offer flexibility for trajectory adjustments and fine control during orbital insertion. In many cases, these systems are paired with solid rocket boosters during the initial stages to provide the high thrust necessary for lifting heavier payloads. Such hybrid propulsion designs enable rockets to achieve the balance of power and precision required for MEO launches.

Payload systems for MEO missions are adapted to support the deployment of large and often clustered satellites, such as those used in navigation constellations like GPS, Galileo, and GLONASS. Rockets are equipped with deployable payload separation mechanisms to ensure smooth and accurate satellite placement into the desired orbital slots.

Prominent rockets used for MEO missions include the United Launch Alliance's Atlas V, frequently employed for GPS satellite launches. The Ariane 5 and its successor Ariane 6 are known for launching global navigation systems, while the Soyuz-ST-B is widely used for deploying GLONASS satellites.

The trajectory for MEO launches involves a longer duration of powered flight compared to Low Earth Orbit missions, reflecting the greater energy needed to achieve higher altitudes. This trajectory requires a precise balance of vertical thrust to overcome gravity and horizontal thrust to achieve the necessary orbital velocity, ensuring an efficient and accurate ascent into MEO.

Launching satellites to Geostationary Orbit (GEO) involves placing payloads at an altitude of 35,786 kilometres, where they remain stationary relative to a fixed point on Earth's surface. This orbit requires achieving a velocity of approximately 3.07 km/s and precise alignment with the equatorial plane. The process typically involves multi-stage rockets, often using three or four stages, to efficiently transport payloads from Earth's surface to a Geostationary Transfer Orbit (GTO). The satellite's onboard propulsion systems then circularize and adjust the orbit to achieve the final GEO position.

Early rocket stages are equipped with high-thrust engines to escape Earth's gravity and provide the initial energy needed for the ascent. Upper stages, which are often cryogenic or hypergolic, play a critical role in achieving precise placement in GTO. Once in GTO, satellites rely on their onboard propulsion systems, which may include chemical thrusters or electric propulsion systems, to perform the final orbital circularization and inclination adjustments.

Payload fairings for GEO missions are designed to protect larger and heavier payloads, such as communication satellites, space observatories, or scientific instruments. These fairings shield the payload from aerodynamic forces during ascent and are jettisoned once the rocket exits the dense atmosphere.

Rockets commonly used for GEO missions include the Ariane 5 and its successor Ariane 6, which are optimized for dual-payload launches. SpaceX's Falcon Heavy is utilized for high-payload missions to GEO and beyond, while the Russian Proton-M rocket has a long history of launching heavy GEO satellites.

The trajectory for GEO missions involves insertion into GTO, where the satellite's onboard propulsion systems take over to circularize the orbit and align it with the equatorial plane. Adjusting the orbit's inclination to achieve equatorial alignment requires significant energy, making GEO missions particularly demanding in terms of propulsion and trajectory design. These features ensure the successful deployment of satellites into this critical orbit, enabling applications such as communication, weather monitoring, and global broadcasting.

Key design considerations for rockets used in satellite launches involve balancing efficiency, structural integrity, launch site selection, and precision to meet the specific demands of each mission and orbit.

Efficiency and adaptability are essential to optimize fuel usage and thrust, reducing costs while maximizing payload capacity [432]. Rocket stages are carefully engineered for quick separation, which minimizes weight during ascent and ensures that each stage operates at peak efficiency [432].

Structural integrity is another critical factor, as rockets experience intense mechanical stress during launch and ascent. To withstand these forces, manufacturers rely on advanced materials such as aluminium alloys and composites, which provide the necessary strength without adding excessive weight [433].

The choice of launch site plays a pivotal role in mission success. Equatorial sites, such as the Kourou Spaceport in French Guiana, offer a significant advantage for geostationary orbit (GEO) missions by leveraging Earth's rotational velocity to reduce the energy required for orbital insertion [434]. Conversely, polar sites like Vandenberg Space Force Base in California are better suited for sun-synchronous and polar orbits, as they allow for launches that pass over Earth's poles [434].

Precision in guidance and navigation is vital for all satellite launches, ensuring accurate orbital insertion. This precision is particularly important for satellite constellations, where tight tolerances are required to maintain orbital alignment, as well as for high-value satellites, where any deviation from the planned trajectory could compromise mission objectives [435].

Satellite Constellations: Starlink, Oneweb, and Their Implications

Satellite constellations have emerged as a transformative technology in global communication and internet access. Two prominent examples are Starlink, developed by SpaceX, and OneWeb, a UK-based company [436, 437]. These constellations aim to provide high-speed, low-latency internet services worldwide, particularly in underserved and remote regions [436, 437].

Starlink operates its satellites in low Earth orbit (LEO) at altitudes ranging from approximately 340 km to 1,200 km, with most operational satellites around 540 km to 570 km [436]. This proximity to Earth enables lower latency compared to traditional geostationary satellites [436]. Each Starlink satellite is equipped with phased-array antennas and laser interlinks, allowing for high-speed communication between satellites and ground terminals [436]. Starlink has already launched thousands of satellites, with plans to expand the network to over 42,000 in the coming years [436]. Its service offers speeds ranging from 50 Mbps to 150 Mbps, with latency as low as 20 milliseconds [436].

In contrast, OneWeb's satellites operate at altitudes of approximately 1,200 km [437]. This configuration reduces the total number of satellites needed for coverage but increases latency slightly compared to Starlink [437]. OneWeb's constellation will consist of around 648 satellites at full capacity, using Ku-band frequencies for communication and designed to provide seamless coverage through overlapping satellite beams [437].

These satellite constellations have significant implications:

1. Global Connectivity: They promise to bridge the digital divide by providing internet access to remote regions where traditional infrastructure is not feasible, potentially improving education, healthcare, and economic opportunities worldwide [436, 437].

2. Economic Opportunities: These constellations represent a significant economic opportunity, both in terms of direct revenue from internet services and the development of satellite manufacturing, launch services, and ground infrastructure industries [436, 437].

3. Military and Strategic Applications: The low latency and global coverage of these constellations make them attractive for military and defence applications, including secure communications, surveillance, and real-time data sharing [436, 437].

4. Astronomical Concerns: The scientific community has raised concerns about the impact of satellite constellations on astronomy, as the brightness of these satellites can interfere with ground-based telescopes, hindering observations and research [436, 437].

5. Space Debris: The rapid deployment of thousands of satellites increases the risk of collisions and contributes to the growing problem of space debris, requiring effective debris mitigation strategies [436, 437].

6. Regulatory and Spectrum Challenges: The allocation of radio frequencies and orbital slots is a complex issue, requiring coordination between international agencies like the International Telecommunication Union (ITU) [436, 437].

As Starlink, OneWeb, and other emerging constellations like Amazon's Project Kuiper and China's Guowang Network continue to shape the future of global communication, addressing the associated challenges will require innovation, collaboration, and careful regulation to ensure that these satellite constellations contribute positively to society and the environment [436, 437].

Chapter 12

Military and Commercial Applications

Defence Applications: Ballistic Missiles and Space Defence Systems

Ballistic missiles are rocket-propelled weapons designed to follow a parabolic trajectory, delivering payloads such as warheads to distant targets. They are categorized based on range, including short-range (SRBM), medium-range (MRBM), intermediate-range (IRBM), and intercontinental (ICBM) ballistic missiles. These missiles consist of several critical components. The launch vehicle provides the initial propulsion to exit the atmosphere, while the re-entry vehicle houses the warhead and is designed to withstand the intense heat generated during atmospheric re-entry. Advanced guidance systems, incorporating inertial navigation, GPS, and star trackers, ensure precise targeting. Propulsion options vary, with solid-fuelled systems offering reliability and low maintenance and liquid-fuelled systems providing higher thrust and greater control.

Ballistic missiles are deployed in various configurations. Land-based systems utilize silos, mobile launchers, or rail-based platforms for strategic flexibility. Sea-based deployments, such as submarine-launched ballistic missiles (SLBMs), offer stealth and mobility. Air-launched systems are deployed from strategic bombers, enabling rapid response capabilities. Modern ballistic missile technology includes features like Multiple Independently Targetable Re-entry Vehicles (MIRVs), which allow a single missile to target multiple locations, and Hypersonic Glide Vehicles (HGVs), capable of high-speed manoeuvres to evade defence systems.

Space defence systems complement missile technologies, addressing threats to satellites and other space-based assets. Anti-Satellite (ASAT) weapons are designed to disable or destroy enemy satellites through kinetic impact, directed energy, or cyberattacks. These systems are typically ground-launched or air-launched. Missile defence systems play a vital role in intercepting and neutralizing incoming ballistic missiles during their boost, midcourse, or terminal phases. Key systems include the Ground-Based Interceptor (GBI) for midcourse interception, Terminal High Altitude Area Defense (THAAD) for terminal-phase destruction, and the Aegis Ballistic Missile Defense (BMD) system, a sea-based platform using SM-3 interceptors.

Figure 86: Terminal High Altitude Area Defense (THAAD). The U.S. Army Ralph Scott/Missile Defense Agency/U.S. Department of Defense, CC BY 2.0, via Wikimedia Commons.

Space-based defence systems enhance capabilities by deploying satellites equipped with sensors and weapons to detect and intercept threats in orbit. Emerging technologies in this domain include directed-energy weapons such as lasers and kinetic kill vehicles. Early warning systems, including ground- and space-based sensors, detect missile launches and track trajectories. Systems like the US Space-Based Infrared System (SBIRS) and Russia's Oko provide real-time data for defence coordination. Electronic warfare and cybersecurity efforts focus on jamming or spoofing enemy satellite communications and navigation systems and conducting cyberattacks on ground control stations to disrupt or disable defence assets.

Rockets play a central role in space-based defence systems, serving as the primary means of deploying, maintaining, and supporting assets critical for national security and strategic deterrence [438]. These systems encompass a range of technologies and capabilities designed to protect and defend against threats originating from space or utilizing space infrastructure [439].

Rockets are essential for placing key defence-related satellites and equipment into orbit [439]. These include early warning satellites that provide real-time data on missile launches [439], anti-satellite (ASAT) weapons capable of neutralizing enemy satellites [440], and reconnaissance satellites for monitoring potential adversaries [439]. Launch vehicles ensure precise placement of these assets in specific orbits, such as geostationary or sun-synchronous orbits, depending on their mission requirements [439].

Rockets are integral to space-based interception systems designed to neutralize threats in space or during their approach to Earth [439]. These include missile defence systems that launch interceptor payloads to destroy incoming ballistic missiles [439], and ASAT interception capabilities that can disrupt enemy satellites [440].

Rockets facilitate the maintenance and resupply of space-based defence systems by launching replacement satellites and delivering supplies, equipment, and crew members to space stations [439].

With advancements in technology, rockets are expanding their role in space defence to include the deployment of space-based weapons, the deployment of satellite constellations for enhanced global coverage, and on-orbit servicing missions to repair or upgrade satellites [439].

Rockets are pivotal in projecting a nation's space defence capabilities, serving as a visible deterrent to potential adversaries [438]. The ability to rapidly launch and deploy defence assets signals preparedness and technological superiority, reinforcing a country's defence posture [438].

The role of rockets in space-based defence systems also introduces challenges, such as the high cost of launching and maintaining space-based infrastructure, the management of space debris generated by defensive actions, and the geopolitical implications of the weaponization of space [439].

Rockets used in space defence systems are meticulously engineered to address diverse and demanding mission requirements, from deploying surveillance satellites to delivering interceptors and anti-satellite weapons. Payload capacities for defence rockets vary based on mission objectives and orbital needs. For example, SpaceX's Falcon 9 can carry up to 22,800 kg to Low Earth Orbit (LEO), making it ideal for deploying reconnaissance satellites and defence-related constellations. Similarly, the Atlas V can transport approximately 20,000 kg to LEO or 8,900 kg to Geostationary Transfer Orbit (GTO), suitable for launching heavier payloads such as missile defence systems.

Advanced propulsion systems are central to these rockets. Liquid engines, such as the Merlin engines in the Falcon 9 or the RD-180 in the Atlas V, provide high thrust and precise control for varied missions. Solid Rocket Boosters (SRBs) are often incorporated to deliver powerful initial thrust during launch, as seen in rockets like the ULA Delta IV Heavy. To ensure precise trajectory alignment, guidance and navigation systems combine inertial navigation with GPS-based augmentation and advanced real-time algorithms.

Ground-based launch vehicles play a significant role in deploying satellites, interceptors, and weapons into specific orbits. The United Launch Alliance (ULA) Atlas V is frequently used to deploy reconnaissance satellites and missile warning systems, such as SBIRS satellites. Its modular design supports diverse payload configurations, while the RD-180 engine and Centaur upper stage offer powerful and precise orbital insertion capabilities. SpaceX's Falcon 9, known for its reusable first stage, is commonly employed for launching defence satellite constellations like GPS III and classified payloads. Its efficient Merlin 1D engines and versatile payload fairings enhance its adaptability for varied missions. The European Ariane 5 and 6 rockets are also noteworthy for their dual-payload capabilities and robust solid boosters, frequently used for geostationary defence satellites.

Sea-based launch platforms like the Trident II (D5) submarine-launched ballistic missile (SLBM) provide stealth and mobility. The Trident II uses a three-stage solid propulsion system with inertial navigation and stellar tracking, making it a critical component of nuclear deterrence.

Interceptors for missile defence, such as the Ground-Based Interceptor (GBI), Terminal High Altitude Area Defense (THAAD), and Aegis Ballistic Missile Defense (BMD), demonstrate the versatility of defence rockets. The GBI employs a three-stage rocket system to intercept threats in midcourse, while THAAD uses a single-stage solid-fuelled rocket for rapid terminal-phase interception. Aegis BMD, a sea-based system, relies on SM-3 interceptors with multi-stage propulsion for high-speed midcourse interceptions.

Anti-satellite (ASAT) rockets are specifically designed to neutralize enemy satellites through kinetic impact or directed energy weapons. India's Mission Shakti successfully demonstrated a ground-launched missile destroying a satellite in LEO, while the USA's ASM-135 ASAT showcased air-launched capabilities targeting low orbit.

Emerging technologies are reshaping defence rockets. Hypersonic Glide Vehicles (HGVs), delivered via rocket stages, operate at speeds exceeding Mach 5, with examples like Russia's Avangard and China's DF-ZF [441]. Reusable rocket systems, such as SpaceX's Starship and Falcon Heavy, are being explored for rapid-response military applications. Hybrid propulsion systems, combining solid and liquid fuels, enhance flexibility and efficiency in defence operations.

Key challenges in defence rockets include rapid deployment readiness, precision in trajectory calculations, and survivability against countermeasures like jamming or cyberattacks. Reusability and modular designs are essential for cost-effective operations. As space becomes an increasingly contested domain, these rocket systems continue to evolve, blending advanced propulsion, guidance, and payload integration to meet complex security demands and safeguard national interests.

Commercial Launch Services: Private Companies and Cost Optimization

Commercial launch services have revolutionized the space industry by enabling greater access to orbit for governments, private enterprises, and academic institutions. These services are provided by private companies that develop, manufacture, and operate rockets for launching satellites, payloads, and even crewed missions into space. Key players in the industry include SpaceX, Blue Origin, Rocket

Lab, United Launch Alliance (ULA), and others, each contributing to a rapidly expanding market through innovation and competitive pricing.

A launch service provider (LSP) is a type of company that utilizes launch vehicles and related services provided by a launch agency, including furnishing the launch vehicles, launch support, equipment, and facilities, for the purpose of launching satellites into orbit or deep space [442]. There are over 100 launch companies worldwide [443], and these companies and their launch vehicles are in various stages of development, with some (such as SpaceX, Rocket Lab, and ULA) already in regular operation, while others are still in the development phase [444].

In 2018, the launch services sector accounted for $5.5 billion out of a total $344.5 billion "global space economy" [443]. An LSP is responsible for the ordering, conversion, or construction of the carrier rocket, assembly and stacking, payload integration, and ultimately conducting the launch itself [442]. Some of these tasks may be delegated or subcontracted to other companies [445].

Private companies offering commercial launch services have brought about significant cost reductions compared to traditional government-led programs. SpaceX, for instance, has led the charge in cost optimization through its groundbreaking development of reusable rocket stages. The Falcon 9 and Falcon Heavy rockets feature reusable first stages, which return to Earth for refurbishment and reuse after each mission. This innovation has drastically reduced the cost per kilogram of payload delivered to orbit, making space more accessible. In addition, streamlined manufacturing processes and vertical integration have allowed companies like SpaceX to control costs by producing critical rocket components in-house.

The launch services sector has seen significant growth in recent years, driven by factors such as reduced launch costs, increased reliance on satellite technologies, and the emergence of private space companies [443]. This has led to a surge in the number of small satellites being launched, with the number of small satellites launched in 2017 exceeding the number of conventional satellites [444].

Blue Origin, another major player, has focused on reusability and innovation to optimize costs. Its New Shepard suborbital vehicle and the forthcoming New Glenn orbital rocket emphasize reusable stages to minimize waste and lower launch expenses. The company's vision of reducing the cost barrier for space access aligns with the broader industry trend toward affordability and sustainability in space operations.

Rocket Lab, with its Electron rocket, has targeted small satellite launches. By designing a dedicated small launch vehicle with efficient production methods and a reusable first stage (Neutron rocket under development), Rocket Lab has created a cost-effective solution for the growing demand in the small satellite sector. The use of 3D-printed components and streamlined supply chains further enhances cost optimization.

Cost optimization in commercial launch services also involves innovative business models. Ride-share programs allow multiple payloads to share the same rocket, dividing the cost among several customers. This approach is widely used for deploying satellite constellations, as seen with SpaceX's Starlink launches and rideshare missions that include CubeSats and other small satellites.

Additionally, dedicated rideshare providers, such as Spaceflight Inc., aggregate payloads to maximize the efficiency of each launch.

Advances in materials, manufacturing techniques, and propulsion systems have also contributed to cost optimization. Lightweight materials, such as carbon composites, reduce the mass of rockets while maintaining structural integrity, allowing for more efficient launches. Propulsion advances, including the development of methalox engines (methane and liquid oxygen) like SpaceX's Raptor, promise lower costs and higher performance for future missions.

Private companies have also fostered competitive pricing by leveraging government contracts and partnerships. For example, SpaceX and ULA regularly win contracts from NASA and the U.S. Department of Defense, which provide stable revenue streams that enable further investments in cost-saving technologies. International collaborations and partnerships, such as those involving European company Arianespace, expand the market reach of commercial launch services, driving down costs through economies of scale.

As the demand for satellite constellations, Earth observation systems, and space exploration missions continues to grow, commercial launch services are poised to play an increasingly critical role. Companies are focusing not only on cost optimization but also on expanding their service offerings, such as rapid launch capabilities and end-to-end mission integration. By continuing to innovate and compete, private companies are democratizing access to space, paving the way for new scientific, commercial, and exploratory opportunities.

A document central to successful launch service provision is the Interface Control Document (ICD), a contract that specifies the integration and mission requirements responsibilities across the service provider and the service solicitor [446]. In some cases, an LSP is not required to launch a rocket, as government organizations such as the military and defence forces may conduct the launch themselves [442].

The adoption of consumer technologies and rapid development cycles by small, agile teams has been a hallmark of the modern small satellite industry [447, 448]. This has enabled the development of more cost-effective and innovative satellite solutions, which in turn has driven the growth of the launch services sector [449].

However, the launch process can be a challenging environment for satellites, with about 45% of satellite faults originating from vibrational damage during the launch process [450]. To address this, the space industry has proposed the concept of whole-spacecraft isolation techniques, which aim to reduce the environmental loads on satellite launches and increase the dynamic performance requirements of satellites and their equipment [450].

Space Tourism and New Frontiers in Commercial Space Travel

Space tourism and the emergence of commercial space travel represent a transformative chapter in humanity's relationship with space. What was once the exclusive domain of government agencies and astronauts is now opening to private individuals and businesses, driven by advancements in technology, innovative business models, and a growing desire for experiences that transcend traditional boundaries.

The history of space tourism is deeply intertwined with humanity's fascination with space exploration, which dates back to the early achievements of the Space Age. The journey began in 1961 when Yuri Gagarin became the first human to venture into space, igniting a global fascination with the possibilities of space travel. While initial space missions were primarily driven by government-funded programs for scientific research and military purposes, the notion of ordinary individuals traveling to space began to take root as advancements in rocket technology matured [451].

In the 1990s, a pivotal shift occurred when private interest in space tourism gained momentum. The Russian Space Agency made a historic leap in 1998 by enabling Dennis Tito, an American businessman, to become the first paying space tourist. Tito travelled to the International Space Station aboard a Soyuz spacecraft, marking a groundbreaking moment in space tourism history. This achievement proved that commercial space travel was feasible, though its accessibility was limited to those with substantial financial resources [451].

The early 2000s saw the space tourism industry take incremental but significant steps forward. Companies like Space Adventures emerged as key players, organizing low Earth orbit trips for private individuals. Despite their exclusivity due to high costs, these missions demonstrated growing interest and demand for personal space experiences. Simultaneously, private aerospace companies began developing technologies to make space tourism safer and more affordable, driven by a vision of democratizing access to space [451].

The launch of the Ansari X Prize in 1996 further catalysed the development of commercial space tourism. This competition, spearheaded by the X Prize Foundation, challenged teams to build a privately funded spacecraft capable of carrying passengers to the edge of space and returning them safely. The success of SpaceShipOne in 2004, which won the prize, showcased the potential for privately funded spaceflight and spurred innovation in the sector [451].

These early efforts were instrumental in laying the groundwork for modern space tourism. The advent of reusable rocket technology, pioneered by companies such as SpaceX and Blue Origin, has dramatically reduced launch costs and enhanced the feasibility of frequent space travel. These advancements have made space tourism a more viable industry and inspired new possibilities for accessible space exploration [451].

As the industry evolves, the initial vision of space tourism is expanding. Early milestones achieved by pioneering individuals and companies have transformed space tourism from a dream to a budding industry poised for mainstream acceptance. Today, with ambitious projects and technological innovations on the horizon, space tourism is becoming an increasingly realistic option for those seeking to venture beyond Earth, heralding a new era of exploration and human achievement [451].

Space tourism is at the forefront of this transformation. Companies like SpaceX, Blue Origin, and Virgin Galactic are pioneering efforts to make suborbital and orbital travel accessible to civilians. Suborbital flights, such as those offered by Blue Origin's New Shepard and Virgin Galactic's

SpaceShipTwo, provide passengers with a brief experience of microgravity and a breathtaking view of Earth from the edge of space. These experiences typically involve a high-altitude ascent, a few minutes of weightlessness, and a safe return to the ground, offering a tantalizing glimpse of space without the prolonged training and risks of orbital missions.

The space tourism industry is witnessing remarkable advancements, led by pioneering companies that are reshaping the way civilians access and experience space. Key players such as Virgin Galactic, Blue Origin, and SpaceX are each making significant contributions, using their unique strategies and technologies to broaden humanity's reach into the cosmos [451].

Virgin Galactic has established itself as a leader in suborbital space tourism with its SpaceShipTwo vehicle. Designed to provide a brief but exhilarating experience in microgravity, this spacecraft offers passengers a chance to witness Earth from the edge of space. Virgin Galactic's operations are centred at Spaceport America in New Mexico, a state-of-the-art facility that offers an immersive pre-flight experience. The company's goal is to make space travel a reality for a broader audience, focusing on individuals seeking a once-in-a-lifetime adventure rather than exclusively scientific or exploratory missions. Its flights, which include a launch from a mothership and a glide back to Earth, represent an innovative approach to commercial space travel [451].

Blue Origin, founded by Jeff Bezos, has also made significant strides in democratizing access to space. The company's New Shepard rocket system is specifically designed for suborbital flights, providing passengers with several minutes of weightlessness and a breathtaking view of Earth. Blue Origin places a strong emphasis on safety and reliability, demonstrated through numerous successful test flights. The company's vision extends beyond tourism to include future possibilities for space colonization and long-term human presence in space. By combining technical excellence with a customer-centric approach, Blue Origin is solidifying its role in the space tourism sector [451].

SpaceX, on the other hand, has taken space tourism to new heights by focusing on orbital flights. Its Crew Dragon spacecraft, initially developed for NASA missions, has been adapted for private ventures, including the groundbreaking Inspiration4 mission, which sent a civilian crew into orbit. SpaceX's ambitious plans for its Starship program promise even greater opportunities, including lunar flybys and, eventually, missions to Mars. By leveraging reusable rocket technology and its proven track record of successful launches, SpaceX is expanding the scope of space tourism far beyond suborbital experiences, making longer-duration and more complex missions a possibility for private individuals [451].

Emerging players are also contributing to the growth of this industry. Axiom Space is developing a commercial space station that will serve as a platform for both research and tourism in low Earth orbit. Space Perspective, another innovative company, is exploring the use of high-altitude balloons to gently lift passengers to the edge of space, offering a more serene and accessible experience. These efforts are diversifying the ways people can experience space, catering to varying levels of ambition and comfort [451].

Collectively, these companies are not only advancing the technical capabilities of space travel but also creating a robust ecosystem that supports the growing demand for space tourism. Their work is laying the foundation for a future where space is no longer the exclusive domain of astronauts and researchers but an accessible frontier for ordinary individuals. As these technologies mature, the dream of experiencing space firsthand is becoming a tangible reality, transforming the very nature of travel and exploration [451].

Orbital tourism is the next frontier, promising longer-duration stays in space. SpaceX has already facilitated private orbital missions, such as the Inspiration4 mission, which sent civilians into orbit aboard the Crew Dragon spacecraft. Plans for missions to the International Space Station (ISS) and even private space stations are underway, with companies envisioning luxurious accommodations and research facilities in low Earth orbit.

Beyond the thrill of space tourism, commercial space travel is expanding into new frontiers that could redefine industries. Private companies are exploring lunar tourism, with missions designed to orbit or land on the Moon. SpaceX's Starship program, for instance, aims to carry private passengers on lunar flybys and, eventually, to Mars. These ambitious projects promise to make deep-space exploration a reality for non-professional astronauts.

Technological advancements are revolutionizing space travel, ushering in an era where commercial space tourism is becoming a tangible reality. These innovations span key domains, including propulsion systems, spacecraft design, life support systems, pre-flight preparation tools, and launch and landing technologies, each playing a pivotal role in making space tourism feasible, safe, and appealing [451].

Rocket propulsion technology has undergone significant advancements, with reusable rocket systems leading the charge. SpaceX's reusable Falcon 9 and Starship rockets are prime examples, drastically reducing the cost of launching payloads and passengers into space. Reusability enables more frequent flights, minimizes production expenses, and lowers ticket prices, making space travel more accessible. Enhanced propulsion capabilities also provide access to a greater variety of orbits and destinations, offering tourists diverse experiences, from suborbital flights to orbital and even lunar adventures [451].

Modern spacecraft design prioritizes not only functionality but also passenger comfort and safety. Companies like Blue Origin and Virgin Galactic have developed spacecraft that emphasize user-friendly interfaces, onboard amenities, and expansive windows that provide breathtaking views of Earth and space. These designs cater to the needs of space tourists by combining cutting-edge safety features with luxurious interiors, creating an unforgettable travel experience in a zero-gravity environment [451].

Life support systems have also seen transformative improvements. Closed-loop systems now enable the efficient recycling of air and water, essential for longer-duration missions and sustainable space travel. These systems integrate advanced monitoring technologies to track passengers' vital signs in real time, ensuring their health and well-being throughout the journey. Such innovations are critical in creating a secure and comfortable environment for tourists venturing beyond Earth's atmosphere [451].

Virtual reality (VR) and augmented reality (AR) are playing an increasingly important role in pre-flight preparation. These technologies simulate the unique aspects of space travel, such as weightlessness and the immersive experience of viewing Earth from space. By offering realistic training scenarios, VR and AR help travellers acclimate to the physical and psychological demands of spaceflight. These tools not only enhance anticipation but also reduce anxiety, ensuring that passengers are fully prepared for their journey [451].

Launch and landing technologies have evolved to ensure safer, more efficient operations. Precision landing systems and vertical landing capabilities enable spacecraft to return to Earth with remarkable accuracy, increasing safety and minimizing environmental impact. These advancements facilitate the development of spaceports in diverse locations, broadening access to space travel and supporting the growing demand for tourism and exploration [451].

Together, these technological advancements are reshaping the landscape of space travel. They are making journeys to the edge of space and beyond safer, more affordable, and more accessible than ever before. By integrating these innovations, the space tourism industry is poised to transform the dream of space travel into a reality for a growing number of individuals, opening new frontiers for adventure and exploration [451].

The commercialization of space travel is also creating opportunities for new ventures in microgravity research, in-orbit manufacturing, and satellite deployment. Microgravity environments offer unique conditions for scientific experiments, leading to breakthroughs in fields such as pharmaceuticals, materials science, and biotechnology. Meanwhile, private companies are developing small satellite deployment platforms that leverage commercial rockets to reduce costs and expand access to space for businesses and researchers.

The economic implications of these advancements are profound. By reducing launch costs through innovations like reusable rockets, companies are not only democratizing access to space but also fostering a burgeoning space economy. This includes the development of spaceports, training facilities, and supporting infrastructure to accommodate the growing demand for commercial space travel.

The emergence of space tourism marks the advent of an industry with far-reaching economic implications and immense market potential. This burgeoning sector is poised to create substantial revenue streams, generate employment opportunities, and drive technological innovation, reshaping the global economic landscape [451].

Projections suggest that the global space tourism market could grow into a multibillion-dollar industry in the coming decade. The target clientele spans a diverse demographic, with affluent individuals and adventurous explorers leading initial demand. This broad base of interest underscores the potential for sustainable growth, encompassing a range of offerings from suborbital flights to full-scale orbital experiences. As interest grows, companies operating in this space can expect steady revenue generation, making the sector a viable investment option [451].

The development of space tourism necessitates significant investment in infrastructure, driving economic activity in multiple domains. Launch facilities, spacecraft manufacturing, and astronaut training centres will require extensive funding and resources. This demand translates into job creation across various sectors, including engineering, design, manufacturing, hospitality, and customer service. As the industry grows, educational institutions and vocational programs will likely expand to meet the need for specialized training, further boosting workforce development [451].

Technological advancements spurred by space tourism will ripple into other industries, amplifying their economic impact. Breakthroughs in propulsion systems, life support technologies, and materials science—essential for commercial space travel—can revolutionize sectors like aviation, telecommunications, and healthcare. These cross-industry synergies will enhance innovation and global competitiveness, demonstrating the far-reaching benefits of investing in space-related technologies [451].

Ancillary markets will also experience significant growth. The fascination with space travel is likely to fuel demand for complementary services, including travel packages, themed merchandise, and space-related media content. Industries such as insurance and legal consulting will find new opportunities as they develop frameworks tailored to the unique risks and challenges of space tourism. Moreover, public interest will drive educational programs, documentaries, and entertainment centred on the exploration of space, expanding cultural and economic horizons [451].

Regulation will play a pivotal role in shaping the market's evolution. Governments and international agencies are working to establish safety standards, liability frameworks, and operational guidelines for commercial space travel. These regulatory developments will influence market entry, competition, and consumer trust. Staying informed about policy changes is crucial for stakeholders aiming to capitalize on the economic opportunities presented by space tourism [451].

Engaging with space tourism offers a chance to be part of a transformative economic and technological movement. Beyond its direct economic contributions, the industry symbolizes the growing intersection of human curiosity, innovation, and entrepreneurship, opening new frontiers for exploration and economic prosperity [451].

While the potential is immense, the challenges remain significant. Safety is paramount, as space tourism must demonstrate a track record of reliability and risk mitigation to gain public trust. Regulatory frameworks are evolving to address liability, environmental impacts, and international cooperation in an increasingly crowded orbital environment. Additionally, ensuring equitable access to space amid high costs will be critical to making space tourism more inclusive.

Navigating the burgeoning field of space tourism requires addressing complex regulatory and safety challenges that are critical to ensuring the industry's long-term viability and public trust. One of the most significant challenges is the establishment of a coherent and unified regulatory framework. Currently, space travel is governed by a patchwork of international treaties, national laws, and regulatory bodies, each addressing different aspects of spaceflight. For example, in the United States, the Federal Aviation Administration (FAA) oversees commercial space launches, while other entities like NASA and the Department of Defense may influence operational guidelines. This fragmented oversight can create legal ambiguities, making it difficult for space tourism companies to navigate compliance requirements across jurisdictions [451].

Safety standards, which are still in their formative stages, present another significant hurdle. The nascent nature of the space tourism industry means there are limited precedents or benchmarks to guide the development of safety protocols. Ensuring the safety of passengers and crew involves extensive testing of spacecraft, propulsion systems, and operational procedures under conditions that replicate real-life scenarios. Companies must account for a range of risks, including equipment malfunctions, human factors like health risks during microgravity exposure, and the potential for catastrophic failures during launch or re-entry. Establishing and adhering to rigorous safety standards is not only a regulatory requirement but also essential for building public confidence in commercial space travel [451].

Obtaining licenses for commercial spaceflights is another critical regulatory challenge. Different countries impose varying requirements for permits and approvals, often tailored to their specific legal, environmental, and security considerations. Companies operating in the space tourism sector must navigate these complex licensing frameworks to ensure compliance while maintaining operational timelines. This often requires significant investment in legal expertise and close collaboration with national and international regulatory bodies [451].

Liability concerns further complicate the landscape. Space tourism inherently carries risks, and clear liability frameworks are necessary to define responsibilities in the event of accidents or malfunctions. Companies must consider how to protect themselves from potential lawsuits while offering adequate protections and recourse for passengers. Developing robust insurance solutions for space tourism remains a challenge, as the insurance industry grapples with quantifying risks associated with this uncharted territory. Without comprehensive coverage options, companies may face significant financial vulnerabilities [451].

Public perception also plays a pivotal role in shaping regulatory developments. As space tourism garners attention, concerns about its environmental impact, such as contributions to atmospheric pollution or space debris, are becoming more pronounced. Engaging with the public, environmental advocates, and policymakers is critical to addressing these concerns and ensuring the industry is seen as responsible and sustainable. Transparency and proactive measures to mitigate environmental impacts will be key to maintaining a positive reputation [451].

Finally, the ethical implications of space tourism add another layer of complexity. Issues such as the equitable use of space resources, the potential for creating additional space debris, and the long-term sustainability of human activities in space demand thoughtful consideration. Companies must collaborate with international organizations and adhere to emerging guidelines to ensure that their operations contribute to the responsible development of space exploration [451].

Addressing these regulatory and safety challenges requires a multifaceted approach that prioritizes compliance, safety, and sustainability. By adopting proactive strategies, investing in legal and technical expertise, and fostering transparent dialogue with stakeholders, space tourism companies can lay the groundwork for a robust and trustworthy industry [451].

Space tourism and commercial space travel are reshaping humanity's understanding of exploration, offering not just a journey into the cosmos but also a glimpse into the future of innovation, collaboration, and ambition. As more companies enter the sector and technology continues to evolve, these new frontiers promise to bring space closer to Earth, making it a tangible destination for adventure, commerce, and discovery.

PART 5

Future of Rocket Design and Space Exploration

Chapter 13
Reusable Rockets and Sustainability

Innovations in Reusability

Rocket reusability has indeed emerged as a transformative innovation within the aerospace industry, significantly reducing the cost of access to space and increasing the frequency of launches. The development of reusable launch systems (RLSs) is widely recognized as a key breakthrough that can potentially lower manufacturing costs by approximately 30% due to reduced refurbishment expenses [452]. This paradigm shift is exemplified by the pioneering efforts of companies like SpaceX and Blue Origin, which have redefined the economics and possibilities of space travel through their innovative reusable systems.

Once recovered, rocket components undergo rigorous inspection and refurbishment to ensure readiness for subsequent launches. This iterative process not only enhances cost-effectiveness but also facilitates the continuous improvement of rocket systems. By analysing flight data and assessing hardware performance after each mission, engineers can make incremental upgrades, ensuring reliability and safety over time.

SpaceX's contributions to reusable rocket technology are most prominently showcased through its Falcon 9 and Falcon Heavy launch systems. The Falcon 9's first stage, powered by nine Merlin engines, is designed for vertical landings, allowing for multiple reuses. After completing its primary mission, the first stage autonomously returns to Earth, landing either on a drone ship at sea or on a ground pad [453]. This capability has been successfully demonstrated in numerous launches, with many boosters completing multiple missions, thus validating the reusability concept [454]. SpaceX's innovations, such as advanced grid fins for precise aerodynamic control during descent and robust thermal protection systems, have significantly lowered launch costs, enabling missions that would have been financially prohibitive with expendable rockets [455].

The Falcon Heavy extends SpaceX's reusable capabilities to heavier payloads, featuring three reusable boosters that often return to land nearly simultaneously, showcasing remarkable precision engineering [453]. Furthermore, SpaceX's ambitious Starship program aims to develop fully reusable two-stage rockets capable of carrying substantial payloads to various destinations, including orbit, the Moon, and Mars. Starship prototypes have conducted high-altitude tests, demonstrating controlled descents and landings, although further development is necessary to ensure reliability for orbital missions [455].

Figure 87: The first launch of the SpaceX Falcon Heavy Rocket on January 6, 2018 from Kennedy Space Center. Daniel Oberhaus, CC BY 4.0, via Wikimedia Commons.

Developing and refining the technologies required for the successful launch and recovery of Falcon 9 and Falcon Heavy first stages, as well as both stages of Starship, has been a critical aspect of SpaceX's innovation journey. Since 2017, the recovery and reuse of Falcon rocket boosters have become routine, marking a significant milestone in reusable rocket technology.

One of the key advancements is the restartable ignition system for the first-stage booster. This system enables multiple engine restarts at different stages of re-entry. The first restart occurs at supersonic velocities in the upper atmosphere to reverse the booster's trajectory and guide it back toward the launch site. Another restart happens at high transonic velocities in the lower atmosphere to slow the terminal descent for a soft landing. An additional burn is required for boosters returning to land shortly after stage separation to reverse their flight direction, totalling four burns for the centre engine during such missions.

SpaceX also developed advanced attitude control technology to manage the booster's descent through various atmospheric conditions. This system provides precise roll control, preventing excessive spinning and ensuring stability. Early tests revealed challenges, such as fuel shifting due to excessive roll, which led to engine shutdowns. The refined technology now handles the transition from the vacuum of space at hypersonic speeds through the dense atmosphere at transonic velocities, ensuring controlled re-entry and successful engine relighting for a stable landing.

Figure 88: Falcon 9 v1.1 with landing legs attached, in stowed position as the rocket is prepared for launch in its hangar. SpaceX, CC0, via Wikimedia Commons.

The addition of hypersonic grid fins has further improved precision landing capabilities. These fins, arranged in an "X" configuration, control the lift vector of the descending rocket, allowing it to target specific landing locations with greater accuracy. Initially introduced in 2014, the grid fins have undergone significant iteration. Larger, forged titanium fins were introduced in 2017 and are now standard on all reusable Block 5 Falcon 9 first stages, offering enhanced durability and performance.

Another critical innovation is the ability to throttle the rocket engine to achieve zero velocity at the exact moment of landing. Since the minimum thrust of a single Merlin 1D engine exceeds the weight of the nearly empty Falcon 9 booster, the rocket cannot hover. Instead, precision timing and control are required to ensure the rocket lands precisely without overshooting or crashing.

SpaceX's advancements also extend to terminal guidance and landing systems, including sophisticated control algorithms, closed-loop thrust vectoring, and throttle control to manage the rocket's thrust-to-weight ratio. Precision navigation sensors further enhance the accuracy of these landings.

To support missions where the first stage lacks the fuel to return to the launch site, SpaceX developed large, floating landing platforms known as Autonomous Spaceport Drone Ships (ASDS). As of 2022,

SpaceX operates three of these drone ships—one on the West Coast and two on the East Coast of the United States—enabling successful recoveries even for offshore launches.

Figure 89: Falcon Heavy — Closeup of Upper Section. Steve Jurvetson from Menlo Park, USA, CC BY 2.0, via Wikimedia Commons.

Blue Origin, founded by Jeff Bezos, has also made significant advancements in rocket reusability. Its New Shepard rocket, designed for suborbital missions, employs a similar vertical landing approach as SpaceX's Falcon 9. The New Shepard's first stage, powered by a BE-3 engine, has successfully completed multiple flights with both cargo and crew, emphasizing Blue Origin's commitment to making space travel accessible and routine [456]. Additionally, Blue Origin's upcoming New Glenn rocket, currently under development, aims to enhance reusability for orbital missions by returning its first stage to a landing platform on an ocean-going vessel, thus optimizing logistical flexibility [456].

The innovations in reusability by both SpaceX and Blue Origin not only enhance the economic viability of space exploration but also create new opportunities for industries reliant on affordable access to orbit. By minimizing waste and enabling frequent launches, these advancements pave the way for ambitious projects, including satellite constellations, lunar bases, and interplanetary missions [452]. The rapid increase in rocket launches, evidenced by a record-breaking 180 successful launches in 2022, underscores the growing significance of reusability in the aerospace sector [457]. Together, these companies are reshaping the future of aerospace, demonstrating how innovation can bring humanity closer to the stars.

The economic impact of reusable rockets is profound. SpaceX has already reduced the cost of launching payloads to Low Earth Orbit (LEO) to approximately $2,700–3,000 per kilogram, compared to the $10,000 per kilogram typical of expendable rockets [458]. Fully reusable systems promise to bring these costs even lower, potentially under $100 per kilogram, enabling a new wave of scientific, commercial, and exploratory missions [458]. Reduced launch costs democratize access to space, fostering innovation and competition, while unlocking new markets in space tourism, in-orbit manufacturing, and asteroid mining [458].

Reusable rockets also offer environmental benefits by reducing resource consumption and minimizing discarded hardware in space and oceans. However, the increased frequency of launches has contributed to the proliferation of space debris, a growing concern for the sustainability of orbital activities. Effective debris mitigation strategies and international cooperation will be critical to addressing this challenge [458].

In terms of reliability and safety, reusability provides opportunities for continuous improvement through iterative testing and analysis. However, maintaining and refurbishing components subjected to extreme conditions, such as engines and turbopumps, remains a challenge. Advances in materials science and inspection techniques are helping to mitigate these risks, ensuring that reusable rockets maintain high safety standards [458].

Beyond the space industry, reusable rocket technology has far-reaching implications. Innovations in propulsion systems, autonomous navigation, and materials science are driving progress in fields such as aviation, automotive engineering, and renewable energy. Additionally, the affordability of reusable systems encourages international collaboration, allowing more nations and organizations to participate in ambitious space missions [458].

Reusable rockets are not merely an innovation in spaceflight; they represent a paradigm shift that democratizes access to space, fosters global collaboration, and drives technological and economic advancements across multiple sectors. As this technology continues to evolve, it promises to unlock unprecedented opportunities for exploration, sustainability, and progress [458].

The technical advancements driven by rocket reusability have sparked transformative changes not only in space exploration but also across a variety of Earth-based industries. These innovations are rooted in materials science, autonomous navigation, and propulsion systems, with far-reaching implications for technology and global collaboration.

In materials science, the need for reusable rockets has catalysed the development of high-performance, lightweight, and durable materials. Components of reusable rockets must withstand extreme conditions, such as intense heat during atmospheric re-entry, high-speed impacts, and the mechanical stress of multiple launches and landings. This has led to innovations in heat-resistant alloys, advanced composites, and thermal protection systems. For example, SpaceX's forged titanium grid fins and heat shields are designed to endure repeated use without significant degradation. These advancements have potential applications in aviation, where lighter and more durable materials can enhance fuel efficiency and aircraft longevity, and in automotive engineering, where similar materials can improve vehicle safety and performance.

Autonomous navigation systems have also seen significant breakthroughs due to reusable rocket technology. Precision landing systems, such as SpaceX's ability to autonomously land boosters on drone ships or landing pads, rely on advanced sensors, real-time data processing, and machine learning algorithms. These systems ensure the precise control of a rocket's trajectory during descent and landing, even in challenging environments. The expertise gained in autonomous navigation is now influencing the development of self-driving cars, drones, and other autonomous vehicles, improving their safety, reliability, and efficiency.

In propulsion systems, reusability has spurred innovations that enhance both performance and efficiency. Engines like the Merlin 1D and BE-3 are designed for rapid restarts and consistent performance over multiple missions. These systems have advanced the understanding of fuel efficiency, throttle control, and thrust-vectoring technologies, which are critical for reusable launch systems. Similar principles are being adapted for renewable energy solutions, such as in the development of efficient turbines for wind and hydroelectric power generation. The ability to optimize energy output while minimizing wear and tear has direct parallels in these sectors.

The economic benefits of reusability also contribute to broader global collaboration. By reducing the cost of launching payloads into space, reusable rockets democratize access to space exploration, enabling smaller nations and organizations to participate in missions that were previously cost-prohibitive. This shift encourages international partnerships, fostering collaborative research, shared technological advancements, and the development of global standards for space exploration. Initiatives such as multinational satellite constellations and joint lunar exploration missions are becoming more feasible, enhancing scientific knowledge and fostering diplomatic relationships.

The technical advancements driven by reusability are reshaping multiple industries and promoting global collaboration. Innovations in materials science, autonomous navigation, and propulsion systems have broad applications, enhancing efficiency and sustainability in aviation, automotive engineering, and renewable energy. Moreover, the reduced costs and increased accessibility of space exploration are creating opportunities for a diverse range of players to contribute to humanity's expanding presence in space. As these advancements continue to evolve, they promise to unlock new possibilities for technological progress and international cooperation.

Reducing Costs and Environmental Impact

The aerospace industry is undergoing a significant transformation driven by the need to reduce both the costs and environmental impacts associated with rocket launches. This dual focus is reshaping the economics of space exploration while addressing sustainability concerns that have long been associated with traditional rocket systems.

One of the most impactful advancements in reducing launch costs has been the development of reusable rockets. Historically, rockets were designed as single-use systems, leading to high costs due to the need for new components for each launch. Companies like SpaceX and Blue Origin have pioneered reusable rocket technology, exemplified by the Falcon 9 and New Shepard, respectively. These systems can land back on Earth, be refurbished, and launched again, resulting in cost reductions of up to 30% for some missions [459]. The economic implications of this technology are profound, as they allow for more frequent launches at a fraction of the previous costs.

Additionally, modular rocket designs are emerging as a viable strategy for cost reduction. By allowing different configurations to accommodate various payloads, modular systems minimize the need for custom designs, thereby streamlining production and reducing manufacturing costs. This standardization is crucial for enhancing the efficiency of rocket production and enabling more cost-effective launch schedules.

Advancements in manufacturing processes, particularly the adoption of 3D printing, have also contributed significantly to cost reductions. This technology enables the rapid production of complex components while minimizing material waste and reducing production time [460]. The integration of these innovative manufacturing techniques is essential for driving down costs in the aerospace sector.

The environmental footprint of rocket launches is a pressing concern, particularly given the reliance on traditional propellants that emit significant pollutants. Conventional fuels, such as kerosene and hypergolic propellants, contribute to greenhouse gas emissions and can damage the ozone layer. In response, the industry is exploring cleaner alternatives, such as liquid hydrogen and liquid oxygen, which produce only water vapor as a byproduct. Furthermore, the development of biofuels and synthetic fuels is gaining traction as a means to minimize the ecological impact of rocket propulsion systems.

The environmental benefits of reusable rockets extend beyond cost savings; they also reduce the demand for raw materials and the energy-intensive processes associated with manufacturing new components. By reusing rocket stages, the overall resource consumption per launch is significantly lowered, contributing to a more sustainable approach to space exploration

Another critical aspect of environmental sustainability in rocketry is the management of space debris. The increasing frequency of launches raises concerns about the accumulation of debris in orbit, which poses risks to operational satellites and future missions. Innovations such as satellite-servicing spacecraft and deorbiting technologies are being developed to address this issue. Reusable rockets, which return to Earth rather than contributing to orbital debris, represent a proactive solution to mitigate this growing problem.

Several innovative technologies are on the horizon that promise to further reduce the environmental impact of rocket launches. Hybrid propulsion systems, which combine solid and liquid fuels, offer a balance between performance and environmental considerations. Additionally, advancements in material science, particularly the development of lightweight and recyclable composites, are reducing the energy required for manufacturing while maintaining the structural integrity necessary for rocket components. These materials not only enhance the efficiency of production but also align with broader sustainability goals.

The future of the aerospace industry will likely see a continued emphasis on integrating cost reduction strategies with environmental sustainability. As the demand for space exploration grows, collaboration among governments, private companies, and international organizations will be essential to promote sustainable practices and technologies. Ongoing innovation in this field is crucial for balancing the ambitious goals of space exploration with the imperative of protecting our planet.

Environmental Concerns and Sustainability Related to Space Tourism

The expansion of space tourism brings with it significant environmental concerns that must be addressed to ensure long-term sustainability. Rocket launches, a cornerstone of the industry, produce emissions that can contribute to atmospheric pollution. Substances such as black carbon and alumina are released into the upper atmosphere, potentially altering climate patterns and depleting the ozone layer. While the scale of these emissions is currently small compared to commercial aviation, the anticipated growth in the frequency of launches could amplify their environmental impact [451].

In addition to emissions, the resource-intensive nature of rocket manufacturing and fuel consumption poses sustainability challenges. The materials used in rocket construction and the vast amounts of propellant required for launches can strain terrestrial ecosystems and energy resources. To mitigate these effects, the space tourism sector is increasingly exploring cleaner propulsion technologies, sustainable fuel alternatives, and recyclable materials that can reduce the overall ecological footprint of space travel [451].

Space debris is another critical issue tied to the commercialization of space. As more satellites, spacecraft, and debris accumulate in orbit, the risk of collisions increases, potentially leading to a cascading effect known as the Kessler Syndrome. This not only jeopardizes future space missions but also threatens operational satellites that provide vital services such as communication and navigation. Effective debris mitigation strategies, including the development of satellites designed for deorbiting or recycling, are essential to preserving the orbital environment [451].

Beyond Earth, the ethical responsibility to protect extraterrestrial environments is becoming increasingly relevant. As missions extend to the Moon, Mars, and beyond, there is a growing need to establish guidelines that prevent the contamination of celestial bodies. Preserving the scientific integrity of these environments is crucial, as they may hold clues about the origins of the solar system or even extraterrestrial life. Ensuring that space tourism operations adhere to planetary protection principles will play a vital role in maintaining these pristine environments for future exploration [451].

Balancing the excitement of space tourism with environmental stewardship requires a collaborative effort from governments, private companies, and international organizations. By prioritizing sustainability in technology development, operations, and regulatory frameworks, the industry can minimize its environmental impact while paving the way for a responsible and thriving future in space tourism [451].

Advances in Fuel Technology and Green Propulsion

Rocket fuel technology is experiencing significant advancements, particularly in the realm of green propulsion, which aims to reduce the environmental impact of rocket launches while maintaining efficiency and reliability. The aerospace industry is increasingly focused on developing alternative fuels that minimize harmful emissions associated with traditional propellants.

Historically, rocket engines have utilized two primary types of propellants: liquid and solid. Liquid propellants, such as kerosene (RP-1) combined with liquid oxygen (LOX), are favoured for their high energy efficiency and controllability, allowing for engine throttling and restarts. In contrast, solid propellants are valued for their simplicity and reliability but lack the same level of control [461]. However, both types of traditional fuels present environmental challenges. Kerosene-based fuels emit substantial amounts of carbon dioxide and other pollutants, while solid propellants often contain toxic compounds like ammonium perchlorate, contributing to air and water pollution [461, 462].

The transition to green propulsion technologies is driven by the need for safer, cleaner alternatives to conventional fuels. One of the most promising advancements is the use of liquid hydrogen and oxygen (LH2/LOX), which produces only water vapor upon combustion. This technology has been employed in NASA's Space Launch System (SLS) and the Space Shuttle, demonstrating high efficiency with minimal environmental impact [461, 462]. However, challenges such as cryogenic storage and the low density of hydrogen necessitate larger fuel tanks, complicating logistics [461].

Methane-based fuels, or Methalox, represent another significant advancement. When burned with LOX, methane produces fewer harmful byproducts compared to kerosene, making it a cleaner alternative. SpaceX's Starship and Blue Origin's BE-4 engines utilize methane-based propellants, highlighting their potential for long-duration missions and in-situ resource utilization (ISRU) on Mars, where methane can be synthesized from local resources [461, 462].

Biofuels and synthetic fuels derived from renewable sources are also being explored. These fuels can potentially reduce carbon emissions during both production and combustion processes. Synthetic fuels, produced through chemical processes that utilize captured carbon dioxide and renewable energy, offer a pathway toward carbon-neutral propulsion systems [461, 462].

Ionic liquids and green monopropellants are emerging as safer alternatives to traditional toxic monopropellants like hydrazine. These non-volatile, non-toxic propellants, such as NASA's Advanced Spacecraft Energetic Non-Toxic (ASCENT) monopropellant, reduce operational risks associated with handling and storage [461, 462]. Additionally, electric propulsion systems, including ion and Hall-effect thrusters, utilize minimal fuel and are highly efficient for long-duration space missions, significantly reducing the environmental impact of propulsion systems [461, 462].

Recent innovations in green propulsion include the development of catalytic and additive technologies that optimize the combustion of green fuels, enhancing efficiency while minimizing emissions [461, 462]. Furthermore, ISRU technologies are being developed to produce rocket fuels on celestial bodies, such as Mars or the Moon, thereby reducing the need to transport fuel from Earth [461, 462]. Advanced methods for producing fuels from captured carbon dioxide align rocket propulsion with global carbon reduction goals, potentially making space exploration carbon-neutral in the future [461, 462].

Despite the advantages of green propulsion, challenges remain regarding scalability, cost, and performance. The development of cryogenic systems requires advanced insulation and handling technologies, while ISRU and carbon-neutral fuels are still in experimental stages, necessitating further investment and testing [461, 462]. Collaborative efforts among governments, private companies, and academia will be crucial in accelerating the adoption of these technologies, alongside regulatory frameworks that ensure alignment with broader sustainability goals [461, 462].

Chapter 14

Next-Generation Propulsion Systems

Nuclear Propulsion for Deep Space Travel

Nuclear Thermal Propulsion (NTP) systems represent a significant advancement in space propulsion technology by utilizing nuclear fission to heat a propellant, typically hydrogen, which is expelled to generate thrust. The fundamental working mechanism of NTP involves a nuclear reactor that generates heat through fission reactions. This heat is transferred to liquid hydrogen, which is stored cryogenically. As the hydrogen absorbs heat, it expands and exits through a nozzle, producing thrust. This process is analogous to traditional chemical rocket engines, but it relies on nuclear reactions for heating rather than chemical combustion [463].

One of the primary advantages of NTP systems is their high specific impulse (Isp), which is approximately double that of conventional chemical rockets. This enhanced efficiency allows for higher speeds or reduced fuel consumption for the same mission profile, making NTP particularly suitable for long-duration missions to destinations such as Mars. The fast transit times associated with NTP can significantly minimize radiation exposure and life support challenges for astronauts, thereby improving mission safety and feasibility [463]. Furthermore, NTP engines can be scaled for various missions, from crewed Mars expeditions to heavy cargo transport, showcasing their versatility in space exploration [463].

However, the development and implementation of NTP systems face several challenges. Reactor safety is a paramount concern, as creating reactor systems that are safe for launch and operation in the harsh environment of space requires rigorous engineering and testing protocols. Additionally, managing cryogenic propellant, particularly maintaining hydrogen in its liquid state over extended missions, poses significant thermal insulation challenges that must be addressed to ensure mission success [463].

Nuclear Electric Propulsion (NEP) systems represent a significant advancement in space exploration technology, converting nuclear energy into electricity to power electric thrusters, such as ion or Hall-effect thrusters. This method of propulsion is particularly advantageous for long-duration missions due to its sustained low-thrust capabilities, which are essential for deep-space travel.

The fundamental mechanism of NEP involves a nuclear reactor that generates heat, which is then converted into electricity. This conversion can be achieved through thermoelectric or thermodynamic processes. The generated electricity is utilized to power electric thrusters that ionize and accelerate propellant, typically xenon, to produce thrust. For instance, the Prometheus project has explored advanced nuclear electric propulsion systems capable of supporting ambitious missions requiring significant changes in velocity (ΔV) [65]. Additionally, studies have indicated that NEP can effectively utilize gravity assists from inner planets to reach outer planets, enhancing mission efficiency [464].

One of the primary advantages of NEP systems is their high efficiency. Compared to chemical propulsion systems, NEP offers significantly higher fuel efficiency, enabling extended mission durations without the need for frequent resupply of propellant [465]. This efficiency is particularly beneficial for missions that require fine trajectory adjustments or prolonged operations in deep space, as highlighted in various studies [466, 467]. Furthermore, the electricity generated by NEP systems can support onboard scientific instruments, communication systems, and life support, making them versatile for various mission requirements [465].

Despite their advantages, NEP systems face several challenges. The continuous but low-thrust nature of NEP makes it unsuitable for quick manoeuvres or rapid acceleration, which can be a limitation during critical mission phases [466]. Additionally, effective thermal management is crucial, as dissipating excess heat from the reactor is essential to maintain operational integrity [468]. The complexity of integrating the reactor, power conversion systems, and thrusters also poses engineering challenges that must be addressed in the design and implementation of NEP systems [469].

Ongoing advancements in materials science, reactor miniaturization, and energy conversion technologies are making nuclear propulsion increasingly viable. Initiatives by NASA, ESA, and private companies indicate a growing commitment to integrating nuclear systems into future space missions. Concepts like NASA's Nuclear Thermal Propulsion Reactor Element project and DARPA's DRACO (Demonstration Rocket for Agile Cislunar Operations) program exemplify the active development of nuclear propulsion technologies.

NASA's NTP Reactor Element project focuses on advancing nuclear thermal propulsion, a technology that uses a nuclear reactor to heat a propellant like liquid hydrogen. The heated propellant expands and is expelled through a nozzle to produce thrust. This method offers significantly higher efficiency than traditional chemical propulsion, allowing for reduced travel times and increased payload capacity for deep space missions. The project involves rigorous testing and validation of reactor components to ensure safety and reliability in space environments. By developing scalable and robust nuclear propulsion systems, NASA aims to support future crewed missions to Mars and beyond, where the enhanced performance of NTP could play a critical role in achieving mission objectives.

DARPA's DRACO program takes a complementary approach by exploring the application of nuclear thermal propulsion for agile and responsive operations in the cislunar space—the region between Earth and the Moon. DRACO's objective is to demonstrate the feasibility of nuclear-powered spacecraft for rapid manoeuvring and extended operational capabilities in this increasingly strategic region. The program highlights the dual-use potential of nuclear propulsion systems, serving both national security interests and broader space exploration goals. By enabling spacecraft to travel faster and operate more efficiently, DRACO seeks to establish a new paradigm for space operations, particularly in environments where traditional propulsion systems are less effective.

Both projects underscore the growing recognition of nuclear propulsion as a transformative technology for future space exploration. By combining high efficiency with substantial thrust capabilities, these systems promise to overcome many of the limitations of current propulsion technologies. The advancements made through NASA's NTP and DARPA's DRACO initiatives not only pave the way for ambitious interplanetary missions but also enhance the strategic capabilities of spacefaring nations in an era of rapidly evolving space exploration.

Solar Sails and Laser Propulsion

Solar sails and laser propulsion are innovative, non-chemical propulsion systems that harness the momentum of photons to generate thrust, offering a sustainable and potentially limitless energy source for deep space exploration. Both technologies aim to overcome the limitations of traditional propulsion methods, such as reliance on finite fuel supplies and limited efficiency over long distances.

Solar sails represent a novel propulsion method that utilizes the momentum of photons emitted by the Sun to generate thrust. The fundamental principle behind solar sails is the transfer of momentum from photons to the sail's surface upon reflection. Although photons are massless, they carry momentum, which can be harnessed to propel a spacecraft. This method is particularly advantageous for long-duration missions, as it does not require onboard fuel and allows for continuous acceleration over extended periods, making it suitable for deep space exploration [470, 471].

The structure and design of solar sails are critical to their functionality. Typically, they are constructed from ultra-thin, lightweight, and highly reflective materials such as aluminized Mylar or Kapton. These materials are chosen for their low weight and high reflectivity, which maximize the momentum transfer from sunlight to the sail [472, 473]. Solar sails can vary significantly in size, often ranging from tens to hundreds of square meters, and they are designed to be compactly folded during launch, deploying once in space [474, 475]. The deployment mechanisms can include innovative designs such as shape memory alloys that facilitate self-deployment, enhancing the feasibility of solar sail missions [476, 477].

The working mechanism of solar sails relies on the pressure exerted by sunlight on the sail's surface, producing a small but constant force. While the initial acceleration may be slow due to the low intensity of solar radiation, the continuous thrust allows the sail to achieve high velocities over time. This characteristic is particularly beneficial for missions that require sustained propulsion without the constraints of fuel limitations [478, 479]. The ability to maintain a trajectory and perform orbital manoeuvres, such as station-keeping at Lagrange points, further underscores the versatility of solar sails in various space applications [480, 481].

Figure 90: LightSail 1 captured this image of its deployed solar sails in Earth orbit on 8 June 2015. The Planetary Society, CC BY-SA 3.0, via Wikimedia Commons.

Solar sails have numerous applications, particularly in interplanetary missions where sunlight is abundant. For instance, NASA's NEA Scout mission aims to utilize solar sails to study near-Earth asteroids, demonstrating their potential for scientific exploration [470, 471]. Additionally, solar sails are being considered for interstellar exploration initiatives, such as the Breakthrough Starshot project, which seeks to send small probes to nearby stars using light propulsion [481, 482]. The advantages of solar sails include reduced mission costs due to the absence of fuel, continuous acceleration capabilities, and scalability for diverse mission profiles [479, 480].

However, challenges remain in the deployment and maintenance of solar sails in the harsh environment of space. The intensity of solar radiation diminishes with distance from the Sun, which can reduce thrust effectiveness. Moreover, the engineering challenges associated with deploying large, thin sails necessitate advanced materials and design strategies to ensure structural integrity and functionality during missions [483, 484].

Laser propulsion represents a cutting-edge technology that utilizes high-powered lasers to propel spacecraft equipped with reflective sails. This method leverages the momentum transfer from laser photons to the sail, generating thrust in a manner akin to solar sails but with significantly enhanced energy density and control capabilities. The underlying mechanics of laser propulsion involve directing focused energy beams at a spacecraft, which can be positioned either on the ground or in orbit. The efficiency of this system is predicated on the ability of the sail to withstand intense laser energy without degradation, necessitating advanced materials and design considerations [485, 486].

The structure and design of laser-propelled spacecraft are critical for their operational success. A reflective sail must be engineered to endure the high-energy laser beams, which can reach intensities of 1 GW/m^2 or more, as indicated by recent studies [486]. The laser system itself comprises arrays of high-powered lasers capable of maintaining precision over long distances, which is essential for effective propulsion and trajectory control [487, 488]. This design allows for a high thrust-to-weight ratio, making laser propulsion particularly advantageous for deep space missions where traditional propulsion methods may falter due to fuel limitations [485, 486].

The working mechanism of laser propulsion is fundamentally based on photon pressure. When the laser beam strikes the sail, it imparts momentum, thereby generating thrust. This mechanism allows for controlled, high-intensity thrust, which is a significant advantage over solar sails that rely solely on sunlight ([489]. The ability to modulate thrust intensity also enables more flexible trajectories, which is crucial for missions requiring precise navigation [488].

Applications of laser propulsion are diverse and promising. One of the most notable projects is Breakthrough Starshot, which aims to send microprobes to Alpha Centauri at speeds reaching 20% of the speed of light [490]. This initiative exemplifies the potential of laser propulsion for interstellar exploration, where traditional propulsion systems would be inadequate. Additionally, laser propulsion can facilitate cargo missions across interplanetary distances, significantly reducing the need for onboard fuel and enhancing mission efficiency [485, 490].

Despite its advantages, laser propulsion faces several challenges. The establishment of a massive ground or orbital infrastructure for laser arrays is a significant hurdle, as is the atmospheric distortion that can diminish laser efficiency when operating from Earth [486, 487]. Furthermore, the materials used for the sails must exhibit exceptional durability to withstand prolonged exposure to laser energy without degrading, which remains an area of active research [486].

Antimatter and Fusion-Based Propulsion Concepts

Antimatter and fusion-based propulsion concepts represent cutting-edge theoretical approaches to achieving revolutionary advancements in space travel, offering unparalleled efficiency and thrust capabilities compared to conventional propulsion systems.

Antimatter Propulsion Concepts

Antimatter propulsion leverages the annihilation reaction that occurs when antimatter and matter collide. This reaction releases an immense amount of energy, approximately 100% of the mass-energy equivalence described by Einstein's equation $E=mc^2$. This efficiency far exceeds that of chemical or even nuclear propulsion, making antimatter one of the most energy-dense substances theoretically available for propulsion.

In an antimatter propulsion system, antimatter (such as positrons or antiprotons) would be stored in magnetic or electric traps to prevent it from coming into contact with normal matter. Upon controlled release, antimatter would interact with matter (usually hydrogen), producing high-energy gamma rays and particles. These products could be directed to produce thrust using magnetic nozzles or other advanced systems.

Antimatter is the counterpart to ordinary matter, comprising antiparticles that share the same mass but have opposite charges and other quantum properties. For example, the antimatter equivalent of an electron, known as a positron, has a positive charge, while an antiproton, the counterpart to the proton, carries a negative charge. Similarly, antineutrons mirror neutrons but with opposite quantum characteristics. When antimatter and matter meet, they annihilate each other, converting their entire mass into energy. This annihilation process is highly efficient, releasing vast amounts of energy, which positions antimatter as a promising candidate for advanced propulsion systems, cutting-edge medical applications, and fundamental scientific exploration.

Antimatter is not naturally abundant on Earth because it annihilates upon contact with matter. However, it can be produced in small quantities through high-energy interactions, both in natural and artificial settings. Naturally, antimatter is generated by cosmic rays interacting with Earth's atmosphere, creating positrons and other antiparticles. Though these processes occur regularly, the quantities produced are minuscule, making them impractical for large-scale collection. Some radioactive isotopes, such as potassium-40, also emit positrons during beta decay, but this process is similarly inefficient for significant antimatter production.

Artificially, antimatter is primarily created in particle accelerators, such as those at CERN. High-energy collisions within these facilities produce antiprotons and other antiparticles as byproducts, which are then captured and stored using electromagnetic fields to prevent contact with regular matter. Positrons can also be generated through nuclear reactions, like the decay of isotopes used in medical imaging technologies such as Positron Emission Tomography (PET) scans. Despite these advancements, the production of antimatter remains limited.

Storing antimatter poses unique challenges since it annihilates upon contact with conventional matter. Specialized containment systems, such as Penning traps, use electromagnetic fields to suspend antimatter particles in a vacuum, ensuring they remain isolated. Facilities like CERN's Antiproton Decelerator are designed to slow down antiprotons for controlled research and storage, but storage durations are still limited.

Producing and storing antimatter is fraught with challenges. The yield from particle accelerators is extremely low, with global production measured in nanograms annually. The high cost of production, estimated at $100 billion per milligram of antiprotons, further restricts large-scale use. Additionally, while storage systems have improved, long-term containment remains technically complex. Current systems allow storage for only about 1,000 seconds under controlled conditions, underscoring the need for advancements in this area.

Future advancements in particle physics, materials science, and energy management hold the potential to make antimatter production and storage more efficient. Dedicated facilities designed for high-yield antimatter production could significantly boost output, while improved magnetic traps might enable the containment of larger quantities for extended periods. There is also growing interest in space-based collection methods, which could tap into naturally occurring antimatter in Earth's magnetic field or other cosmic sources.

Despite its current limitations, antimatter remains a focus of research due to its immense potential. Its applications in energy generation, space propulsion, and medicine promise revolutionary breakthroughs, driving continued innovation and investment in this field. As technology advances, the barriers to antimatter production and use may eventually be overcome, opening new frontiers in science and industry.

The primary advantage of antimatter propulsion is its extraordinary energy density, enabling spacecraft to achieve extremely high speeds, potentially reaching significant fractions of the speed of light. This makes it an ideal candidate for interstellar travel. However, technical challenges remain daunting. These include the production and storage of antimatter, which currently exists only in minute quantities in laboratory settings, and the development of systems to safely harness its annihilation energy.

Antimatter-Catalysed Nuclear Pulse Propulsion

Antimatter-catalysed nuclear pulse propulsion (ACNPP) represents a significant advancement in space propulsion technology, leveraging the energy released from antimatter-matter annihilation to initiate nuclear reactions. This innovative approach differs from traditional nuclear pulse propulsion by utilizing small quantities of antimatter, such as antiprotons, to catalyse fission and fusion reactions in subcritical masses of nuclear fuel, including plutonium and uranium. The mechanism involves injecting antiprotons into a mass of nuclear fuel, where they annihilate upon contact with the positively charged nuclei, producing high-energy gamma rays and kinetic energy that facilitate fission and potentially fusion reactions, thereby amplifying the energy yield significantly [491, 492].

The theoretical framework for ACNPP suggests that a microgram of antihydrogen could replace the conventional fission trigger in thermonuclear devices. In this design, the antimatter is magnetically levitated and cooled, isolated from the surrounding fuel until a desired reaction is initiated through compression with high-explosive lenses [492]. This method allows for a highly efficient propulsion system that can be tailored to mission requirements, emphasizing either high-thrust microfission reactions for rapid missions or a combination of fission and fusion for sustained thrust over longer durations [492, 493].

Despite its potential, the development of ACNPP faces substantial challenges, particularly in the areas of antimatter production and storage. Currently, the global production of antimatter is limited to nanograms annually, with costs for producing even a milligram estimated at around $100 billion [494]. Moreover, the safe and reliable storage of antimatter is critical, as it annihilates upon contact with normal matter, necessitating advanced containment technologies [494]. Noteworthy advancements, such as the successful storage of antiprotons for over 1,000 seconds at CERN, indicate progress but also highlight the nascent stage of this technology [494].

Research into ACNPP has been ongoing since its initial proposal by researchers at Pennsylvania State University in the early 1990s, with various studies exploring its feasibility for space propulsion [491, 492, 494]. Institutions like Lawrence Livermore National Laboratory have investigated antiproton-initiated fusion for inertial confinement systems, demonstrating the potential of antimatter as a viable energy driver that could significantly reduce the mass and complexity of traditional propulsion systems [492]. However, practical implementation will require breakthroughs in antimatter production, containment, and injection technologies to make ACNPP a reality [493].

The implications of successfully developing ACNPP for space exploration are profound. Its high energy efficiency could enable rapid travel to distant planets and even interstellar destinations, drastically reducing mission durations and expanding the scope of human exploration [492]. Furthermore, the reduced size and weight of propulsion systems could facilitate more compact spacecraft designs, enhancing payload capacities and reducing costs [494]. Nevertheless, realizing these benefits hinges on overcoming significant technological barriers and making antimatter production economically viable [493, 494].

Fusion-Based Propulsion Concepts

Fusion-based propulsion systems harness the energy released from nuclear fusion reactions, the same process that powers stars. In a fusion reaction, light atomic nuclei, such as isotopes of hydrogen (deuterium and tritium), combine under extreme conditions of pressure and temperature, releasing energy in the form of high-speed particles. This energy can be effectively converted into thrust, making fusion propulsion a promising avenue for advanced space exploration [67, 495, 496].

Several concepts for fusion propulsion have been proposed, each utilizing different mechanisms to achieve the necessary conditions for fusion. One prominent method is Inertial Confinement Fusion (ICF), which employs lasers or ion beams to compress a small pellet of fusion fuel, achieving the conditions required for fusion. The energy released from this reaction can be directed through a nozzle to produce thrust [67, 496]. Another approach is Magnetic Confinement Fusion (MCF), which utilizes magnetic fields to contain and control hot plasma. Systems such as tokamaks or stellarators could potentially be adapted for propulsion, directing the energy from the fusion reaction into a controlled exhaust [495, 497]. The Direct Fusion Drive (DFD) concept combines magnetic confinement with direct conversion of fusion energy into thrust, promising high efficiency and specific impulse [498].

Fusion propulsion offers several advantages over traditional chemical propulsion systems. It has a significantly higher energy density than chemical propulsion and even nuclear fission, enabling faster travel to distant destinations within the solar system and potentially beyond [495, 499, 500]. Additionally, fusion reactions produce fewer harmful byproducts, making them a cleaner propulsion option [67, 496]. The ability to eliminate mission window constraints allows for more flexible mission planning, which is particularly beneficial for interplanetary exploration [500].

However, fusion-based propulsion systems face considerable challenges. Achieving and sustaining fusion reactions in a compact, spacecraft-friendly system remains a major engineering hurdle [492, 497]. Issues related to fusion fuel storage and management, as well as heat dissipation and energy conversion systems, must be addressed to make these systems viable for practical use [495, 500]. Despite these challenges, ongoing research continues to explore the feasibility of various fusion propulsion concepts, with the potential to revolutionize space travel in the future [67, 492, 498].

The use of advanced fuels like helium-3 makes this approach even more viable. Helium-3, an isotope abundant on the Moon, can be fused with deuterium to produce energy, propelling spacecraft without significant radioactive byproducts. However, the deuterium-tritium reaction commonly used in fusion releases a large portion of its energy as neutrons, making direct thrust generation less efficient and necessitating heavy shielding. Helium-3's advantages, coupled with its potential lunar abundance, make it a promising candidate for future propulsion systems.

Even if self-sustaining fusion cannot be achieved, hybrid systems may leverage fusion to enhance the performance of other propulsion technologies, such as the Variable Specific Impulse Magnetoplasma Rocket (VASIMR). These hybrid systems would combine the benefits of fusion's high energy output with existing efficient propulsion methods.

Fusion confinement approaches also shape propulsion system designs. Magnetic confinement, as in tokamaks, traps plasma in magnetic fields to sustain fusion reactions. While tokamaks are currently heavy and unsuitable for spacecraft, smaller designs like NASA's proposed spherical torus reactor for the "Discovery II" vehicle could revolutionize this field. This concept envisions delivering a substantial payload to destinations like Jupiter or Saturn using fusion-powered hydrogen heating for thrust, achieving impressive exhaust velocities and reduced propellant mass.

Inertial confinement fusion (ICF), exemplified by Project Daedalus, uses high-energy beams or lasers to ignite small fusion fuel pellets. This approach could employ helium-3 or aneutronic fusion to maximize charged particle energy output while minimizing radiation. Concepts like the VISTA spacecraft demonstrate the potential of ICF-powered vehicles to carry substantial payloads to planetary orbits, although they require massive quantities of propellant and fusion fuel.

Magnetized target fusion (MTF) blends the strengths of magnetic and inertial confinement, using plasma guns to compress magnetically confined fuel to achieve fusion. This method offers compact, cost-effective reactors and has been explored in designs like NASA's HOPE spacecraft, which could deliver payloads to Jupiter with high fuel efficiency and exhaust velocities.

Inertial electrostatic confinement (IEC) represents another innovative approach, employing electric fields to trap and compress fusion fuel. Designs like the "Fusion Ship II" propose using D-He3 IEC reactors to power ion thrusters, offering efficient, long-duration missions to the outer planets. Such concepts promise significant advances in propulsion capability, although they remain in the conceptual stage.

Antimatter-catalysed propulsion represents a more speculative but potentially transformative technology. By using antimatter to initiate fission and fusion reactions, this method could enable smaller, more efficient fusion explosions. While research, such as the AIMStar project, has explored this concept, technical and production challenges related to antimatter remain formidable, preventing its near-term feasibility.

Together, these diverse propulsion technologies represent the forefront of space exploration innovation, each addressing unique challenges and promising unprecedented capabilities for future interplanetary and interstellar missions.

Implications for Space Exploration

Both antimatter and fusion propulsion concepts hold the promise of enabling humanity to venture far beyond the limitations of current propulsion technologies. They could reduce travel times to distant planets and even make interstellar travel feasible. For example, a fusion-powered spacecraft might reach Mars in weeks rather than months, while an antimatter-powered vehicle could reach the nearest stars within decades.

As research and technology development progress, these propulsion concepts may revolutionize how we approach space exploration, unlocking the potential for faster, more efficient, and more sustainable space missions. However, significant scientific and engineering breakthroughs are required before these systems can transition from theoretical concepts to practical applications.

Chapter 15
The Role of AI and Robotics in Rocketry

Autonomous Spacecraft Systems

Autonomous spacecraft systems represent a transformative advancement in space exploration, allowing spacecraft to operate independently from ground control stations. This capability is crucial for future missions, particularly those involving collaborative operations among multiple spacecraft or those venturing into remote areas of space where real-time communication is impractical. The integration of autonomy in space missions enhances operational efficiency and mission reliability, enabling spacecraft to make decisions based on real-time data without waiting for instructions from Earth [501].

One prominent example of such autonomous systems is NASA's Distributed Spacecraft Autonomy (DSA) project. This initiative aims to enable multiple spacecraft to function cohesively as a unit, integrating advanced capabilities such as scalable communication networks, distributed coordination, and human-swarm interaction. These features empower spacecraft to share information, coordinate actions, and adapt to changing mission requirements autonomously. The DSA project is particularly significant for missions that require large constellations or swarms of satellites, which are essential for Earth observation, planetary exploration, and interstellar research [501].

The Distributed Spacecraft Autonomy (DSA) project, initiated by NASA, is a groundbreaking initiative aimed at developing advanced autonomy capabilities for spacecraft operating within distributed architectures. This project focuses on enabling multiple spacecraft to function collaboratively as a unified system, reducing reliance on ground control and significantly enhancing mission efficiency and reliability. It addresses the challenges posed by complex missions involving multiple spacecraft, such as swarm formations, constellation management, and collaborative sensing.

The primary objectives of the DSA project include the development of scalable communication systems that allow seamless interaction between spacecraft, enabling efficient data sharing and task coordination. Additionally, it aims to facilitate distributed coordination, allowing spacecraft to collaborate autonomously in real time to achieve mission objectives. Another key aspect of the project is the design of intuitive interfaces for operators to interact effectively with a network of spacecraft, thereby improving mission oversight and decision-making.

The technical features of the DSA project are built on cutting-edge advancements in autonomy and artificial intelligence. The project employs AI algorithms that enable spacecraft to make decisions and adapt to changing mission environments or unexpected conditions. Through task allocation, AI autonomously assigns responsibilities among spacecraft based on mission priorities and available resources. Machine learning models further enhance behavioural adaptation, allowing spacecraft to respond to anomalies or system failures dynamically. Collaborative learning among spacecraft ensures continuous improvement by sharing insights and experiences.

Distributed sensing and coordination form the backbone of DSA's capabilities. Spacecraft in distributed missions can integrate data from multiple sources through advanced data fusion techniques, providing a comprehensive understanding of their environment. Relative positioning algorithms ensure precise coordination for tasks such as imaging or scientific experiments. Moreover, the system allows dynamic reconfiguration of the constellation to optimize coverage or avoid potential hazards, enhancing mission flexibility.

Communication architecture in DSA is designed to be scalable and resilient. Inter-satellite links enable high-bandwidth communication directly between spacecraft, minimizing dependency on ground relays. The network is scalable, supporting the addition of new spacecraft without compromising efficiency. Fault tolerance is another critical feature, with systems designed to handle communication failures and reroute data through alternate paths to maintain uninterrupted operations.

Resilience and fault management are integral to the DSA project. Onboard systems equipped with anomaly detection and recovery capabilities ensure continuous functionality by isolating and addressing faults autonomously. The distributed nature of the system provides inherent redundancy, allowing the network to continue operations even if individual spacecraft fail. Autonomous energy management further optimizes power usage across the constellation, extending mission lifespans.

The DSA project has diverse applications, including Earth observation, where multiple satellites can simultaneously focus on different regions or phenomena for comprehensive analysis. In asteroid and planetary exploration, swarms of small spacecraft equipped with DSA capabilities can perform distributed sensing and coordinated manoeuvres. Space weather monitoring is another key application, with networks of spacecraft providing real-time updates and improving predictive models. Additionally, DSA can play a vital role in space traffic management by tracking and managing objects in orbit, reducing collision risks, and ensuring sustainable space use.

Despite its promise, the DSA project faces challenges. High-performance onboard computing is essential for AI and autonomy, but spacecraft power and thermal constraints remain significant hurdles. Communication latency, particularly in deep-space missions, poses an ongoing technical challenge. Integration with legacy systems requires substantial modifications, and ensuring robust cybersecurity is critical to protect the network from potential threats.

Looking ahead, NASA envisions expanding the capabilities of the DSA project by integrating quantum communication for enhanced security and exploring swarm intelligence for more complex missions. Efforts are also underway to develop self-healing mechanisms, further improving system resilience and reliability.

The Distributed Spacecraft Autonomy project represents a transformative approach to managing and deploying spacecraft constellations. By enabling autonomous operations, collaborative decision-making, and efficient communication, DSA is set to revolutionize fields such as Earth observation, planetary exploration, and space traffic management. Its capabilities position NASA at the forefront of distributed space system technology, paving the way for a new era in space exploration.

Another noteworthy project is SCARLET, which focuses on enhancing AI-based autonomy for next-generation space systems. SCARLET aims to minimize communication delays and operational costs by equipping spacecraft with sophisticated decision-making algorithms. This initiative is vital for missions that require immediate responses to environmental changes or anomalies, thereby improving spacecraft efficiency and responsiveness. By integrating artificial intelligence into mission planning and execution, SCARLET addresses challenges such as data latency and limited access to ground stations, which are critical for the success of autonomous operations in space.

The SCARLET (Spacecraft Autonomy, Resilience, and Learning Enhanced Technology) project is a groundbreaking initiative aimed at advancing AI-based autonomy for the next generation of space systems. It focuses on enhancing the operational independence, reliability, and efficiency of spacecraft, especially in scenarios where real-time communication with ground control is limited or impractical. By integrating advanced machine learning, onboard decision-making algorithms, and adaptive systems, SCARLET addresses critical challenges for modern and future space missions.

SCARLET seeks to minimize reliance on ground-based control by enabling spacecraft to make real-time decisions and adapt to evolving mission parameters or unexpected conditions. The project emphasizes resilience by improving system reliability and fault tolerance, ensuring mission success even in the presence of anomalies. SCARLET also aims to optimize mission operations by reducing communication and data latency, a crucial capability for deep-space and distributed satellite systems. Additionally, the project is designed to lower mission costs by automating processes traditionally handled by ground control, reducing the need for extensive human oversight.

At its core, SCARLET leverages advanced AI techniques to empower spacecraft with the ability to perform complex tasks independently. This includes the use of machine learning models trained to recognize patterns, predict system behaviour, and adapt to new conditions. Reinforcement learning allows spacecraft to identify optimal strategies through trial and error, while neural networks are deployed for tasks such as image recognition, anomaly detection, and navigation.

The project incorporates powerful onboard data processing systems to handle large volumes of information in real time. High-performance processors are custom-designed to handle AI workloads with low power consumption, while edge computing processes data locally, reducing latency and dependency on ground-based resources. Data compression and prioritization ensure that only critical information is transmitted back to Earth, conserving bandwidth and power.

Resilience is a cornerstone of SCARLET's design, with advanced anomaly detection systems identifying deviations from expected behaviour and triggering automated recovery protocols. Self-healing systems enable spacecraft to isolate and repair faults autonomously, maintaining operational continuity. Redundant architectures provide backup systems to ensure continued functionality in case of hardware or software failures.

SCARLET also facilitates seamless collaboration in missions involving multiple spacecraft. Swarm intelligence allows satellites to communicate and coordinate their actions, achieving collective objectives efficiently. Distributed ledger technology ensures secure and reliable data exchange, while scalable communication protocols adapt to the complexity of tasks and the number of satellites in the network.

SCARLET has a wide range of applications, including deep-space missions where communication delays make real-time control impractical. Its autonomy is crucial for navigating asteroid belts, conducting scientific experiments on distant planets or moons, and adapting to unforeseen challenges like dust storms or equipment malfunctions. SCARLET also enhances the functionality of distributed satellite networks by enabling autonomous reconfiguration to maintain coverage and avoid collisions, as well as coordinated data collection for Earth observation and space situational awareness. In time-critical missions, SCARLET's capabilities allow spacecraft to respond in real time, monitoring and reacting to transient phenomena like solar flares and supporting defence applications such as reconnaissance and missile detection.

Despite its promise, SCARLET faces technical challenges, including balancing AI performance with the constraints of onboard hardware, such as power and thermal limits. Ensuring the reliability of AI systems across all scenarios, including unforeseen edge cases, remains a priority. Additionally, integrating SCARLET with legacy systems and diverse mission architectures presents further complexities.

Future advancements in SCARLET may include the integration of quantum computing to enhance AI capabilities, expansion into fully autonomous mission planning and execution, and deployment in emerging space applications like space tourism and asteroid mining. SCARLET represents a significant leap forward in spacecraft autonomy, combining cutting-edge AI, robust resilience mechanisms, and efficient data management. By reducing reliance on ground control and enhancing operational efficiency, SCARLET is poised to transform the landscape of space exploration, making missions more capable, cost-effective, and adaptable to the challenges of the final frontier.

The DARPA Blackjack program exemplifies the application of autonomous systems in defence and security missions. This experimental satellite constellation program seeks to establish a network of small satellites with autonomous capabilities for orbital operations. The Pit Boss subsystem within Blackjack emphasizes autonomy, system integration, and cybersecurity, ensuring that the constellation can perform complex tasks while remaining resilient against cyber threats [502]. This program represents a significant advancement toward creating robust, decentralized satellite networks capable of reconnaissance, communication, and navigation without heavy reliance on ground control.

The DARPA Blackjack program is an innovative initiative aimed at revolutionizing defence satellite architecture by deploying a constellation of small, low-cost satellites in Low Earth Orbit (LEO). This project leverages commercial space technology to create a resilient and distributed satellite network that supports national security missions. The program's primary goals include enhancing mission resilience by replacing traditional monolithic satellites with distributed constellations that are more difficult to target and disable. It also emphasizes cost efficiency by utilizing commercially available satellite buses and payloads to reduce development and launch expenses significantly. Another critical objective is to develop advanced autonomous capabilities, reducing reliance on ground control and enabling rapid, independent decision-making in orbit. Additionally, Blackjack aims to integrate military and commercial data streams seamlessly to provide enhanced situational awareness.

The architecture of the Blackjack satellite constellation is designed to maximize efficiency and resilience. The program envisions deploying between 20 to 200 small satellites in LEO at altitudes ranging from 500 to 1,200 kilometres, ensuring low-latency communication and high revisit rates over target areas. The satellites use standard commercial satellite buses, which reduce production costs and capitalize on existing manufacturing expertise. This distributed network ensures that the failure of one satellite does not compromise the entire system, as satellites communicate with each other using inter-satellite links to create a self-healing mesh network.

A cornerstone of the program is the Pit Boss subsystem, an advanced autonomy processor that enables satellites to operate independently or in coordination with minimal ground intervention. Pit Boss processes and prioritizes data onboard, coordinates distributed operations among satellites, and implements robust cybersecurity protocols to safeguard operations. This system integrates cutting-edge technologies such as onboard artificial intelligence (AI) for real-time decision-making and distributed ledger technology for secure data sharing and command verification.

The program's payloads include a mix of military and commercial capabilities, such as Intelligence, Surveillance, and Reconnaissance (ISR) sensors for real-time imaging and data collection, communication payloads for secure data relay, and missile detection sensors for early warning systems. These payloads are modular, allowing for a flexible approach to mission requirements. Launch and deployment are facilitated through commercially available small satellite launch vehicles like SpaceX's Falcon 9 and Rocket Lab's Electron, enabling quick and cost-effective satellite deployment. Satellites can be launched in batches to scale up the constellation or replace damaged units efficiently.

The Blackjack program focuses on cost-effectiveness, with DARPA targeting a unit cost of $6 million per satellite, significantly lower than traditional military satellites that cost hundreds of millions. This cost reduction is achieved by relying on commercial off-the-shelf (COTS) components, minimizing custom development, and reducing lead times. The modular and scalable design of the constellation allows for the seamless integration of new satellites or upgrades without disrupting existing operations.

Despite its benefits, the program faces challenges, including integrating commercial and military technologies to meet strict reliability and security standards. Developing high levels of autonomy while maintaining trust and transparency in system decisions is another critical hurdle. Moreover, cybersecurity concerns must be addressed to protect the constellation from potential cyberattacks, particularly given its reliance on interconnected networks and commercial platforms.

The advantages of autonomous spacecraft systems are manifold. They enhance mission reliability by enabling spacecraft to swiftly respond to unexpected conditions, such as avoiding collisions with space debris or adapting to equipment failures. Furthermore, autonomy increases productivity by allowing spacecraft to execute tasks independently, which is particularly beneficial for time-sensitive operations. Additionally, reducing the need for constant communication lowers operational risks and costs, making space missions more sustainable and scalable.

However, the development of autonomous spacecraft systems is not without challenges. A significant hurdle is the integration of various systems, as ensuring seamless coordination between software, hardware, and mission objectives necessitates sophisticated engineering. Moreover, autonomous systems require substantial onboard computing power, which is often constrained by weight, energy, and space limitations on spacecraft. Addressing these challenges will demand advancements in computing technology, system architecture, and AI algorithms.

As space exploration ambitions grow, autonomous spacecraft systems will play a crucial role in enhancing mission capabilities and expanding humanity's reach into the cosmos. By enabling spacecraft to operate independently, these systems facilitate more complex, efficient, and cost-effective missions, heralding a new era of exploration.

AI-Driven Design and Optimization

Artificial intelligence (AI) and machine learning (ML) techniques are being employed to improve every aspect of rocket development, from conceptual design to testing, manufacturing, and operational performance.

AI algorithms play a crucial role in the early stages of rocket development by exploring design spaces more efficiently than traditional methods. Machine learning models can analyse vast datasets to identify optimal design configurations based on mission requirements. AI-driven tools allow engineers to simulate aerodynamics, propulsion, and structural integrity, reducing the need for time-consuming and costly physical prototypes.

Generative design algorithms, a subset of AI, enable the creation of innovative and unconventional rocket components by exploring thousands of potential geometries. These algorithms balance competing factors like weight, strength, and thermal properties, resulting in highly optimized designs tailored to specific missions.

AI systems are being used to identify and develop advanced materials for rocket construction. By analysing material properties and testing data, AI can recommend lightweight, durable, and heat-resistant materials. Additionally, AI-driven tools can simulate material behaviour under extreme conditions, such as high temperatures and pressures, ensuring reliability in flight.

In additive manufacturing (3D printing), AI enhances the process by optimizing material usage and identifying potential flaws before production. This capability reduces waste and accelerates the development of complex rocket components.

AI is transforming propulsion system design by optimizing engine configurations and improving fuel efficiency. Machine learning models can analyse historical engine performance data to predict how design changes will impact thrust, fuel consumption, and heat generation. These insights enable engineers to refine propulsion systems for specific mission profiles, such as low-Earth orbit (LEO) launches or interplanetary travel.

AI also facilitates the development of advanced propulsion systems like electric or nuclear engines by modelling complex interactions between physical and chemical processes. This capability speeds up innovation and helps engineers overcome challenges in developing next-generation propulsion technologies.

Rocket testing is a critical phase where AI can significantly improve efficiency and accuracy. AI-powered systems monitor test data in real time, identifying anomalies and providing actionable insights. Predictive analytics allow engineers to foresee potential failures and address them before they occur, reducing the risk of costly test failures.

In manufacturing, AI enhances quality assurance by using computer vision and sensor data to detect defects in rocket components. These systems can identify microscopic cracks, inconsistencies, or misalignments that might otherwise go unnoticed, ensuring the reliability of each component.

AI-driven optimization extends to the development of autonomous flight systems, which manage rocket navigation, trajectory adjustments, and landing operations. For reusable rockets, AI algorithms process sensor data to guide precise landings on drone ships or landing pads. These systems rely on deep learning models trained on vast datasets of flight simulations and real-world launches.

Autonomous systems also enable rockets to adapt to unexpected conditions during flight, such as weather changes or component failures, ensuring mission success and safety.

One of the most significant advantages of AI-driven design is the reduction in development costs and timelines. AI accelerates the iterative design process by identifying optimal solutions faster than traditional engineering methods. By automating routine tasks and streamlining workflows, AI allows engineers to focus on innovation and problem-solving.

AI's ability to process and analyse data in real time enables rockets to optimize their performance during flight. For example, AI can adjust fuel flow, engine thrust, and trajectory to account for unexpected variables like wind conditions or payload imbalances. This dynamic optimization enhances efficiency and ensures mission objectives are met.

The future of AI-driven rocket design is poised to be transformed by groundbreaking advancements in technology, including quantum computing, digital twins, and collaborative AI systems. These trends promise to significantly enhance the efficiency, accuracy, and innovation within the aerospace industry.

Quantum computing offers the potential to revolutionize rocket design by exponentially accelerating simulations and optimizations. Traditional computing methods, while powerful, struggle with the immense complexity of rocket systems, which involve intricate interactions of physics, material science, and engineering principles. Quantum computers, with their ability to process vast amounts of data and solve problems involving multiple variables simultaneously, could enable engineers to explore design configurations and propulsion models that were previously unattainable. This technology holds promise for optimizing fuel efficiency, structural integrity, and thermal management at a scale and speed unmatched by current computational methods.

Digital twins represent another transformative trend, leveraging AI to create virtual replicas of rockets that simulate and predict their real-world performance. These AI-driven models incorporate data from every stage of a rocket's lifecycle, from design and testing to launch and mission execution. By analysing this data, digital twins provide invaluable insights into potential issues, allowing for pre-emptive troubleshooting and iterative improvements. This approach enhances reliability, extends the operational lifespan of rocket systems, and reduces costs associated with physical testing and maintenance. Digital twins also enable ongoing adjustments and optimizations even during flight, ensuring that rockets perform optimally under dynamic conditions.

Collaborative AI systems are set to redefine the interaction between human engineers and intelligent technologies. Unlike conventional AI tools that function independently, collaborative systems are designed to work alongside human designers, offering suggestions, conducting analyses, and even co-creating innovative solutions. These systems enable a more intuitive and effective design process, where AI can augment human creativity and expertise with data-driven insights and predictive capabilities. By fostering seamless collaboration, these AI systems can accelerate the development timeline, enhance precision, and foster innovation in rocket design and construction.

Together, these future trends signify a shift toward more intelligent, integrated, and efficient methods in rocket engineering. By combining the computational power of quantum systems, the predictive accuracy of digital twins, and the intuitive interaction of collaborative AI, the aerospace industry is on the cusp of a new era of exploration and technological advancement. These innovations promise not only to enhance the capabilities of rockets but also to make space exploration more accessible, sustainable, and impactful for humanity.

Several AI tools are currently revolutionizing rocket design by optimizing processes, enhancing efficiency, and improving overall performance. These tools span a variety of applications, from conceptual design to post-launch analytics, and integrate advanced machine learning, data analytics, and simulation capabilities to streamline the aerospace development lifecycle.

Generative design tools use AI algorithms to generate multiple design options based on specific parameters, such as weight, strength, and material properties. Software like Autodesk Fusion 360 with generative design capabilities allows engineers to explore innovative structures for rocket components, such as fuel tanks or fuselages, while reducing weight and maintaining structural integrity. These tools are invaluable in optimizing designs for both manufacturing and performance.

AI-enhanced CFD tools are widely used to simulate aerodynamics, heat transfer, and flow dynamics. Tools such as ANSYS Fluent and Siemens Simcenter integrate AI to speed up simulations by predicting outcomes based on historical data and machine learning models. This enables rapid evaluation of rocket designs, particularly for aerodynamic optimization and thermal management systems.

AI-driven tools like Altair OptiStruct and Dassault Systèmes' Abaqus use advanced algorithms to analyse stress, strain, and structural performance under various conditions. These tools help in designing lightweight yet robust rocket structures by identifying potential failure points and recommending material optimizations.

AI tools such as AGI's Systems Tool Kit (STK) use machine learning to optimize mission parameters, including launch windows, orbital insertion trajectories, and payload deployment. These tools reduce mission planning time by automating complex calculations and suggesting optimal configurations based on mission objectives and environmental constraints.

AI tools like Model-Based Systems Engineering (MBSE) platforms, including MATLAB and Simulink with AI integration, help design and optimize propulsion systems. These tools simulate engine performance, fuel efficiency, and thrust vectors, allowing engineers to test various configurations digitally before physical prototyping.

Platforms like Materials Studio and Quantum Espresso use AI and machine learning to discover and evaluate new materials for rocket components. These tools accelerate the development of heat-resistant alloys, lightweight composites, and high-performance fuels, all critical for improving rocket efficiency and durability.

Digital twin platforms, such as those offered by Siemens and GE Digital, create virtual replicas of rockets to simulate their performance throughout their lifecycle. These AI-powered systems enable real-time monitoring and predictive maintenance, ensuring optimal performance and reducing downtime for critical components.

AI tools like Autodesk Netfabb and Siemens NX are used for additive manufacturing (3D printing) of rocket parts. These tools optimize the printing process, reduce material waste, and ensure precision in producing complex geometries, such as fuel injectors or cooling channels in engines.

AI-based platforms like TensorFlow and PyTorch are increasingly being used to design autonomous navigation and control systems for rockets. These tools enable the development of intelligent algorithms for real-time decision-making, ensuring stable flight paths and precise orbital insertions.

Platforms like AWS SageMaker and IBM Watson are employed to analyse vast amounts of telemetry data collected during and after rocket launches. These tools identify performance anomalies, improve future launch reliability, and enhance predictive maintenance protocols.

The integration of these AI tools in rocket design is transforming the aerospace industry by enabling faster innovation, reducing costs, and improving the safety and efficiency of space exploration. As AI technologies continue to advance, their role in rocket development will become even more pivotal, pushing the boundaries of what is possible in the field of aerospace engineering.

Robots in Construction, Maintenance, and Space Exploration

Robots play a transformative role in rocket construction, maintenance, and space exploration, contributing to efficiency, precision, and the ability to perform tasks in environments that are hazardous or inaccessible to humans. From automating production processes to enabling deep-space missions, robotics has become indispensable in the aerospace industry.

Robots are widely used in the construction of rockets to improve precision, enhance safety, and reduce costs.

Automated Assembly: Robotic arms and automated systems are integral in assembling rocket components. Tasks such as welding, fastening, and material handling require high levels of accuracy, which robots can achieve consistently. For instance, robotic systems are used to assemble fuel tanks, attach thermal protection systems, and integrate payloads into rockets.

Composite Material Manufacturing: Rockets often use lightweight yet strong composite materials. Robots are employed in filament winding and automated fibre placement processes to build these components with precision, ensuring uniformity and structural integrity.

3D Printing: Additive manufacturing, facilitated by robotic systems, is revolutionizing rocket part production. Robots use 3D printing technology to create complex components, such as engine nozzles and fuel injectors, with reduced waste and faster production times. Companies like SpaceX and Rocket Lab use 3D printing extensively in rocket construction.

Inspection and Quality Control: Robotic systems equipped with advanced sensors, cameras, and AI algorithms perform non-destructive testing (NDT) to identify defects in rocket components. Robots ensure that parts meet stringent safety and performance standards, minimizing the risk of failure during launch or operation.

Robots also enable efficient and precise maintenance of rockets, both on the ground and in space.

Launch Pad Maintenance: Robots are deployed to inspect and maintain rocket launch pads, which endure extreme conditions during liftoff. Autonomous robots can identify wear and tear, clean surfaces, and perform repairs, reducing downtime between launches.

Reusable Rocket Inspection: For reusable rockets, such as SpaceX's Falcon 9, robots inspect engines, landing legs, and thermal protection systems after recovery. This ensures that components are safe and functional for subsequent missions, streamlining the refurbishment process.

Cryogenic System Monitoring: Robots monitor and maintain cryogenic fuel systems, which involve handling extremely cold temperatures and volatile substances. They ensure proper sealing, leak detection, and pressure regulation in fuel storage and transfer systems.

Robots are further essential for space exploration missions, performing tasks that are beyond human capability in remote and hostile environments.

Planetary Rovers: Robotic rovers, such as NASA's Perseverance and Curiosity on Mars, explore planetary surfaces, conduct scientific experiments, and gather data. These robots are equipped with advanced instruments, cameras, and mobility systems to navigate rugged terrain and extreme conditions.

Robotic Arms and Manipulators: Robotic arms, like those on the International Space Station (e.g., Canadarm2), are used to capture satellites, conduct repairs, and assemble structures in space. They provide astronauts with the ability to manipulate objects with precision in microgravity.

Figure 91: Unlike the Canadarm on the shuttle, Canadarm2 is not permanently anchored at one end. This design gives Canadarm2 the ability to "walk" around the International Space Station on its own. Canadarm2 can move end-over-end to fixtures placed around the station exterior. Each fixture provides the arm with power and a computer and video link to astronaut controllers inside. Defense Visual Information Distribution Service, Public Domain, via National Archives and Defense Visual Information Distribution Service.

Autonomous Spacecraft: Autonomous robotic spacecraft, such as the European Space Agency's Rosetta and NASA's OSIRIS-REx, perform complex missions, including asteroid sampling and comet exploration. These robots operate independently, making decisions based on pre-programmed algorithms and real-time data.

In-Situ Resource Utilization (ISRU): Future missions aim to use robots for in-situ resource utilization, such as mining lunar regolith for oxygen or water and constructing habitats on Mars. Robotic systems will play a critical role in making long-term human presence in space sustainable.

Satellite Servicing: Robots are increasingly used for on-orbit servicing of satellites, including refuelling, repairing, and upgrading. NASA's Robotic Refuelling Mission and Northrop Grumman's Mission Extension Vehicle demonstrate the potential of robotic systems to extend satellite lifespans and reduce space debris.

Robots offer significant advantages across various stages of aerospace development, particularly in terms of safety, precision, efficiency, and autonomy. Their ability to operate in extreme environments greatly reduces risks to human workers, making them indispensable for tasks in hazardous conditions such as deep-space exploration or handling volatile materials during rocket manufacturing and maintenance.

Robotic systems are renowned for their unmatched accuracy, which is critical in aerospace applications where even the smallest error can have catastrophic consequences. From assembling intricate components to performing delicate repairs in space, robots provide the precision needed to meet stringent quality and performance standards.

Efficiency is another key benefit of robotic systems. By streamlining processes in manufacturing, maintenance, and exploration, robots save time and reduce costs, enabling faster turnaround times and increased productivity. Their capacity to handle repetitive and labour-intensive tasks frees up human workers to focus on higher-level problem-solving and innovation.

The autonomy of robots is particularly transformative for deep-space missions, where real-time human control is often impossible due to communication delays. Autonomous robots can make independent decisions, adapt to unexpected conditions, and carry out complex tasks, ensuring mission success even in challenging and unpredictable environments.

As robotic technologies continue to advance, their role in the aerospace industry is set to expand further. Robots are reshaping every stage of rocket development, operation, and exploration, paving the way for new possibilities in human-robot collaboration and unlocking the potential for groundbreaking achievements in space exploration and beyond.

PART 6

Challenges and Opportunities

Chapter 16

Regulations and Ethics in Space Exploration

International Space Laws and Treaties

International space laws and treaties are essential frameworks that promote the peaceful, responsible, and equitable use of outer space. These legal agreements address key issues such as weaponization, exploration, satellite deployment, liability for damage, environmental protection, and the equitable sharing of space benefits. They provide a structured approach to managing the complexities of space exploration and operations, particularly regarding the use of rockets.

The Outer Space Treaty (OST), adopted by the United Nations in 1967, is the cornerstone of international space law. This treaty establishes foundational principles for space exploration and rocket use. It mandates that outer space, including the Moon and other celestial bodies, must be used exclusively for peaceful purposes, prohibiting the deployment of weapons of mass destruction. The treaty also asserts that no nation can claim sovereignty over outer space or celestial bodies, ensuring that space remains free from territorial disputes. Furthermore, it holds nations accountable for their space activities, including those carried out by private entities, and emphasizes the need to prevent harmful contamination of space and celestial environments.

Building on the OST, the Rescue Agreement (1968) establishes protocols for assisting astronauts in distress and recovering space objects such as rockets or satellites that re-enter Earth's atmosphere. This agreement obligates nations to provide rescue assistance and encourages international cooperation to return recovered space objects to their launching state.

The Liability Convention (1972) defines the liability rules for damage caused by space objects, including rockets. Under this convention, launching states are held absolutely liable for any damage their rockets or space objects cause on Earth or within its atmosphere. In space, liability is determined based on fault, ensuring accountability and financial compensation for damages resulting from space missions.

The Registration Convention (1976) requires states to register their space objects, including rockets, with the United Nations. This centralized registry records key details such as orbit parameters and the purpose of the space object. Registration promotes transparency and helps track space objects, including rocket stages that may remain in orbit or re-enter Earth's atmosphere.

The Moon Agreement (1984) expands on principles related to the use of the Moon and other celestial bodies. It prohibits the militarization of the Moon and emphasizes the equitable sharing of resources derived from lunar exploration. While less widely adopted than the OST, it highlights important considerations for activities involving rockets used in resource extraction or other operations on celestial bodies.

The Guidelines for the Long-Term Sustainability of Outer Space Activities, issued by the UN Committee on the Peaceful Uses of Outer Space (COPUOS), promote sustainable practices for space operations, including rocket launches. These guidelines focus on minimizing space debris, developing debris mitigation technologies, and fostering international coordination to prevent collisions and ensure safe rocket operations.

The Missile Technology Control Regime (MTCR), an informal political agreement, aims to prevent the proliferation of missile technology capable of delivering weapons of mass destruction. Although focused on missile systems, its principles are directly relevant to rockets, particularly those with dual-use capabilities for both civilian and military applications.

As rocket technology and space exploration evolve, these international laws face several challenges. The growing involvement of private companies in rocket launches and satellite deployment raises questions about regulatory oversight and liability. Space debris management remains a critical issue, as rockets contribute significantly to orbital debris, and binding agreements on debris mitigation are lacking. Concerns about the weaponization of space also persist, as the OST prohibits weapons of mass destruction but does not comprehensively address other military uses of rockets. Additionally, equitable access to space resources and technologies for developing nations remains a pressing issue.

International space laws and treaties form the foundation of cooperative and peaceful space exploration, particularly regarding rocket use. As technology advances and space activities increase, these frameworks must adapt to address emerging challenges. By doing so, they will continue to ensure that the exploration and utilization of space benefit all humanity.

Space law is an evolving body of international and domestic legal frameworks that governs activities related to outer space. This legal domain encompasses a wide range of issues, including space exploration, liability for damages, the use of weapons, rescue operations, environmental preservation, the regulation of new technologies, and ethical considerations. The integration of various disciplines such as administrative law, intellectual property law, arms control law, insurance law, and environmental law reflects the complexity of space-related activities [503-506].

The origins of space law can be traced back to the early 20th century, particularly with the recognition of national sovereignty over airspace, which was established in 1919 and reinforced by the 1944 Chicago Convention. The Cold War era significantly accelerated the development of comprehensive space laws, particularly following the launch of Sputnik 1 in 1957, which prompted the United States to establish NASA through the Space Act. This period marked the evolution of space law into an independent field, distinct from traditional aerospace law, to address the unique challenges associated with international boundaries in space exploration [507, 508].

The cornerstone of space law is the 1967 Outer Space Treaty (OST), which lays down foundational principles for the exploration and use of outer space. It mandates that space must be utilized for peaceful purposes and prohibits the placement of weapons of mass destruction in space. The treaty also establishes the principle of non-appropriation, ensuring that no nation can claim sovereignty over celestial bodies. Furthermore, the OST holds states accountable for their space activities, whether conducted by governmental or private entities, and emphasizes the importance of preventing harmful contamination of outer space [504, 505, 507].

In addition to the OST, several treaties and agreements have been established to further govern space activities. The Rescue Agreement (1968) facilitates international cooperation in rescuing astronauts and recovering space objects, while the Liability Convention (1972) delineates liability rules for damages caused by space objects, holding launching states accountable for damages both on Earth and in space. The Registration Convention (1976) requires states to register their space objects with the United Nations, promoting transparency and accountability. The Moon Treaty (1979) elaborates on the use of the Moon and other celestial bodies, emphasizing their use for peaceful purposes and equitable sharing of resources, although it has seen limited adoption [503, 505, 507].

The governance of space activities is also shaped by international principles and declarations issued by the United Nations, which include provisions for the use of artificial Earth satellites, remote sensing, nuclear power in space, and international cooperation in space exploration. The concept of outer space as the "province of all mankind," established by the OST, has guided international space governance, emphasizing that the exploration and utilization of space should benefit all humanity [503].

Furthermore, collaborative agreements such as the 1998 International Space Station (ISS) Agreement exemplify international cooperation in space exploration. This agreement outlines the roles and responsibilities of participating nations, including jurisdiction over their respective modules and protection of intellectual property rights, serving as a potential model for future agreements governing lunar or Martian bases [503-505].

Space law faces several challenges, including defining airspace boundaries, allocating geostationary orbital slots, and managing space debris. These issues underscore the necessity for updated legal frameworks to address the increasing complexity of space activities. The growing involvement of private entities in space exploration raises questions about regulatory oversight, resource appropriation, and the ethical implications of commercial space activities [503, 505, 506].

Environmental concerns are also central to modern space law, with efforts to mitigate space debris, protect celestial environments, and regulate the use of nuclear power in space being critical for the sustainability of space activities. The integration of bioengineering measures to prevent contamination and the consideration of ethical frameworks for space exploration further highlight the need for comprehensive governance [503].

The academic and legal communities play a vital role in shaping space law. Institutions such as McGill University and the University of Mississippi are advancing research and education in this field, while international projects like the McGill Manual on International Law Applicable to Military Uses of Outer Space (MILAMOS) aim to clarify the application of existing laws. Collaborative initiatives such as the Open Lunar Foundation are exploring innovative approaches to space governance, including the preservation of space heritage and the responsible use of lunar resources [503-505].

As humanity ventures further into space, the urgency for robust legal frameworks becomes increasingly apparent. International cooperation, ethical considerations, and sustainable practices will be essential to ensure that space exploration and utilization remain a shared endeavour for the benefit of all. Space law continues to evolve, reflecting the dynamic nature of this frontier and the challenges and opportunities it presents [503-505].

Balancing Exploration with Environmental Responsibility

Balancing space exploration with environmental responsibility is an increasingly critical challenge as humanity extends its reach beyond Earth. The dual objectives of advancing scientific knowledge and preserving the integrity of both terrestrial and extraterrestrial environments necessitate meticulous planning, regulation, and innovation. Space exploration presents significant opportunities for scientific discovery and technological advancement; however, it also raises pressing concerns regarding pollution, resource exploitation, and contamination that must be effectively addressed to ensure sustainable practices [27, 509, 510].

One of the foremost environmental concerns associated with space exploration is the generation of space debris. Each rocket launch contributes to the accumulation of spent rocket stages, defunct satellites, and fragments from collisions, creating a hazardous environment in low Earth orbit. These debris pose risks to operational satellites, space stations, and future missions, necessitating the development of debris mitigation strategies. Such strategies include designing rockets and satellites for deorbiting after use, employing active debris removal technologies, and establishing international agreements for shared responsibility in managing orbital space [27, 510]. The United Nations' guidelines on long-term space sustainability highlight the importance of collaborative action to reduce orbital pollution, underscoring the need for a concerted global effort [104].

Rocket launches also contribute to atmospheric pollution and climate change. The combustion of rocket propellants releases greenhouse gases, particulates, and other pollutants into the atmosphere, which can have detrimental effects on climate systems. Although the current scale of emissions from space activities is relatively small compared to other industries, the anticipated growth in launches due to commercial space ventures and space tourism could significantly amplify their impact [27, 510, 511]. Developing cleaner propulsion technologies, such as methane-based engines or electric and hybrid systems, is crucial for reducing the environmental footprint of space exploration [509, 512, 513]. Research indicates that the atmospheric consequences of rocket emissions, including perturbations of ozone chemistry and radiative forcing, are significant and warrant further investigation [27, 511].

Another critical aspect of environmental responsibility involves the protection of celestial bodies from contamination. Planetary protection protocols, established by organizations like NASA and the Committee on Space Research (COSPAR), aim to prevent the biological contamination of other planets and moons by Earth-based organisms. These measures are essential for preserving the natural state of celestial bodies and ensuring the integrity of future scientific studies, particularly in the search for extraterrestrial life [104, 509]. Additionally, preventing the return of alien microbes to Earth is vital for safeguarding our biosphere from potential biological risks [509].

The extraction of extraterrestrial resources presents both opportunities and challenges. Mining asteroids, the Moon, or other celestial bodies for valuable materials could reduce the need for resource-intensive launches from Earth and support the development of in-space infrastructure. However, such activities must be guided by ethical and environmental considerations to avoid over-exploitation, habitat disruption, or geopolitical tensions over resource ownership. Transparent international agreements, inspired by principles like those in the Outer Space Treaty, are essential to ensure that resource utilization is conducted equitably and responsibly [27, 509, 510].

On Earth, space exploration projects also have significant environmental impacts. The construction and operation of launch sites, tracking stations, and testing facilities can disrupt local ecosystems and wildlife habitats. To mitigate these effects, space agencies and private companies must adopt sustainable practices, such as minimizing land use, preserving biodiversity, and utilizing renewable energy sources for their infrastructure [27, 509, 510]. Innovations in green propulsion technologies, recyclable satellite components, and autonomous deorbiting systems demonstrate the potential to align technological progress with ecological stewardship [509].

Public awareness and policy frameworks are equally critical. Educating stakeholders about the environmental impacts of space exploration and incorporating sustainability into national and international space policies can drive meaningful change. Transparency in operations and adherence to global standards for environmental protection will strengthen trust and cooperation among spacefaring nations and organizations [27, 509, 510].

Financial and Logistical Hurdles

The design and construction of rockets are among the most complex engineering feats, requiring significant financial investment and sophisticated logistical management. These challenges stem from the intricate nature of rocket systems, the precision required, and the scale of operations involved in their development and deployment.

Rocket development is characterized by exceptionally high research and development (R&D) costs, often amounting to billions of dollars. This financial burden arises from the need for advanced materials, cutting-edge propulsion technologies, and extensive testing protocols designed to ensure safety and optimal performance. The iterative nature of the design process further compounds these costs, as each prototype failure necessitates redesigns, which are both resource-intensive and time-consuming.

The manufacturing process for rockets also adds to the financial strain. It involves the production of specialized components such as propulsion systems, guidance mechanisms, and thermal protection systems. These components often require custom fabrication using expensive materials like lightweight composites and heat-resistant alloys. Moreover, the necessity for precision machining, assembly in cleanroom environments, and adherence to stringent quality control standards significantly increases production expenses.

Regulatory compliance introduces another layer of financial complexity. Adhering to international and national regulations, including environmental standards and safety certifications, demands considerable resources. Costs associated with obtaining launch licenses, conducting environmental impact assessments, and complying with export controls for sensitive technologies further elevate the financial outlay required for rocket development.

Insurance costs also weigh heavily on the budget. Given the high stakes of rocket launches, comprehensive insurance coverage is indispensable. This includes protection against launch failures, payload loss, or damage to third-party property, all of which can result in claims amounting to hundreds of millions of dollars. Consequently, insurance premiums for rocket projects are exceptionally high.

Finally, the financial returns on investments in rocket design and construction are often long-term and uncertain. Companies typically face extended waiting periods before they can recoup their expenditures. This delayed return on investment makes the industry less attractive to risk-averse investors, leaving many rocket ventures reliant on government contracts or subsidies to sustain their operations.

The cost of producing a rocket from design to launch is influenced by various factors, including the type of rocket, payload capacity, mission objectives, and the technology employed. This variability is evident across different categories of launch vehicles, which can be broadly classified into small, medium-lift, heavy-lift, and super heavy-lift rockets.

Small Launch Vehicles: Small rockets, such as Rocket Lab's Electron and Firefly Aerospace's Alpha, are designed to launch lightweight payloads, typically costing between $2 million and $15 million for production and launch (all amounts in USD). The initial research and development (R&D) costs for these vehicles range from $10 million to $50 million, while the per-unit production cost is approximately $1 million to $5 million. Launch costs, including logistics and ground support, can add another $1 million to $10 million to the total. The demand for low-cost access to space is primarily driven by commercial applications, which further influences the pricing strategies of small launch vehicles [514].

Medium-Lift Rockets: Medium-lift rockets, like SpaceX's Falcon 9, are capable of carrying payloads between 5 and 20 tons to low Earth orbit (LEO). The costs associated with these rockets range from $60 million to $100 million per launch. The design and development phase can incur expenses between $500 million and $1 billion, with production costs per unit estimated at $30 million to $40 million. Launch costs for medium-lift rockets typically fall between $30 million and $60 million per mission. The advancements in reusability, particularly with the Falcon 9, have significantly reduced launch costs, demonstrating a trend towards more cost-effective solutions in the aerospace sector [515].

Heavy-Lift Rockets: Heavy-lift rockets, such as the Falcon Heavy or United Launch Alliance's Delta IV Heavy, can transport over 20 tons to LEO, with launch costs ranging from $100 million to $300 million. The R&D expenses for these rockets can reach between $1 billion and $2 billion, while the production costs per unit are estimated at $90 million to $150 million. Launch costs for heavy-lift missions can be substantial, reflecting the complexity and scale of these operations [173]. The financial implications of heavy-lift capabilities are critical for missions that require significant payloads, such as interplanetary exploration.

Super Heavy-Lift Rockets: The most advanced category includes super heavy-lift rockets like NASA's Space Launch System (SLS) and SpaceX's Starship, designed for interplanetary missions. The costs for these launches can range from $500 million to over $4 billion. R&D expenditures can vary widely, from $2 billion to $20 billion, depending on the project's scope and complexity. The production costs for the SLS are estimated between $300 million and $500 million per unit, while SpaceX aims to reduce its launch costs to between $10 million and $50 million through full reusability strategies. This shift towards reusable systems is expected to significantly lower overall launch costs in the future [104].

Factors Influencing Costs: Several factors contribute to the overall costs of rocket production and launch. Innovations in reusability, such as those seen with SpaceX's Falcon 9 and Starship, have been pivotal in reducing launch expenses by 30-40% per mission. Additionally, custom payload integration for specialized missions can increase costs substantially, as can the development and maintenance of launch infrastructure, including launch pads and mission control facilities. Regulatory compliance also adds to the financial burden, as meeting environmental and safety standards can incur millions in additional costs [516].

The landscape of rocket production and launch costs is evolving, with advancements in technology and increased competition among private aerospace companies. As companies like SpaceX continue to innovate, the expectation is that launch costs will decrease significantly in the coming decades. For instance, SpaceX's goal for the Starship is to achieve launch costs below $10 million per mission, which would represent a dramatic reduction compared to current heavy-lift rocket costs [517].

Logistical challenges in rocket design and construction are immense, starting with the complexities of supply chains. Rocket manufacturing depends on a global network of suppliers to deliver high-precision parts and rare materials, often sourced from specialized industries. Managing this intricate supply chain involves ensuring timely deliveries and consistent quality across multiple tiers. Any delay or defect in a single component can cascade through the production timeline, causing significant disruptions.

Infrastructure requirements also pose substantial logistical hurdles. Building rockets demands specialized facilities such as test stands, cleanrooms, and launch pads, each tailored to handle the specific needs of high-precision engineering and large-scale testing. These facilities are not only costly to construct but also expensive to maintain. Furthermore, transporting rockets from manufacturing sites to often-remote launch locations requires heavy-lift transport vehicles and meticulous handling procedures to prevent damage during transit.

Testing and iteration are another critical aspect of rocket development. Comprehensive testing, from engine firings to structural evaluations and full-scale launches, is essential to validate design and performance. Coordinating these tests requires precise planning and the mobilization of large teams, specialized equipment, and carefully controlled environments. Test failures can lead to timeline delays and additional costs, highlighting the need for rigorous logistical management.

Workforce management adds to the logistical complexities. The aerospace industry demands highly skilled engineers, scientists, and technicians with expertise in diverse fields. Recruiting, training, and retaining such talent is challenging, especially in a competitive market where qualified professionals are in high demand. Effective workforce planning is essential to ensure the seamless progression of rocket development projects.

Environmental concerns further complicate logistical planning. Rocket launches and operations must comply with stringent environmental regulations, requiring manufacturers to minimize emissions and address the issue of space debris. Developing sustainable practices and technologies to meet these requirements introduces additional design and operational constraints that must be managed effectively.

The integration of rockets with their payloads adds another layer of logistical difficulty. Rockets must be designed to accommodate a wide range of payloads, from communication satellites to scientific instruments and crewed modules. This necessitates close collaboration with payload manufacturers and the incorporation of adaptable design features to meet diverse and evolving requirements. Ensuring compatibility and smooth integration adds significant complexity to the development process, further emphasizing the multifaceted logistical challenges inherent in rocket construction.

The logistical challenges in rocket design and construction are multifaceted, necessitating a vast and specialized infrastructure that incurs significant costs and operational considerations. The complexity of rocket manufacturing is underscored by the reliance on a global supply chain that delivers high-precision components and rare materials, such as heat-resistant alloys and advanced composites. Effective management of this intricate supply chain requires robust tracking systems and quality assurance protocols, which can cost tens to hundreds of millions of dollars annually, depending on the scale of operations [145, 518, 519]. Delays or defects in any component can lead to substantial ripple effects across the production timeline, resulting in increased costs associated with storage, reordering, or redesigning components [137, 520].

The infrastructure required for rocket manufacturing includes specialized facilities such as cleanrooms, test stands, and assembly bays. Cleanrooms are essential for assembling sensitive components like guidance systems and avionics, with construction costs ranging from $1,000 to $5,000 per square foot, alongside ongoing operational expenses for air filtration and contamination control [163, 519]. Test stands, crucial for static fire tests and structural evaluations, can cost between $20 million and $50 million each, while state-of-the-art assembly bays may reach costs of $50 million to $100 million [519, 521]. The investment in these facilities reflects the high stakes involved in ensuring the reliability and performance of rocket systems.

Launch infrastructure represents another significant financial burden, with the development of launch pads and associated facilities costing between $500 million and $1 billion for a single site [519]. This includes the necessary fuelling systems, telemetry stations, and mission control facilities, all of which require ongoing maintenance and upgrades. The logistics of transporting rockets from manufacturing hubs to remote launch locations further complicate operations, necessitating investment in heavy-lift transporters and custom containers, which can also amount to tens of millions of dollars [519, 520].

Testing and iteration costs are integral to the rocket development process. Comprehensive testing, including engine firings and full-scale launches, is essential for validating the integration of all rocket systems. Operating a test facility can cost several million dollars per test, while full-scale launch tests may run into the tens of millions [519, 521]. Failures during testing can lead to additional costs for redesign, repairs, and subsequent tests, emphasizing the need for meticulous planning and execution in the rocket development lifecycle.

The workforce required for rocket design and construction is highly specialized, with salaries for aerospace engineers and other professionals often exceeding $100,000 annually. The recruitment and retention of such talent necessitate significant investments in training programs and benefits, with labour costs for large-scale projects potentially reaching $200 million annually [163, 519].

Moreover, compliance with environmental regulations adds another layer of complexity and cost. The aerospace industry must invest in systems to minimize emissions and address space debris, with innovations like deorbit systems potentially adding $10 million to $50 million per mission [519, 521]. The cumulative costs associated with building and maintaining the infrastructure for rocket design, manufacturing, and launching can range from $2 billion to $5 billion for medium-scale programs, with large-scale initiatives exceeding $10 billion [519, 520].

Competition and Collaboration among Nations and Private Entities

The growing demand for space exploration, satellite deployment, and commercial space ventures has brought about a dynamic interplay of competition and collaboration among nations and private entities in rocket development. As space technology evolves, both government-led programs and private companies are shaping the aerospace industry's future. This dual dynamic of rivalry and partnership is fostering innovation while presenting unique challenges.

Competition among nations has been a cornerstone of space exploration since the Cold War, marked by the historic space race between the United States and the Soviet Union. Today, national space programs such as NASA, ESA, CNSA, ISRO, and Roscosmos continue to compete for scientific achievements, economic gains, and geopolitical influence. Technological leadership remains a primary focus, with advancements in reusable rockets, heavy-lift capabilities, and propulsion systems driving nations to strive for independent access to space and leadership in exploration missions targeting the Moon and Mars. Geopolitical rivalries further fuel this competition as countries vie for strategic dominance through space-based technologies for defense, communication, and exploration. Economically, nations are working to secure their positions in the lucrative commercial satellite launch market by investing heavily in infrastructure and providing subsidies to their space industries.

Private entities are also intensely competitive, driving rapid advancements in the aerospace sector. Companies such as SpaceX, Blue Origin, Rocket Lab, and Relativity Space are leading a commercial space race focused on satellite launches, crewed missions, and space tourism. Innovations in cost reduction, reusable systems, and rapid prototyping provide these companies with significant competitive advantages. Market segmentation highlights competition across areas such as small satellite launches, interplanetary missions, and human spaceflight, with startups disrupting traditional aerospace markets by offering lower-cost solutions. Additionally, competition extends to intellectual property and trade secrets, with firms fiercely protecting their proprietary technologies in propulsion, materials, and manufacturing processes.

Despite the competitive landscape, collaboration among nations remains a vital aspect of space exploration. International partnerships, such as the International Space Station (ISS), showcase joint efforts by agencies like NASA, ESA, Roscosmos, and JAXA. Collaborative missions, including the James Webb Space Telescope and Mars rovers, underline the benefits of shared resources and expertise. Global agreements like the Outer Space Treaty and Artemis Accords provide frameworks promoting peaceful exploration, while organizations such as the United Nations Committee on the Peaceful Uses of Outer Space (COPUOS) facilitate coordination. Collaborative research in propulsion, robotics, and sustainability, as well as educational exchanges, further highlight the importance of cooperation in addressing complex challenges.

The integration of public and private efforts has also gained momentum through public-private partnerships (PPPs). Initiatives like NASA's Commercial Crew Program exemplify how government contracts can foster private innovation, enabling companies like SpaceX and Boeing to contribute to satellite launches, lunar landers, and cargo resupply missions. Shared infrastructure, such as national facilities and launch sites, facilitates collaboration, while regulatory frameworks and incentives encourage private companies to participate in national space programs.

However, this interplay of competition and collaboration is not without challenges. Political tensions and geopolitical conflicts often restrict international partnerships and impose export controls on technology sharing. Balancing national interests with private profitability raises concerns about monopolies and fair competition within the sector. Additionally, addressing sustainability and ethical concerns, such as space debris, environmental impact, and equitable access to resources, requires coordinated efforts and the establishment of international norms.

Future trends in rocket development are likely to involve hybrid models that combine competition and collaboration. Programs like NASA's Artemis initiative already exemplify such efforts by including international and private partners. The rise of new spacefaring nations, such as the UAE and South Korea, will further diversify the global space ecosystem. Meanwhile, private companies are expected to expand their roles in interplanetary exploration, space mining, and orbital infrastructure development, paving the way for unprecedented achievements in space exploration.

Ultimately, the intertwined dynamics of competition and collaboration are driving significant progress in rocket development. Striking a balance between strategic interests and cooperative efforts will be crucial in shaping the future of space exploration and ensuring that its benefits are shared globally.

Emerging Markets in Space Technology

The emergence of new markets in space technology signifies a transformative shift in the industry, driven by advancements in technology, decreasing costs, and increasing accessibility. These developments are creating opportunities for innovation, economic growth, and international collaboration, while also presenting challenges that necessitate careful consideration of regulatory, technological, and ethical aspects.

One of the most significant developments in the space sector is the deployment of satellite constellations aimed at providing global internet services. Companies such as SpaceX with its Starlink project, Amazon's Project Kuiper, and OneWeb are at the forefront of this initiative, deploying low Earth orbit (LEO) satellites to deliver high-speed internet access to underserved regions. This initiative has the potential to bridge the digital divide and enhance global connectivity, particularly in remote areas [522]. However, the rapid expansion of these constellations raises critical concerns regarding space debris and orbital congestion, which could jeopardize the sustainability of space operations [522, 523]. The need for effective debris mitigation strategies is underscored by the increasing number of satellites, which poses risks to both operational spacecraft and future missions [524, 525].

Space tourism is transitioning from a theoretical concept to a burgeoning industry, with companies like Blue Origin, Virgin Galactic, and SpaceX successfully conducting commercial spaceflights. This market is expected to grow as technological advancements reduce costs and make space travel more accessible to a broader audience [526]. While space tourism can inspire public interest in space exploration, it also raises significant challenges, including safety concerns, environmental impacts, and ethical questions regarding resource allocation [525]. The industry must navigate these complexities to ensure sustainable growth and public trust in commercial space endeavours.

The increasing number of satellites has catalysed the development of on-orbit services, including satellite servicing, refuelling, and debris removal. Companies like Northrop Grumman and Astroscale are pioneering technologies to extend satellite lifespans and mitigate space debris, which poses severe risks to operational spacecraft [527, 528]. The necessity for sustainable space operations is becoming increasingly apparent, as the accumulation of debris threatens the viability of future space activities [529, 530]. Innovations in robotics and artificial intelligence are critical to advancing these on-orbit servicing capabilities, which are essential for maintaining the safety and functionality of space infrastructure [531, 532].

The concept of space mining is gaining traction as a potential solution to resource scarcity on Earth. Celestial bodies such as asteroids and the Moon are believed to contain valuable materials, including rare metals and water, which could be utilized for various industrial applications [533, 534]. Companies like Planetary Resources and Deep Space Industries are exploring the feasibility of extracting these resources, which could revolutionize industries like energy and manufacturing. However, the development of international legal frameworks is crucial to address ownership and environmental concerns associated with space resource utilization [535].

Microgravity environments present unique opportunities for scientific research and manufacturing processes. The International Space Station (ISS) has facilitated significant advancements in fields such as pharmaceuticals and material science [536]. Emerging markets in microgravity manufacturing aim to produce high-value products that are challenging to create on Earth, leveraging the unique conditions of space [537]. Investment in infrastructure, such as space habitats and orbital labs, is essential to scale these operations and unlock their economic potential [538].

The proliferation of small satellites and advanced sensors is driving growth in Earth observation and data analytics. Companies like Planet Labs and Maxar Technologies are providing high-resolution imagery and real-time data to support various industries, including agriculture and urban planning [539]. The ability to harness space-derived data for actionable insights represents a transformative opportunity, fostering innovation in artificial intelligence and big data analytics [540, 541]. This capability is crucial for addressing global challenges, such as climate change and disaster management [542].

The militarization of space and the increasing reliance on satellite technologies for defense and national security have spurred significant investment in this sector. Nations are developing advanced capabilities for missile detection, secure communications, and space-based surveillance [543]. The competitive market for dual-use technologies underscores the strategic importance of space in geopolitics and global security [544]. As emerging markets in space technology continue to evolve, collaboration between governments and private entities will be essential to navigate the complexities of this landscape.

The demand for satellite launches has led to the development of new spaceports and launch services worldwide. Emerging markets in this area focus on providing cost-effective and flexible access to space, with companies like Rocket Lab and Relativity Space driving innovation in launch technologies [545]. Countries such as the UAE and Australia are investing in spaceport infrastructure to position themselves as regional hubs, reflecting the growing global interest in commercial space activities.

Emerging nations are increasingly contributing to the space technology market, with countries like India, Brazil, South Korea, and the UAE investing in their space programs. These nations are fostering innovation and creating opportunities for international partnerships, enhancing the diversity and inclusivity of the global space industry. Their involvement is crucial for the sustainable development of space technologies and for addressing global challenges collaboratively.

Despite the immense opportunities presented by emerging markets in space technology, challenges such as high entry costs, regulatory complexities, and environmental concerns persist. Collaboration among governments, private entities, and international organizations will be essential to address these issues effectively. As technology continues to advance and commercial activities expand, the space sector is poised to become a cornerstone of the global economy, driving innovation and shaping humanity's future in space.

References

1. Neufeld, M.J., *Spaceflight.* 2018.
2. MacDonald, A.E., *The Long Space Age.* 2017.
3. Benson, T., *Brief History of Rockets*. 2021, National Aeronautics and Space Administration.
4. Brake, M. and N. Hook, *Rocketry, Film and Fiction: The Road to Sputnik.* Physics Education, 2007. **42**(4): p. 345-350.
5. Herman, D.A. and A.D. Gallimore, *Discharge Chamber Plasma Potential Mapping of a 40-Cm NEXT-type Ion Engine.* 2005.
6. Zahari, A.R. and F.I. Romli, *Potential of Commercial Human Spaceflight.* International Review of Aerospace Engineering (Irease), 2017. **10**(5): p. 277.
7. Motoki, A., et al., *SATELLITE-DERIVED GRAVIMETRY FOR ABROLHOS CONTINENTAL SHELF, STATES OF ESPÍ RITO SANTO AND BAHIA, BRAZIL, AND ITS RELATION TO TECTONIC GENESIS OF SEDIMENTARY BASINS.* Brazilian Journal of Geophysics, 2014. **32**(4): p. 735-751.
8. Lee, W.-C., K.-S. Kim, and Y.H. Kwon, *Review of the history of animals that helped human life and safety for aerospace medical research and space exploration.* 항공우주의학회지, 2020. **30**(1): p. 18-24.
9. Nelson, S., *Giant Leaps for Knowledge.* Physics World, 2019. **32**(7): p. 38-43.
10. Wood, C.A., *Scientific Knowledge of the Moon, 1609 to 1969.* Geosciences, 2018. **9**(1): p. 5.
11. Burnett, D.S., *Lunar Science: The Apollo Legacy.* Reviews of Geophysics, 1975. **13**(3): p. 13-34.
12. Holland, T., *From Apollo to the ISS: The Televisual Image in Human Spaceflight.* Television & New Media, 2023. **25**(1): p. 57-73.
13. Kawamura, T., et al., *Lunar Surface Gravimeter as a Lunar Seismometer: Investigation of a New Source of Seismic Information on the Moon.* Journal of Geophysical Research Planets, 2015. **120**(2): p. 343-358.
14. Ming, D.W., *Lunar Sourcebook. A User's Guide to the Moon.* Endeavour, 1992. **16**(2): p. 96.
15. Kumar, K., *Space Exploration Technologies Corporation Aka SpaceX's Amazing Accomplishments: A Complete Analysis.* Interantional Journal of Scientific Research in Engineering and Management, 2023. **07**(06).
16. Cai, J., et al., *SpaceX's Network Effects and Innovation Strategy Analysis.* Highlights in Business Economics and Management, 2024. **30**: p. 234-238.
17. Sagar, R., *The SpaceX Effect.* New Space, 2018. **6**(2): p. 125-134.
18. Shammas, V.L. and T.B. Holen, *One Giant Leap for Capitalistkind: Private Enterprise in Outer Space.* Palgrave Communications, 2019. **5**(1).
19. Kim, M.J., et al., *Effects of Value-Belief-Norm Theory, ESG, and AI on Space Tourist Behavior for Sustainability With Three Types of Space Tourism.* Journal of Travel Research, 2023. **63**(6): p. 1395-1410.

20. Ryan, R.G., et al., *Impact of Rocket Launch and Space Debris Air Pollutant Emissions on Stratospheric Ozone and Global Climate.* Earth S Future, 2022. **10**(6).
21. Cao, Y., *Analyzing SpaceX's International Collaborations: Conflicts and Coordination Mechanisms With Space Partners.* Highlights in Business Economics and Management, 2024. **24**: p. 602-607.
22. Beavers, L., *National Space Policy: International Comparison of Policy and the 'Gray' Area.* 2021. **2**(1).
23. Hodkinson, P.D., et al., *An Overview of Space Medicine.* British Journal of Anaesthesia, 2017. **119**: p. i143-i153.
24. Wissehr, C., J. Concannon, and L.H. Barrow, *Looking Back at the Sputnik Era and Its Impact on Science Education.* School Science and Mathematics, 2011. **111**(7): p. 368-375.
25. Weiss, S.I. and A.R. Amir, *Secondary and tertiary aerospace systems*, in *Britannica*. 2024.
26. Li, X., et al., *Short-Period Concentric Traveling Ionospheric Disturbances Excited by the Launch of China's Long March 4B Rocket Detected by 1 Hz GNSS Data.* Space Weather, 2022. **20**(6).
27. Sirieys, E., et al., *Space Sustainability Isn't Just About Space Debris: On the Atmospheric Impact of Space Launches.* 2022. **3**: p. 143-151.
28. Ross, M.N. and D.W. Toohey, *The Coming Surge of Rocket Emissions.* Eos, 2019. **100**.
29. Trzun, Z., M. Vrdoljak, and H. Cajner, *The Effect of Manufacturing Quality on Rocket Precision.* Aerospace, 2021. **8**(6): p. 160.
30. Johnson, D.L. and W.W. Vaughan, *The Role of Terrestrial and Space Environments in Launch Vehicle Development.* Journal of Aerospace Technology and Management, 2019.
31. Zosimovych, N., et al., *Integrated Guidance System of a Commercial Launch Vehicle.* Matec Web of Conferences, 2018. **179**: p. 03002.
32. Guo, Z., J.X. Liu, and W.C. Luo, *Parametric Modeling and Simulation for Aerodynamic Design of Launch Vehicle.* Applied Mechanics and Materials, 2011. **101-102**: p. 697-701.
33. Duan, L., et al., *Data-Driven Model-Free Adaptive Attitude Control Approach for Launch Vehicle With Virtual Reference Feedback Parameters Tuning Method.* Ieee Access, 2019. **7**: p. 54106-54116.
34. Johnson, D.L. and W.W. Vaughan, *The Wind Environment Interactions Relative to Launch Vehicle Design.* Journal of Aerospace Technology and Management, 2020(12).
35. Sim, C.-H., et al., *Experimental and Computational Modal Analyses for Launch Vehicle Models Considering Liquid Propellant and Flange Joints.* International Journal of Aerospace Engineering, 2018. **2018**: p. 1-12.
36. Pu, P. and Y. Jiang, *Assessing Turbulence Models on the Simulation of Launch Vehicle Base Heating.* International Journal of Aerospace Engineering, 2019. **2019**: p. 1-14.
37. Sünör, E., *Current Trends in the Aerospace Industry.* 2024, StartupBlink.
38. National Aeronautics and Space Administration, *Rocket Principles.* 2021, National Aeronautics and Space Administration.
39. De Curtò, J. and I. De Zarzà, *Optimizing Propellant Distribution for Interorbital Transfers.* Mathematics, 2024. **12**(6): p. 900.
40. Bruce, A.L., *A General Quadrature Solution for Relativistic, Non-Relativistic, and Weakly-Relativistic Rocket Equations.* 2015.
41. Lee, H.-J., C.-H. Chiu, and W.-K. Hsia, *Integrated Energy Method for Propulsion Dynamics Analysis of Air-Pressurized Waterjet Rocket.* Transactions of the Japan Society for Aeronautical and Space Sciences, 2001. **44**(143): p. 1-7.

42. Hippke, M., *Spaceflight From Super-Earths Is Difficult.* International Journal of Astrobiology, 2018. **18**(05): p. 393-395.
43. Krzysiak, A., et al., *Experimental Study of the Boosters Impact on the Rocket Aerodynamic Characteristics.* Aircraft Engineering and Aerospace Technology, 2022. **95**(2): p. 193-200.
44. Lee, I. and J. Koo, *Break-Up Characteristics of Gelled Propellant Simulants With Various Gelling Agent Contents.* Journal of Thermal Science, 2010. **19**(6): p. 545-552.
45. Glenn Research Center, *Chemical Propulsion Systems: Designing and testing chemical propulsion systems and nuclear thermal engines for satellites and spacecraft, in support of NASA's space exploration missions.* 2024, National Aeronautics and Space Administration.
46. Shafaee, M., P.M. Zadeh, and H. Fallah, *Design Optimization of a Thrust Chamber Using a Mass-Based Model to Improve the Geometrical and Performance Parameters of Low-Thrust Space Propulsion Systems.* Proceedings of the Institution of Mechanical Engineers Part G Journal of Aerospace Engineering, 2018. **233**(5): p. 1820-1837.
47. Chen, L., et al., *Numerical Simulation of Gas-Liquid Interface in a Blade Type Propellant Tank.* Journal of Physics Conference Series, 2024. **2752**(1): p. 012190.
48. Motooka, N., et al., *Microgravity Evaluation of Advantages of Porous Metal in the Gas-Liquid Equilibrium Thruster for Small Spacecraft.* Transactions of the Japan Society for Aeronautical and Space Sciences Aerospace Technology Japan, 2012. **10**(ists28): p. Pb_19-Pb_23.
49. Pettersson, G.M., O. Jia-Richards, and P. Lozano, *Development and Laboratory Testing of a CubeSat-Compatible Staged Ionic-Liquid Electrospray Propulsion System.* 2022.
50. Gao, H., et al., *Experimental Study on Thermal and Catalytic Decomposition of a Dual-Mode Ionic Liquid Propellant.* E3s Web of Conferences, 2021. **257**: p. 01041.
51. Rafalskyi, D. and A. Aanesland, *Brief Review on Plasma Propulsion With Neutralizer-Free Systems.* Plasma Sources Science and Technology, 2016. **25**(4): p. 043001.
52. Mazouffre, S., *Electric Propulsion for Satellites and Spacecraft: Established Technologies and Novel Approaches.* Plasma Sources Science and Technology, 2016. **25**(3): p. 033002.
53. Leomanni, M., et al., *All-Electric Spacecraft Precision Pointing Using Model Predictive Control.* Journal of Guidance Control and Dynamics, 2015. **38**(1): p. 161-168.
54. Takahashi, K., *Helicon-Type Radiofrequency Plasma Thrusters and Magnetic Plasma Nozzles.* Reviews of Modern Plasma Physics, 2019. **3**(1).
55. Kiss'ovski, Z., et al., *Microwave Electrothermal Thruster With Surface Wave Plasma.* Contributions to Plasma Physics, 2024. **64**(3).
56. Yildiz, M.S. and M. Çelik, *Numerical Investigation of the Electric Field Distribution and the Power Deposition in the Resonant Cavity of a Microwave Electrothermal Thruster.* Aip Advances, 2017. **7**(4).
57. Burke, L.M., M.C. Martini, and S.R. Oleson, *A High Power Solar Electric Propulsion - Chemical Mission for Human Exploration of Mars.* 2014.
58. Keidar, M., et al., *Electric Propulsion for Small Satellites.* Plasma Physics and Controlled Fusion, 2014. **57**(1): p. 014005.
59. Nassersharif, B. and D. Thomas, *Nuclear Propulsion.* 2023.
60. Summerer, L. and K. Stephenson, *Nuclear Power Sources: A Key Enabling Technology for Planetary Exploration.* Proceedings of the Institution of Mechanical Engineers Part G Journal of Aerospace Engineering, 2010. **225**(2): p. 129-143.
61. Petitgenet, V., et al., *A Coupled Approach to the Design Space Exploration of Nuclear Thermal Propulsion Systems.* 2020.

62. Gabrielli, R. and G. Herdrich, *Review of Nuclear Thermal Propulsion Systems.* Progress in Aerospace Sciences, 2015. **79**: p. 92-113.
63. Khatry, J. and F. Aydoğan, *Modeling Loss-of-Flow Accidents and Their Impact on Radiation Heat Transfer.* Science and Technology of Nuclear Installations, 2017. **2017**: p. 1-15.
64. Song, J., et al., *Neutronics and Thermal Hydraulics Analysis of a Conceptual Ultra-High Temperature MHD Cermet Fuel Core for Nuclear Electric Propulsion.* Frontiers in Energy Research, 2018. **6**.
65. Randolph, T., et al., *The Prometheus 1 Spacecraft Preliminary Electric Propulsion System Design.* 2005.
66. Glenn Research Center, *Nuclear Thermal Propulsion Systems: Leading research, testing and analysis to support the development of nuclear thermal propulsion for spacecraft and vehicles*. 2024, National Aeronautics and Space Administration.
67. Petkow, D., et al., *Comparative Investigation of Fusion Reactions for Space Propulsion Applications.* Transactions of the Japan Society for Aeronautical and Space Sciences Space Technology Japan, 2009. **7**(ists26): p. Pb_59-Pb_63.
68. Dujarric, C., A. Santovincenzo, and L. Summerer, *The Nuclear Thermal Electric Rocket: A Proposed Innovative Propulsion Concept for Manned Interplanetary Missions.* 2013.
69. Pizarro-Chong, A., et al., *Development of Space Nuclear Reactors for Lunar Purposes: Overview of Technical and Non-Technical Issues.* 2010.
70. Houts, M.G., et al., *The Nuclear Cryogenic Propulsion Stage.* 2014.
71. Leishman, J.G., *Introduction to Aerospace Flight Vehicles*. 2024: Embry-Riddle Aeronautical University.
72. Creech, S., *NASA's Space Launch System: An Enabling Capability for Discovery.* 2014: p. 1-11.
73. Creech, S., *Game Changing: NASA's Space Launch System and Science Mission Design.* 2013.
74. Ahmed, M.M.Z., et al., *Friction Stir Welding of Aluminum in the Aerospace Industry: The Current Progress and State-of-the-Art Review.* Materials, 2023. **16**(8): p. 2971.
75. Singer, J., J. Pelfrey, and G. Norris, *Enabling Science and Deep Space Exploration Through Space Launch System Secondary Payload Opportunities.* 2016.
76. Robinson, K.F., A.A. Schorr, and D. Hitt, *NASA's Space Launch System: Exceptional Opportunities for Secondary Payloads to Deep Space.* 2018.
77. Bramon, C., et al., *NASA Space Rocket Logistics Challenges.* 2014.
78. Trotta, D., et al., *Optimal Tuning of Adaptive Augmenting Controller for Launch Vehicles in Atmospheric Flight.* Journal of Guidance Control and Dynamics, 2020. **43**(11): p. 2133-2140.
79. Pei, J. and P.M. Rothhaar, *Demonstration of the Space Launch System Augmenting Adaptive Control Algorithm on Pole-Cart Platform.* 2018.
80. Crocker, A., et al., *Update on Risk Reduction Activities for an F-1-Based Advanced Booster for NASA's Space Launch System.* 2014.
81. Crocker, A., et al., *The Benefits of an Advanced Booster Competition for NASA's Space Launch System.* 2013.
82. National Aeronautics and Space Administration, *State-of-the-Art of Small Spacecraft Technology* 2024: Hanover, MD.
83. Strojny-Nędza, A., K. Pietrzak, and W. Węglewski, *The Influence of Al2O3 Powder Morphology on the Properties of Cu-Al2O3 Composites Designed for Functionally Graded*

Materials (FGM). Journal of Materials Engineering and Performance, 2016. **25**(8): p. 3173-3184.
84. Nkhasi, N., W. du Preez, and H. Bissett, *Plasma Spheroidisation and Characterisation of Commercial Titanium Grade 5 Powder for Use in Metal Additive Manufacturing.* Matec Web of Conferences, 2023. **388**: p. 03004.
85. Gomez-Gallegos, A., et al., *Studies on Titanium Alloys for Aerospace Application.* Defect and Diffusion Forum, 2018. **385**: p. 419-423.
86. Peters, M., et al., *Titanium Alloys for Aerospace Applications.* Advanced Engineering Materials, 2003. **5**(6): p. 419-427.
87. Bach, C., F. Wehner, and J. Sieder-Katzmann, *Investigations on an All-Oxide Ceramic Composites Based on Al2O3 Fibres and Alumina–Zirconia Matrix for Application in Liquid Rocket Engines.* Aerospace, 2022. **9**(11): p. 684.
88. Schmidt, S., et al., *Ceramic Matrix Composites: A Challenge in Space-Propulsion Technology Applications.* International Journal of Applied Ceramic Technology, 2005. **2**(2): p. 85-96.
89. Panakarajupally, R.P., et al., *Fatigue Characterization of SiC/SiC Ceramic Matrix Composites in Combustion Environment.* Journal of Engineering for Gas Turbines and Power, 2020. **142**(12).
90. Kumaran, S.S., et al., *Fabrication of Al₂O₃ Based Ceramic Matrix Composite by Conventional Sintering and Sol-Gel Process.* Advanced Materials Research, 2011. **335-336**: p. 856-860.
91. Wessels, W., *The Different Materials Used To Make Orbital Rockets*. 2024, Headed for Space.
92. Yee, S.V., et al., *The Influence of ECAP Pass Through Bc Route on Mechanical Properties of Aluminum Alloy 6061.* Advanced Materials Research, 2014. **1024**: p. 219-222.
93. Abtan, N.S., A.H. Jassim, and M.S.M. Al-Janabi, *Tensile Strength, Micro-Hardness and Microstructure of Friction-Stir-Welding AA6061-T4 Joints.* Tikrit Journal of Engineering Sciences, 2018. **25**(4): p. 51-56.
94. Ma, X., et al., *Effect of Ultrasonic Surface Rolling Process on Surface Properties and Microstructure of 6061 Aluminum Alloy.* Materials Research, 2023. **26**.
95. Wang, G., et al., *Research on Corrosion Performance of 6061 aluminum Alloy in Salt Spray Environment.* Materialwissenschaft Und Werkstofftechnik, 2020. **51**(12): p. 1686-1699.
96. Zhang, S.X., G. Cai, and H.H. Wu, *Effects of Solution and Two-Stage Ageing Treatment Process on Microstructure and Properties of 6061 Aluminum Alloy.* Advanced Materials Research, 2011. **335-336**: p. 822-825.
97. Pajaroen, N., et al., *Influence of Solution Heat Treatment Temperature and Time on the Microstructure and Mechanical Properties of Gas Induced Semi-Solid (GISS) 6061 Aluminum Alloy.* Applied Mechanics and Materials, 2013. **313-314**: p. 67-71.
98. Bujuru, K., *Maraging Steel as a Material Choice for Rocket Motor Casings.* 2020.
99. Han, J.A., L. Feng, and H. Xia, *Matching Principle of Material Selection in Product Design.* Applied Mechanics and Materials, 2014. **496-500**: p. 414-417.
100. Mayyas, A. and M. Omar, *Eco-Material Selection for Lightweight Vehicle Design.* 2020.
101. Kanazaki, M., et al., *Conceptual Design of Single-Stage Rocket Using Hybrid Rocket byMeans of Genetic Algorithm.* Procedia Engineering, 2015. **99**: p. 198-207.
102. Tikul, N., *Environmental and Economic of Flooring Building Materials.* Applied Environmental Research, 2014: p. 47-59.

103. Louis-Charles, H.M., et al., *Emergency Management and the Final Frontier: Preparing Local Communities for Falling Space Debris.* Risk Hazards & Crisis in Public Policy, 2023. **14**(3): p. 247-266.
104. Byers, M., et al., *Unnecessary Risks Created by Uncontrolled Rocket Reentries.* Nature Astronomy, 2022. **6**(9): p. 1093-1097.
105. Ragul, M.S., et al., *Theoretical Model Study on Chemical Compositions Affecting the Space Launch Vehicles.* 2022. **8**(1): p. 35-38.
106. Chang, I.S. and E.J. Tomei, *Solid Rocket Failures in World Space Launches.* 2005.
107. Hidayah, Q., U. Salamah, and M. Sasono, *Analisis Uji Peluncuran Roket Air Berbasis Carbon Fiber Menggunakan Sistem Telemetri.* Jurnal Teori Dan Aplikasi Fisika, 2022. **10**(1): p. 81.
108. Ismail, I.I., et al., *Metals and Alloys Additives as Enhancer for Rocket Propulsion: A Review.* Journal of Advanced Research in Fluid Mechanics and Thermal Sciences, 2021. **90**(1): p. 1-9.
109. Burke, W., *Engineer's Guide to Lightweight Part Design*. 2024, Five Flute In.
110. Gao, X., et al., *Fused Deposition Modeling With Polyamide 1012.* Rapid Prototyping Journal, 2019. **25**(7): p. 1145-1154.
111. Tlegenov, Y., W.F. Lu, and G.Y. Hong, *A Dynamic Model for Current-Based Nozzle Condition Monitoring in Fused Deposition Modelling.* Progress in Additive Manufacturing, 2019. **4**(3): p. 211-223.
112. Bai, W., et al., *Academic Insights and Perspectives in 3D Printing: A Bibliometric Review.* Applied Sciences, 2021. **11**(18): p. 8298.
113. Mendenhall, R. and B. Eslami, *Experimental Investigation on Effect of Temperature on FDM 3D Printing Polymers: ABS, PETG, and PLA.* Applied Sciences, 2023. **13**(20): p. 11503.
114. Rojek, I., et al., *Bulletin of the Polish Academy of Sciences: Technical Sciences.* 2021.
115. Reis, R.I., W.K. Shimote, and L.C. Pardini, *Anomalous Behavior of a Solid Rocket Motor Nozzle Insert During Static Firing Test.* Journal of Aerospace Technology and Management, 2016. **8**(4): p. 483-490.
116. Mastura, M.T., et al., *Concurrent Material Selection of Natural Fibre Filament for Fused Deposition Modeling Using Integration of Analytic Hierarchy Process/Analytic Network Process.* Journal of Renewable Materials, 2022. **10**(5): p. 1221-1238.
117. Alafaghani, A.a., A. Qattawi, and M.A. Ablat, *Design Consideration for Additive Manufacturing: Fused Deposition Modelling.* Open Journal of Applied Sciences, 2017. **07**(06): p. 291-318.
118. Wałpuski, B. and M. Słoma, *Accelerated Testing and Reliability of FDM-Based Structural Electronics.* Applied Sciences, 2022. **12**(3): p. 1110.
119. Martinez, D.W.C., et al., *A Comprehensive Review on the Application of 3D Printing in the Aerospace Industry.* Key Engineering Materials, 2022. **913**: p. 27-34.
120. Laudante, E., et al., *Human–Robot Interaction for Improving Fuselage Assembly Tasks: A Case Study.* Applied Sciences, 2020. **10**(17): p. 5757.
121. Betts, E.M. and R.A. Frederick, *A Historical Systems Study of Liquid Rocket Engine Throttling Capabilities.* 2010.
122. Ziegler, B., J. Mosędrżny, and N. Lewandowska, *Swirled Injector Modeling for Cavitating Multiphase Flow.* E3s Web of Conferences, 2019. **128**: p. 06007.
123. Farrokhi, A., *Welding Properties of Titanium Alloys Grade 5.* 2023.
124. Cican, G., et al., *Design, Manufacturing, and Testing Process of a Lab Scale Test Bench Hybrid Rocket Engine.* Engineering Technology & Applied Science Research, 2023. **13**(6): p. 12039-12046.

125. Mahottamananda, S.N., N.P. Kadiresh, and Y. Pal, *Regression Rate Characterization of HTPB-Paraffin Based Solid Fuels for Hybrid Rocket.* Propellants Explosives Pyrotechnics, 2020. **45**(11): p. 1755-1763.
126. Kaneko, Y., et al., *Fuel Regression Rate Behavior of CAMUI Hybrid Rocket.* Transactions of the Japan Society for Aeronautical and Space Sciences Space Technology Japan, 2009. **7**(ists26): p. Pa_77-Pa_80.
127. Ismail, I.I., et al., *Modelling of Hybrid Rocket Flow-Fields With Computational Fluid Dynamics.* CFD Letters, 2022. **14**(3): p. 53-67.
128. Chen, P., et al., *The Effects of Non-Uniform Distribution of Oxidizer Flow on High-Frequency Combustion Instability.* Matec Web of Conferences, 2019. **257**: p. 01005.
129. Glenn Research Center, *Rocket Control.* 2024, National Aeronautics and Space Administration.
130. Ran, C. and Z. Deng, *Two Average Weighted Measurement Fusion Kalman Filtering Algorithms in Sensor Networks.* 2008: p. 2387-2391.
131. Feng, B., et al., *Real-time State Estimator Without Noise Covariance Matrices Knowledge – Fast Minimum Norm Filtering Algorithm.* Iet Control Theory and Applications, 2015. **9**(9): p. 1422-1432.
132. Imron, I., B. Satria, and M.D.B. Barus, *Implementation of Angklung Beat Density With Arduino and Piezoelectric Sensor Using Kalman Filter Applied for Reduce Noise Sensor.* International Journal of Economic Technology and Social Sciences (Injects), 2023. **3**(2): p. 346-355.
133. Schultz, J. and T.D. Murphey, *Extending Filter Performance Through Structured Integration.* 2014: p. 430-436.
134. Bae, J. and Y. Kim, *Attitude Estimation for Satellite Fault Tolerant System Using Federated Unscented Kalman Filter.* International Journal of Aeronautical and Space Sciences, 2010. **11**(2): p. 80-86.
135. Sreekantamurthy, V., R.M. Narayanan, and A.F. Martone, *Combined Kalman and Kalman-Levy Filter for Maneuvering Target Tracking.* 2024: p. 18.
136. Jiang, C.-H., S. Zhang, and Q. Zhang, *A Novel Robust Interval Kalman Filter Algorithm for GPS/INS Integrated Navigation.* Journal of Sensors, 2016. **2016**: p. 1-7.
137. Liang, X., *Principles of Multistage Rocket Vehicle and Concepts of Propulsion Methods for Rocket Applications.* Highlights in Science Engineering and Technology, 2022. **27**: p. 858-865.
138. Blanco, P.R., *Learning About Rockets, in Stages.* Physics Education, 2022. **57**(4): p. 045035.
139. Baldieri, F., E. Martelli, and A. Riccio, *A Numerical Study on Carbon-Fiber-Reinforced Composite Cylindrical Skirts for Solid Propeller Rockets.* Polymers, 2023. **15**(4): p. 908.
140. Dinesh, M. and R. Kumar, *Effect of Protrusion on Combustion Stability of Hybrid Rocket Motor.* Propellants Explosives Pyrotechnics, 2022. **47**(4).
141. Cui, S., et al., *Overall Parameters Design of Air-Launched Rockets Using Surrogate Based Optimization Method.* Aerospace, 2021. **9**(1): p. 15.
142. Mirshams, M., et al., *Liquid Propellant Engine Conceptual Design by Using a Fuzzy-Multi-Objective Genetic Algorithm (MOGA) Optimization Method.* Proceedings of the Institution of Mechanical Engineers Part G Journal of Aerospace Engineering, 2014. **228**(14): p. 2587-2603.
143. Khan, S.S., et al., *Comparison of Optimization Techniques and Objective Functions Using Gas Generator and Staged Combustion LPRE Cycles.* Applied Sciences, 2022. **12**(20): p. 10462.

144. Hernandez, R.N., et al., *Design and Performance of Modular 3-D Printed Solid-Propellant Rocket Airframes.* Aerospace, 2017. **4**(2): p. 17.
145. Okoli, B.I., O.S. Sholiyi, and R.O. Durojaye, *Design, Analysis and Simulation of a Single Stage Rocket (Launch Vehicle) Using RockSim.* International Journal of Science and Engineering Applications, 2021. **10**(04): p. 034-039.
146. Wenzhi, H., et al., *Design Optimization of a Low-Cost Three-Stage Launch Vehicle With Modular Hybrid Rocket Motors.* Journal of Physics Conference Series, 2024. **2764**(1): p. 012026.
147. Palharini, R.C., T. Scanlon, and J.M. Reese, *Effects of Angle of Attack on the Behaviour of Imperfections in Thermal Protection Systems of Re-Entry Vehicles.* 2015: p. 551-556.
148. Zhang, S., et al., *Thermally Insulating Polybenzoxazine/Nanosilica Aerogel Ablation Resistant to 1100 °C for Re-Entry Capsules.* Acs Applied Polymer Materials, 2023. **5**(10): p. 8223-8234.
149. Rold, G.D., et al., *Barriers of Oxidation Ans Ageing of Space Shuttle Material.* Advanced Materials Research, 2010. **89-91**: p. 136-141.
150. Pinto, J.R.A., et al., *Development of Asbestos-Free and Environment-Friendly Thermal Protection for Aerospace Application.* Materials Research, 2018. **21**(6).
151. Nagata, M., *A Space-Flight Ship Travelling by a Plasma Rocket Engine From the Earth Ground to the Moon.* Journal of Modern Physics, 2023. **14**(12): p. 1578-1586.
152. Goel, C. and G. Srinivas, *Mechanisms and Applications of Vibration Energy Harvesting in Solid Rocket Motors.* Microsystem Technologies, 2021. **27**(10): p. 3927-3933.
153. Biryukov, V.I. and R.A. Tsarapkin, *Damping Decrements in the Combustion Chambers of Liquid-Propellant Rocket Engines.* Russian Engineering Research, 2019. **39**(1): p. 6-12.
154. Zhan, Z.-H., et al., *Design of Active Vibration Control for Launcher of Multiple Launch Rocket System.* Proceedings of the Institution of Mechanical Engineers Part K Journal of Multi-Body Dynamics, 2011. **225**(3): p. 280-293.
155. Makihara, K. and S. Shimose, *Supersonic Flutter Utilization for Effective Energy-Harvesting Based on Piezoelectric Switching Control.* Smart Materials Research, 2012. **2012**: p. 1-10.
156. Dąbrowski, A. and S.R. Krawczuk, *Analysis of a Mechanical Vibration Filter and Amplifier for Sounding Rocket Applications.* 2021.
157. Wang, H., L. Liu, and W. Zhang, *Optimization Analysis of Dynamic Modal Characteristics of Large Draw Ratio Rocket Body Structure.* 2017.
158. Duan, X.L. and X.Y. Liu, *Noise Prediction of Liquid Rocket Engine by the Software AutoSEA2.* Advanced Materials Research, 2012. **466-467**: p. 794-798.
159. Nikolayev, D., *Forecasting of the Spacecraft Dynamic Loading Under the Rocket Engine Thrust Oscillations of the Launch Vehicle.* 2024.
160. Pinalia, A., et al., *Design of Propellant Composite Thermodynamic Properties Using Rocket Propulsion Analysis (RPA) Software.* Reaktor, 2022. **22**(1): p. 1-6.
161. ÖZel, C., C.K. MacİT, and M. ÖZel, *Investigation of Flight Performance of Notched Delta Wing Rockets on Different Types of Nose Cones.* Turkish Journal of Science and Technology, 2023. **18**(2): p. 435-447.
162. Srivastava, N., P.T. Tkacik, and R.G. Keanini, *Influence of Nozzle Random Side Loads on Launch Vehicle Dynamics.* Journal of Applied Physics, 2010. **108**(4).
163. Buchanan, G., et al., *The Development of Rocketry Capability in New Zealand—World Record Rocket and First of Its Kind Rocketry Course.* Aerospace, 2015. **2**(1): p. 91-117.
164. Shoyama, T., et al., *Air-Launch Experiment Using Suspended Rail Launcher for Rockoon.* Aerospace, 2021. **8**(10): p. 289.

165. Sarigul-Klijn, N. and M. Sarigul-Klijn, *A Comparative Analysis of Methods for Air-Launching Vehicles From Earth to Sub-Orbit or Orbit.* Proceedings of the Institution of Mechanical Engineers Part G Journal of Aerospace Engineering, 2006. **220**(5): p. 439-452.
166. Kelly, J.W., et al., *Motivation for Air-Launch: Past, Present, and Future.* 2017.
167. Tartabini, P.V., et al., *A Multidisciplinary Performance Analysis of a Lifting-Body Single-Stage-to-Orbit Vehicle.* 2000.
168. Kovač, M., et al., *Multi-Stage Micro Rockets for Robotic Insects.* 2012.
169. Ledsinger, L.A. and J.R. Olds, *Optimized Solutions for the Kistler K-1 Branching Trajectory Using MDO Techniques.* 2000.
170. Yoshida, H., et al., *Integrated Optimization for Single-Stage-to-Orbit Using a Pulse Detonation Engine.* Journal of Spacecraft and Rockets, 2019. **56**(4): p. 983-989.
171. Xue, R., et al., *A Survey on the Conceptual Design of Hypersonic Aircraft Powered by RBCC Engine.* Proceedings of the Institution of Mechanical Engineers Part C Journal of Mechanical Engineering Science, 2021. **237**(18): p. 4213-4245.
172. Bayley, D., et al., *Design Optimization of a Space Launch Vehicle Using a Genetic Algorithm.* 2007.
173. Fujikawa, T., T. Tsuchiya, and S. Tomioka, *Multi-Objective, Multidisciplinary Design Optimization of TSTO Space Planes With RBCC Engines.* 2015.
174. Lockwood, M.K., *Overview of Conceptual Design of Early VentureStar Configurations.* 2000.
175. Webber, H.L., A. Bond, and C.M. Hempsell, *Sensitivity of Pre-Cooled Air-Breathing Engine Performance to Heat Exchanger Design Parameters.* 2006.
176. Landis, G.A. and V. Denis, *High Altitude Launch for a Practical SSTO.* 2003. **654**: p. 290-295.
177. Han, P., R. Mu, and N. Cui, *Effective Fault Diagnosis Based on Strong Tracking UKF.* Aircraft Engineering and Aerospace Technology, 2011. **83**(5): p. 275-282.
178. Sato, M., et al., *Development of Main Propulsion System for Reusable Sounding Rocket: Design Considerations and Technology Demonstration.* Transactions of the Japan Society for Aeronautical and Space Sciences Aerospace Technology Japan, 2014. **12**(ists29): p. Tm_1-Tm_6.
179. Bhavana, Y., *Reusable Launch Vehicles: Evolution Redefined.* Journal of Aeronautics & Aerospace Engineering, 2013. **02**(02).
180. Torres, A.I., *Reusable Rockets and the Environment.* Uc Merced Undergraduate Research Journal, 2020. **12**(2).
181. Tománek, R. and J. Hospodka, *Reusable Launch Space Systems.* Mad - Magazine of Aviation Development, 2018. **6**(2): p. 10-13.
182. Hariharan, R., L.R. N, and G. Ravi, *Reusable Rockets and Multi-Planetary Human Life.* International Journal of Engineering and Advanced Technology, 2019. **8**(6s2): p. 576-579.
183. Wuilbercq, R., et al., *Robust Multidisciplinary Design and Optimisation of a Reusable Launch Vehicle.* 2014.
184. Dresia, K., et al., *Multidisciplinary Design Optimization of Reusable Launch Vehicles for Different Propellants and Objectives.* 2020.
185. Morrell, B., et al., *Development of a Hypersonic Aircraft Design Optimization Tool.* Applied Mechanics and Materials, 2014. **553**: p. 847-852.
186. Billingsley, M., et al., *Extent and Impacts of Hydrocarbon Fuel Compositional Variability for Aerospace Propulsion Systems.* 2010.
187. Cheng, T., *Review of Novel Energetic Polymers and Binders – High Energy Propellant Ingredients for the New Space Race.* Designed Monomers & Polymers, 2019. **22**(1): p. 54-65.

188. Vala, M.M., Y. Bayat, and M. Bayat, *Synthesis and Thermal Decomposition Kinetics of Epoxy Poly Glycidyl Nitrate as an Energetic Binder.* Defence Science Journal, 2020. **70**(4): p. 461-468.
189. Kondo, K., et al., *Vacuum Test of a Micro-Solid Propellant Rocket Array Thruster.* Ieice Electronics Express, 2004. **1**(8): p. 222-227.
190. Tanaka, S., et al., *MEMS-Based Solid Propellant Rocket Array Thruster With Electrical Feedthroughs.* Transactions of the Japan Society for Aeronautical and Space Sciences, 2003. **46**(151): p. 47-51.
191. Wu, C., et al., *Study on Mechanical Properties and Failure Mechanisms of Highly Filled Hydroxy-Terminated Polybutadiene Propellant Under Different Tensile Loading Conditions.* Polymers, 2023. **15**(19): p. 3869.
192. Wen, Z., et al., *Molecular Dynamics Simulation of the Pyrolysis and Oxidation of NEPE Propellant.* Propellants Explosives Pyrotechnics, 2022. **47**(12).
193. Pang, W., et al., *Effect of Metal Nanopowders on the Performance of Solid Rocket Propellants: A Review.* Nanomaterials, 2021. **11**(10): p. 2749.
194. Wang, Y., *Effect of Molecular Perovskite Energetic Materials DAP-4 on Energy Performances of Solid Propellants.* 2023.
195. Sharma, J. and A. Miglani, *Time-Varying Oscillatory Response of Burning Gel Fuel Droplets.* Gels, 2023. **9**(4): p. 309.
196. Chauhan, D., et al., *Studies on the Processing of HTPB-based Fast-burning Propellant With Trimodal Oxidiser Distribution and Its Rheological Behaviour.* Asia-Pacific Journal of Chemical Engineering, 2022. **17**(3).
197. El-Dakhakhny, M., et al., *Comparative Study Between Different Fillers Used as Reinforcements of Rubber Thermal Insulators.* International Conference on Aerospace Sciences and Aviation Technology, 2015. **16**(AEROSPACE SCIENCES): p. 1-15.
198. Bhuvaneswari, C.M., et al., *Ethylene-Propylene Diene Rubber as a Futuristic Elastomer for Insulation of Solid Rocket Motors.* Defence Science Journal, 2006. **56**(3): p. 309-320.
199. Ahmed, A.K.W. and S.V. Hoa, *Improvement of the Properties of Insulating Polymers Using Aramid Fiber for Solid Rocket Motor Insulation.* The International Conference on Chemical and Environmental Engineering, 2008. **4**(6): p. 467-478.
200. George, K., et al., *Recent Developments in Elastomeric Heat Shielding Materials for Solid Rocket Motor Casing Application for Future Perspective.* Polymers for Advanced Technologies, 2017. **29**(1): p. 8-21.
201. Yuan, W., et al., *Designing High-Performance Hypergolic Propellants Based on Materials Genome.* Science Advances, 2020. **6**(49).
202. Niwa, M., A. Santana, and K. Kessaev, *Development of a Resonance Igniter for GO/Kerosene Ignition.* Journal of Propulsion and Power, 2001. **17**(5): p. 995-997.
203. Kapitonova, T. and S. Bullock, *What Is the Optimal Fuel for Space Flight? Efficiency, Cost, and Environmental Impact.* 2023.
204. Borovik, I., et al., *Influence of Polyisobutylene Kerosene Additive on Combustion Efficiency in a Liquid Propellant Rocket Engine.* Aerospace, 2019. **6**(12): p. 129.
205. Xu, J., et al., *Energy Estimation and Testing Verification on Ignition of a Torch Ignition System.* Journal of Physics Conference Series, 2022. **2235**(1): p. 012052.
206. Fareghi-Alamdari, R., N. Zohari, and N. Sheibani, *Reliable Evaluation of Ignition Delay Time of Imidazolium Ionic Liquids as Green Hypergolic Propellants by a Novel Theoretical Approach.* Propellants Explosives Pyrotechnics, 2019. **44**(9): p. 1147-1153.
207. Badakhshan, A., et al., *Nano-Ignition Torch Applied to Cryogenic H2/O2 Coaxial Jet.* 2016.

208. Owio, F., *Design and Testing of the Ignition System for Hybrid Rocket Motor.* International Conference on Aerospace Sciences and Aviation Technology, 2011. **11**(ASAT CONFERENCE): p. 1-14.

209. Schneider, S., et al., *Liquid Azide Salts and Their Reactions With Common Oxidizers IRFNA and N₂O₄.* Inorganic Chemistry, 2008. **47**(13): p. 6082-6089.

210. Li, S., H. Gao, and J.n.M. Shreeve, *Borohydride Ionic Liquids and Borane/Ionic-Liquid Solutions as Hypergolic Fuels With Superior Low Ignition-Delay Times.* Angewandte Chemie, 2014. **126**(11): p. 3013-3016.

211. Bhosale, V.K., S.G. Kulkarni, and P.S. Kulkarni, *Ionic Liquid and Biofuel Blend: A Low–cost and High Performance Hypergolic Fuel for Propulsion Application.* Chemistryselect, 2016. **1**(9): p. 1921-1925.

212. Bhosale, V.K. and P.S. Kulkarni, *Ultrafast Igniting, Imidazolium Based Hypergolic Ionic Liquids With Enhanced Hydrophobicity.* New Journal of Chemistry, 2017. **41**(3): p. 1250-1258.

213. Marothiya, G., et al., *Development of H₂O₂ Based Mixed Hybrid Rocket.* Propellants Explosives Pyrotechnics, 2021. **46**(11): p. 1687-1695.

214. Kara, O. and A. Karabeyoğlu, *Hybrid Propulsion System: Novel Propellant Design for Mars Ascent Vehicles.* 2021.

215. Huh, J.W., et al., *Preliminary Assessment of Hydrogen Peroxide Gel as an Oxidizer in a Catalyst Ignited Hybrid Thruster.* International Journal of Aerospace Engineering, 2018. **2018**: p. 1-14.

216. Surmacz, P., *Green Rocket Propulsion Research and Development at the Institute of Aviation: Problems and Perspectives.* Journal of Kones Powertrain and Transport, 2016. **23**(1): p. 337-344.

217. Gligorijević, N., et al., *Influence of the Ignition Mixture Particle Size on Rocket Motor Igniter Pressure Gradient.* Scientific Technical Review, 2018. **68**(1): p. 33-39.

218. Whitmore, S.A., D.P. Merkley, and N.R. Inkley, *Development of a Power Efficient, Restart-Capable Arc Ignitor for Hybrid Rockets.* 2014.

219. Boiron, A.J. and B. Cantwell, *Hybrid Rocket Propulsion and in-Situ Propellant Production for Future Mars Missions.* 2013.

220. Paccagnella, E., et al., *Scaling Parameters of Swirling Oxidizer Injection in Hybrid Rocket Motors.* Journal of Propulsion and Power, 2017. **33**(6): p. 1378-1394.

221. Zilliac, G.G. and M.A. Karabeyoğlu, *Hybrid Rocket Fuel Regression Rate Data and Modeling.* 2006.

222. Morita, T., et al., *Solid-Fuel Regression Rate for Standard-Flow Hybrid Rocket Motors.* Journal of Thermal Science and Technology, 2012. **7**(2): p. 387-398.

223. Chiaverini, M.J., et al., *Regression Rate Behavior of Hybrid Rocket Solid Fuels.* Journal of Propulsion and Power, 2000. **16**(1): p. 125-132.

224. Chiaverini, M.J., et al., *Regression-Rate and Heat-Transfer Correlations for Hybrid Rocket Combustion.* Journal of Propulsion and Power, 2001. **17**(1): p. 99-110.

225. Barato, F., *Challenges of Ablatively Cooled Hybrid Rockets for Satellites or Upper Stages.* Aerospace, 2021. **8**(7): p. 190.

226. Kafafy, R., M.H. Azami, and M. Idres, *Hybrid Rocket Performance With Varying Additive Concentrations.* 2013.

227. Alsaidi, S.B., J. Huh, and M.Y.E. Selim, *Combustion of Date Stone and Jojoba Solid Waste in a Hybrid Rocket-Like Combustion Chamber.* Aerospace, 2024. **11**(3): p. 181.

228. Holste, K., et al., *Ion thrusters for electric propulsion: Scientific issues developing a niche technology into a game changer.* Review of Scientific Instruments, 2020. **91**(6).
229. Koda, D. and H. Kuninaka, *Demonstration of Negative Fullerene Ion Thruster Combined With Positive Xenon Ion Thruster.* Transactions of the Japan Society for Aeronautical and Space Sciences Aerospace Technology Japan, 2016. **14**(ists30): p. Pb_203-Pb_208.
230. Diamant, K.D., et al., *Ionization, Plume Properties, and Performance of Cylindrical Hall Thrusters.* Ieee Transactions on Plasma Science, 2010. **38**(4): p. 1052-1057.
231. Roibás, E., et al., *Characterization of the Ion Beam Neutralization of Plasma Thrusters Using Collecting and Emissive Langmuir Probes.* Contributions to Plasma Physics, 2013. **53**(1): p. 57-62.
232. Nakayama, Y., I. Funaki, and H. Kuninaka, *Sub-Milli-Newton Class Miniature Microwave Ion Thruster.* Journal of Propulsion and Power, 2007. **23**(2): p. 495-499.
233. Ekholm, J., et al., *Plume Characteristics of the Busek 600 W Hall Thruster.* 2006.
234. Anderson, J.R., I. Katz, and D.M. Goebel, *Numerical Simulation of Two-Grid Ion Optics Using a 3D Code.* 2004.
235. Takao, Y., et al., *Three-Dimensional Particle-in-Cell Simulation of a Miniature Plasma Source for a Microwave Discharge Ion Thruster.* Plasma Sources Science and Technology, 2014. **23**(6): p. 064004.
236. Charles, C., *Plasmas for Spacecraft Propulsion.* Journal of Physics D Applied Physics, 2009. **42**(16): p. 163001.
237. Bundesmann, C., et al., *In Situ Thermal Characterization of the Accelerator Grid of an Ion Thruster.* Journal of Propulsion and Power, 2011. **27**(3): p. 532-537.
238. Schneider, R., et al., *Particle-in-Cell Simulations for Ion Thrusters.* Contributions to Plasma Physics, 2009. **49**(9): p. 655-661.
239. Patino, M.I., L. Chu, and R.E. Wirz, *Ion-Neutral Collision Analysis for a Well-Characterized Plasma Experiment.* 2012.
240. Neumann, H., et al., *Broad Beam Ion Sources for Electrostatic Space Propulsion and Surface Modification Processes: From Roots to Present Applications.* Contributions to Plasma Physics, 2007. **47**(7): p. 487-497.
241. Greig, A.D., C. Charles, and R. Boswell, *Simulation of Main Plasma Parameters of a Cylindrical Asymmetric Capacitively Coupled Plasma Micro-Thruster Using Computational Fluid Dynamics.* Frontiers in Physics, 2015. **2**.
242. Wirz, R.E., J.R. Anderson, and I. Katz, *Time-Dependent Erosion of Ion Optics.* Journal of Propulsion and Power, 2011. **27**(1): p. 211-217.
243. Nakayama, Y., *Experimental Visualization of Ion Thruster Discharge and Beam Extraction.* Transactions of the Japan Society for Aeronautical and Space Sciences Space Technology Japan, 2009. **7**(ists26): p. Pb_29-Pb_34.
244. Aanesland, A., et al., *Development and Test of the Negative and Positive Ion Thruster PEGASES.* 2014.
245. Nakayama, Y. and F. Tanaka, *Experimental Evaluation of Neutralization Phenomena With Visualized Ion Thruster.* Transactions of the Japan Society for Aeronautical and Space Sciences Aerospace Technology Japan, 2014. **12**(ists29): p. Pb_53-Pb_58.
246. Criado, E., et al., *Ion Beam Neutralization and Properties of Plasmas From Low Power Ring Cusp Ion Thrusters.* Physics of Plasmas, 2012. **19**(2).
247. Lafleur, T., et al., *Electron Dynamics and Ion Acceleration in Expanding-Plasma Thrusters.* Plasma Sources Science and Technology, 2015. **24**(6): p. 065013.

248. Singh, L.A., et al., *Operation of a Carbon Nanotube Field Emitter Array in a Hall Effect Thruster Plume Environment.* Ieee Transactions on Plasma Science, 2015. **43**(1): p. 95-102.
249. Nakagawa, Y., et al., *Water and Xenon ECR Ion Thruster—comparison in Global Model and Experiment.* Plasma Sources Science and Technology, 2020. **29**(10): p. 105003.
250. Singhal, N., et al., *3D-Printed Multilayered Reinforced Material System for Gas Supply in CubeSats and Small Satellites.* Advanced Engineering Materials, 2019. **21**(11).
251. Souhair, N., *Numerical Suite for the Design, Simulation and Optimization of Cathode-Less Plasma Thrusters.* 2023.
252. Charles, C., R. Boswell, and K. Takahashi, *Investigation of Radiofrequency Plasma Sources for Space Travel.* Plasma Physics and Controlled Fusion, 2012. **54**(12): p. 124021.
253. Takahashi, K., et al., *Magnetic Nozzle Radiofrequency Plasma Systems for Space Propulsion, Industry, and Fusion Plasmas.* Plasma and Fusion Research, 2023. **18**(0): p. 2501050-2501050.
254. Otsuka, S., et al., *Study on Plasma Acceleration in Completely Electrodeless Electric Propulsion System.* Plasma and Fusion Research, 2015. **10**(0): p. 3401026-3401026.
255. Yang, B., J. Miao, and Y. Yang, *Terminal Sliding Mode Control of a Lunar Lander With Electric Propulsion.* Applied Mechanics and Materials, 2014. **494-495**: p. 1195-1201.
256. Song, H., et al., *Experimental Study of Electromagnetically Enhanced Laser Propulsion Performance.* 2022: p. 50.
257. Zhang, D.X., et al., *Discharge Characteristics of a Laser-Electromagnetic Coupling Plasma Thruster for Spacecraft Propulsion.* Applied Mechanics and Materials, 2012. **232**: p. 337-341.
258. Cannat, F., et al., *Optimization of a Coaxial Electron Cyclotron Resonance Plasma Thruster With an Analytical Model.* Physics of Plasmas, 2015. **22**(5): p. 053503.
259. Kajimura, Y., et al., *Numerical Simulation of Dipolar Magnetic Field Inflation Due to Equatorial Ring-Current.* Plasma and Fusion Research, 2014. **9**(0): p. 2405008-2405008.
260. Moritaka, T., et al., *Full Particle-in-Cell Simulation on a Small-Scale Magnetosphere Using Uniform and Nested Grid Systems.* Plasma and Fusion Research, 2011. **6**: p. 2401101-2401101.
261. Merino, M. and E. Ahedo, *Space Plasma Thrusters: Magnetic Nozzles For.* 2016: p. 1329-1351.
262. Bevilacqua, R., et al., *Guidance Navigation and Control for Autonomous Multiple Spacecraft Assembly: Analysis and Experimentation.* International Journal of Aerospace Engineering, 2011. **2011**: p. 1-18.
263. Nabeel I. Abdulbaki, Y., *Enhancement of Guidance System Using Kalman Filter.* Engineering and Technology Journal, 2010. **28**(3): p. 445-454.
264. Li, K., H.S. Shin, and A. Tsourdos, *Capturability of a Sliding-Mode Guidance Law With Finite-Time Convergence.* Ieee Transactions on Aerospace and Electronic Systems, 2020. **56**(3): p. 2312-2325.
265. He, S. and C.-H. Lee, *Optimality of Error Dynamics in Missile Guidance Problems.* Journal of Guidance Control and Dynamics, 2018. **41**(7): p. 1624-1633.
266. de Celis, R., et al., *Neural Network-Based Ambiguity Resolution for Precise Attitude Estimation With GNSS Sensors.* Ieee Transactions on Aerospace and Electronic Systems, 2024. **60**(5): p. 6702-6716.
267. Haytham, A., et al., *System Design and Realization of an Autonomous Unmanned Ground Vehicle Using GPS-Based Navigation.* International Conference on Aerospace Sciences and Aviation Technology, 2013. **15**(AEROSPACE SCIENCES): p. 1-13.

268. Gavilan, F., R. Vázquez, and S. Esteban, *Trajectory Tracking for Fixed-Wing UAV Using Model Predictive Control and Adaptive Backstepping.* Ifac-Papersonline, 2015. **48**(9): p. 132-137.
269. Schulte, P. and D.A. Spencer, *Development of an Integrated Spacecraft Guidance, Navigation, &Amp; Control Subsystem for Automated Proximity Operations.* Acta Astronautica, 2016. **118**: p. 168-186.
270. Chowdhary, G., et al., *Fully Autonomous Indoor Flight Relying on Only Five Very Low-Cost Range Sensors.* Journal of Aerospace Information Systems, 2013. **10**(1): p. 21-31.
271. Pu, Y. and C. Bei, *Angle-Only Autonomous Terminal Guidance and Navigation Algorithm for Asteroid Defense Based on Meta-Reinforcement Learning.* Journal of Physics Conference Series, 2023. **2632**(1): p. 012032.
272. Li, S., et al., *Image processing algorithms for deep-space autonomous optical navigation.* The Journal of Navigation, 2013. **66**(4): p. 605-623.
273. Downes, L.M., T.J. Steiner, and J.P. How. *Lunar terrain relative navigation using a convolutional neural network for visual crater detection*. in *2020 American Control Conference (ACC)*. 2020. IEEE.
274. Zhang, L., et al., *Relative attitude and position estimation for a tumbling spacecraft.* Aerospace Science and Technology, 2015. **42**: p. 97-105.
275. Xiong, K. and C. Wei, *Integrated celestial navigation for spacecraft using interferometer and earth sensor.* Proceedings of the Institution of Mechanical Engineers, Part G: Journal of Aerospace Engineering, 2020. **234**(16): p. 2248-2262.
276. Christian, J.A., *StarNAV: Autonomous optical navigation of a spacecraft by the relativistic perturbation of starlight.* Sensors, 2019. **19**(19): p. 4064.
277. HE, F.-p., S.-y. ZHU, and P.-y. CUI. *Autonomous Optical Navigation for Spacecraft in Earth Departure Phase*. in *International Conference on Computer Networks and Communication Technology (CNCT 2016)*. 2016. Atlantis Press.
278. de Gioia, F., et al., *A robust RANSAC-based planet radius estimation for onboard visual based navigation.* Sensors, 2020. **20**(14): p. 4041.
279. Bowman, A., *5.0 Guidance, Navigation, and Control*. 2024, NASA.
280. Keanini, R.G., et al., *Stochastic Rocket Dynamics Under Random Nozzle Side Loads: Ornstein-Uhlenbeck Boundary Layer Separation and Its Coarse Grained Connection to Side Loading and Rocket Response.* Annalen Der Physik, 2011. **523**(6): p. 459-487.
281. Zhou, B., H. Wang, and W. Ruan, *Numerical Analysis of 3-D Inner Flow Field for Ladder-Shaped Multiple Propellant Rocket Motor.* Journal of Physics Conference Series, 2022. **2336**(1): p. 012025.
282. Eerland, W., et al., *An Open-Source, Stochastic, Six-Degrees-of-Freedom Rocket Flight Simulator, With a Probabilistic Trajectory Analysis Approach.* 2017.
283. Mingireanu, F., et al., *Solid Rocket Motors Internal Ballistic Model With Erosive and Condensed Phase Considerations.* The International Conference on Applied Mechanics and Mechanical Engineering, 2018. **18**(18): p. 1-13.
284. Celano, M.P., et al., *Injector Characterization for a Gaseous Oxygen-Methane Single Element Combustion Chamber.* 2016.
285. Marquardt, T. and J. Majdalani, *Review of Classical Diffusion-Limited Regression Rate Models in Hybrid Rockets.* Aerospace, 2019. **6**(6): p. 75.
286. Li, Z., et al., *Numerical Simulation of the Effect of Multiple Factors on the Ignition Process of a Solid Rocket Motor.* Propellants Explosives Pyrotechnics, 2024. **49**(5).
287. Blanco, P.R., *A Discrete, Energetic Approach to Rocket Propulsion.* Physics Education, 2019. **54**(6): p. 065001.

288. Ryazantsev, A., A. Shirokozhukhova, and S.M. Kovalev, *Application of Combined Processing Methods for High-Tech Products Manufacturing*. Key Engineering Materials, 2022. **910**: p. 61-66.
289. Lin, Y., et al., *Architecture Design and Timing Analysis of GNC System Based on Time-Triggered Architecture*. 2014.
290. Liu, H., et al., *Modeling and Simulation for Safety Redundant Architecture in Train Control System*. 2016.
291. Peng, W., et al., *A High-Precision Dynamic Model of a Sounding Rocket and Rapid Wind Compensation Method Research*. Advances in Mechanical Engineering, 2017. **9**(7): p. 168781401771394.
292. Nie, Y., Y. Cheng, and J. Wu, *Liquid-Propellant Rocket Engine Online Health Condition Monitoring Base on Multi-Algorithm Parallel Integrated Decision-Making*. Proceedings of the Institution of Mechanical Engineers Part G Journal of Aerospace Engineering, 2016. **231**(9): p. 1621-1633.
293. Guram, S., et al., *Review Study on Thermal Characteristics of Bell Nozzle Used in Supersonic Engine*. 2023. **2**(1): p. 4-14.
294. Atygayev, T., V.P. Ivel, and Y.V. Gerasimova, *Development of a Hardware and Software Model of a Rocket Motion Correction System*. Eastern-European Journal of Enterprise Technologies, 2021. **3**(3 (111)): p. 15-23.
295. Schwarz, R. and S. Theil, *A Fault-Tolerant on-Board Computing and Data Handling Architecture Incorporating a Concept for Failure Detection, Isolation, and Recovery for the SHEFEX III Navigation System*. 2014.
296. Huh, J.W. and S. Kwon, *A Practical Design Approach for a Single-Stage Sounding Rocket to Reach a Target Altitude*. The Aeronautical Journal, 2022. **126**(1301): p. 1084-1100.
297. Chen, L., et al., *Rocket Recovery Carbin Section Passive Location Based on Whale Optimization Algorithm*. Journal of Physics Conference Series, 2022. **2364**(1): p. 012068.
298. Wu, T.-Y., et al., *TPAD: Hardware Trojan Prevention and Detection for Trusted Integrated Circuits*. Ieee Transactions on Computer-Aided Design of Integrated Circuits and Systems, 2016. **35**(4): p. 521-534.
299. Li, W., et al., *Powered Landing Control of Reusable Rockets Based on Softmax Double DDPG*. Aerospace, 2023. **10**(7): p. 590.
300. Han, Y., H. Sun, and H. Guo, *Research on Rocket Laser Scattering Characteristic Simulation Software*. Laser Physics, 2013. **23**(5): p. 056007.
301. Harari, Z., *Derivation of a Revised Tsiolkovsky Rocket Equation That Predicts Combustion Oscillations*. Advances in Aerospace Science and Technology, 2024. **09**(01): p. 10-27.
302. Pylypenko, O.V., et al., *Mathematical Modelling of Start-Up Transients at Clustered Propulsion System With POGO-suppressors for CYCLON-4M Launch Vehicle*. Kosmìčna Nauka ì Tehnologìâ, 2021. **27**(6): p. 3-15.
303. Fink, W., T.L. Huntsberger, and H. Aghazarian, *Dynamic Optimization of <i>N</I>-joint Robotic Limb Deployments*. Journal of Field Robotics, 2009. **27**(3): p. 268-280.
304. Berger, R.W., et al., *The RAD6000MC System-on-Chip Microcontroller for Spacecraft Avionics and Instrument Control*. 2008.
305. Nakamura, Y., et al., *Exploration of Energization and Radiation in Geospace (ERG): Challenges, Development, and Operation of Satellite Systems*. Earth Planets and Space, 2018. **70**(1).
306. Lubin, P., A.N. Cohen, and J. Erlikhman, *Radiation Effects From ISM and Cosmic Ray Particle Impacts on Relativistic Spacecraft*. 2022.

307. Bonet, M.S. and L. Kosmidis, *SPARROW: A Low-Cost Hardware/Software Co-Designed SIMD Microarchitecture for AI Operations in Space Processors*. 2022.
308. Lindsay, P., K. Winter, and N. Yatapanage, *Safety Assessment Using Behavior Trees and Model Checking*. 2010: p. 181-190.
309. Sreekumar, A., et al., *Enhanced Performance Capability in a Dual Redundant Avionics Platform – Fault Tolerant Scheduling With Comparative Evaluation*. Procedia Computer Science, 2015. **46**: p. 921-932.
310. Gavriluţ, V., D. Tămaş-Selicean, and P. Pop, *Fault-Tolerant Topology Selection for TTEthernet Networks*. 2015: p. 4001-4009.
311. Chu, J., et al., *Optimal Design of Configuration Scheme for Integrated Modular Avionics Systems With Functional Redundancy Requirements*. Ieee Systems Journal, 2021. **15**(2): p. 2665-2676.
312. Zhao, C., et al., *Research on Resource Allocation Method of Integrated Avionics System Considering Fault Propagation Risk*. International Journal of Aerospace Engineering, 2022. **2022**: p. 1-19.
313. Lerro, A., et al., *Experimental Analysis of Neural Approaches for Synthetic Angle-of-Attack Estimation*. International Journal of Aerospace Engineering, 2021. **2021**: p. 1-13.
314. Horváth, Á., et al., *Hardware-Software Allocation Specification of IMA Systems for Early Simulation*. 2014.
315. Li, W. and G. Shi, *Redundancy Management Strategy for Electro-Hydraulic Actuators Based on Intelligent Algorithms*. Advances in Mechanical Engineering, 2014. **12**(6).
316. Hoffmann, M., et al., *Effectiveness of Fault Detection Mechanisms in Static and Dynamic Operating System Designs*. 2014: p. 230-237.
317. Wang, H., et al., *Integrated Modular Avionics System Safety Analysis Based on Model Checking*. 2017: p. 1-6.
318. Hickey, C., et al., *The Legacy of Space Shuttle Flight Software*. 2011.
319. Ichikawa, T., et al., *Development of Redundant Integrated Navigation System (RINS) for Launch Vehicle*. 2023.
320. Zhang, C., X. Shi, and D. Chen, *Safety Analysis and Optimization for Networked Avionics System*. 2014.
321. Edwards, B., et al., *Original Research by Young Twinkle Students (ORBYTS): Ephemeris Refinement of Transiting Exoplanets III*. Astronomy Theory Observation and Methods, 2021. **2**(1).
322. Porter, S.B., et al., *Orbits and Occultation Opportunities of 15 TNOs Observed by New Horizons*. The Planetary Science Journal, 2022. **3**(1): p. 23.
323. Stern, S.A., et al., *Initial Results From the New Horizons Exploration of 2014 MU ₆₉, a Small Kuiper Belt Object*. Science, 2019. **364**(6441).
324. Sababha, B.H., O. Rawashdeh, and W. Sadeh, *A Real-Time Gracefully Degrading Avionics System for Unmanned Aerial Vehicles*. 2012.
325. Ellis, S.J., et al., *Runtime Fault Detection in Programmed Molecular Systems*. Acm Transactions on Software Engineering and Methodology, 2019. **28**(2): p. 1-20.
326. Wessels, W., *How Rockets Are Made – Where And How Orbital Launch Vehicles Are Build*. 2024, Headed For Space.
327. Kehayas, N., *Earth-to-Space and High-Speed "Air" Transportation: An Aerospaceplane Design*. Aircraft Engineering and Aerospace Technology, 2019. **91**(2): p. 381-403.
328. Su, Y., et al., *Simulations and Software Development for the Hard X-Ray Imager Onboard ASO-S*. Research in Astronomy and Astrophysics, 2019. **19**(11): p. 163.

329. Feng, L., et al., *Space Weather Related to Solar Eruptions With the ASO-S Mission.* Frontiers in Physics, 2020. **8**.
330. Montopoli, M., et al., *Remote Sensing of the Moon's Subsurface With Multifrequency Microwave Radiometers: A Numerical Study.* Radio Science, 2011. **46**(1).
331. Siemes, C., et al., *CASPA-ADM: A Mission Concept for Observing Thermospheric Mass Density.* Ceas Space Journal, 2022. **14**(4): p. 637-653.
332. Scaccabarozzi, D., et al., *MicroMED, Design of a Particle Analyzer for Mars.* Measurement, 2018. **122**: p. 466-472.
333. Ferrando, P., *The COSPIX Mission: Focusing on the Energetic and Obscured Universe.* 2011.
334. Dunwoody, R., et al., *Development, Description, and Validation of the Operations Manual for EIRSAT-1, a 2U CubeSat With a Gamma-Ray Burst Detector.* Journal of Astronomical Telescopes Instruments and Systems, 2023. **9**(03).
335. Mudarris, M., M.R. Basirung, and I. Sumariyanto, *Rocket Load Test Based on Inertial Measurement Unit Sensor in Supporting National Air Defense.* Jurnal Pertahanan Media Informasi TTG Kajian & Strategi Pertahanan Yang Mengedepankan Identity Nasionalism & Integrity, 2022. **8**(1): p. 1.
336. Pütz, P., *Space Robotics.* Reports on Progress in Physics, 2002. **65**(3): p. 421-463.
337. Pursiainen, S. and M. Kaasalainen, *Electromagnetic 3D Subsurface Imaging With Source Sparsity for a Synthetic Object.* Inverse Problems, 2015. **31**(12): p. 125004.
338. Grundmann, J.T., et al., *Capabilities of Gossamer-1 Derived Small Spacecraft Solar Sails Carrying Mascot-Derived Nanolanders for in-Situ Surveying of NEAs.* Acta Astronautica, 2019. **156**: p. 330-362.
339. Ji, X. and Y.-G. Zhao, *Architecture Design for Unmanned Aerial Vehicle Mission Planning System.* 2019.
340. Yang, S., et al., *Autonomous Attitude Reconstruction Analysis for Propulsion System With Typical Thrust Drop Fault.* Aerospace, 2022. **9**(8): p. 409.
341. Yoshioka, H., et al., *Façade Tests on Fire Propagation Along Combustible Exterior Wall Systems.* Fire Science and Technology, 2014. **33**(1): p. 1-15.
342. Gooranorimi, O., et al., *Residual Mechanical Properties of Fire Exposed GFRP Reinforcement in Concrete Elements.* 2016. **2**: p. 985-994.
343. Ellis, D.S., H. Tabatabai, and A. Nabizadeh, *Residual Tensile Strength and Bond Properties of GFRP Bars After Exposure to Elevated Temperatures.* Materials, 2018. **11**(3): p. 346.
344. Raouffard, M.M. and M. Nishiyama, *Residual Load Bearing Capacity of Reinforced Concrete Frames After Fire.* Journal of Advanced Concrete Technology, 2016. **14**(10): p. 625-633.
345. Li, L.Z., et al., *Experimental Study on Seismic Performance of Post-Fire Reinforced Concrete Frames.* Engineering Structures, 2019. **179**: p. 161-173.
346. Zhang, X., et al., *Experimental Study on Fire Resistance of Reinforced Concrete Frame Structure.* 2014.
347. Li, Z., et al., *Numerical Simulation of Blast Resistance of Reinforced Concrete Slabs Under Fire Conditions.* Advances in Structural Engineering, 2023. **26**(10): p. 1877-1894.
348. Ünal, A., et al., *Design and Implementation of a Thrust Vector Control (TVC) Test System.* Journal of Polytechnic, 2018.
349. Koltunov, A., et al., *The Development and First Validation of the GOES Early Fire Detection (GOES-EFD) Algorithm.* Remote Sensing of Environment, 2016. **184**: p. 436-453.
350. Ruchała, P., et al., *Wind Tunnel Tests of Influence of Boosters and Fins on Aerodynamic Characteristics of the Experimental Rocket Platform.* Transactions on Aerospace Research, 2017. **2017**(4): p. 82-102.

351. Bryson, H., et al., *Vertical Wind Tunnel for Prediction of Rocket Flight Dynamics.* Aerospace, 2016. **3**(2): p. 10.
352. Ocokoljić, G., B. Rašuo, and A. Bengin, *Aerodynamic Shape Optimization of Guided Missile Based on Wind Tunnel Testing and Computational Fluid Dynamics Simulation.* Thermal Science, 2017. **21**(3): p. 1543-1554.
353. Camussi, R., et al., *Wind Tunnel Measurements of the Surface Pressure Fluctuations on the New VEGA-C Space Launcher.* Aerospace Science and Technology, 2020. **99**: p. 105772.
354. Hu, B., et al., *Engineering Calculation and Analysis of Aerodynamic Heating for Hypersonic Rocket Sled.* Journal of Physics Conference Series, 2023. **2478**(8): p. 082009.
355. Kimmel, R.L., et al., *Hypersonic International Flight Research Experimentation-5b Flight Overview.* Journal of Spacecraft and Rockets, 2018. **55**(6): p. 1303-1314.
356. Zilliac, G.G., et al., *A Comparison of the Measured and Computed Skin Friction Distribution on the Common Research Model.* 2011.
357. Kwiek, A., et al., *Results of Simulation and Scaled Flight Tests Performed on a Rocket-Plane at High Angles of Attack.* Aircraft Engineering and Aerospace Technology, 2021. **93**(9): p. 1445-1459.
358. Goetzendorf-Grabowski, T. and A. Kwiek, *Study of the Impact of Aerodynamic Model Fidelity on the Flight Characteristics of Unconventional Aircraft.* Applied Sciences, 2023. **13**(22): p. 12522.
359. Ramamurthy, B., E. Horowitz, and J.R. Fragola, *Physical Simulation in Space Launcher Engine Risk Assessment.* 2010.
360. Zhang, J.H. and S.S. Jiang, *Rigid-Flexible Coupling Model and Dynamic Analysis of Rocket Sled.* Advanced Materials Research, 2011. **346**: p. 447-454.
361. Venkateswarlu, V., et al., *Failure Analysis of Ball-Bearing of Turbo-Pump Used in Liquid Rocket Engine.* Materials Science Forum, 2015. **830-831**: p. 709-712.
362. Deswandri, N., et al., *Risk Assessment of Solid Propellant Rocket Motor Using a Combination of HAZOP and FMEA Methods.* Journal of Advanced Research in Fluid Mechanics and Thermal Sciences, 2023. **110**(1): p. 63-78.
363. Ćatić, D. and J. Glišović, *Failure Mode, Effects and Criticality Analysis of Mechanical Systems' Elements.* Mobility and Vehicle Mechanics, 2019. **45**(3): p. 25-39.
364. Guixiang, S., et al., *System Failure Analysis Based on DEMATEL-ISM and FMECA.* Journal of Central South University, 2014. **21**(12): p. 4518-4525.
365. Zhu, M., et al., *A New Fault Injection Method for Liquid Rocket Pressurization and Feed System.* 2015.
366. Nagashima, F., et al., *Improvement in Identification Accuracy of a Failure Diagnostic System for a Reusable Rocket Engine.* 2023. **4**(1).
367. Nagashima, F., et al., *Development of Failure Diagnostic System for a Reusable Rocket Engine Using Simulation.* 2022: p. 734-739.
368. Korzun, A.M. and L.A. Cassel, *Scaling and Similitude in Single Nozzle Supersonic Retropropulsion Aerodynamics Interference.* 2020.
369. Braun, R.D., B. Sforzo, and C.H. Campbell, *Advancing Supersonic Retropropulsion Using Mars-Relevant Flight Data: An Overview.* 2017.
370. Korzun, A.M., J.R. Cruz, and R.D. Braun, *A Survey of Supersonic Retropropulsion Technology for Mars Entry, Descent, and Landing.* 2008.
371. DeHart, M.D., S. Schunert, and V. Labouré, *Nuclear Thermal Propulsion.* 2022.

372. Myers, R., M. DeHart, and D. Kotlyar, *Integrated Steady-State System Package for Nuclear Thermal Propulsion Analysis Using Multi-Dimensional Thermal Hydraulics and Dimensionless Turbopump Treatment.* Energies, 2024. **17**(13): p. 3068.
373. Clough, J., et al., *Integrated Propulsion and Power Modeling for Bimodal Nuclear Thermal Rockets.* 2007.
374. Sessim, M. and M. Tonks, *Multiscale Simulations of Thermal Transport in W-UO₂ CERMET Fuel for Nuclear Thermal Propulsion.* Nuclear Technology, 2021. **207**(7): p. 1004-1014.
375. Hickman, R., J. Broadway, and O.R. Mireles, *Fabrication and Testing of CERMET Fuel Materials for Nuclear Thermal Propulsion.* 2012.
376. Taylor, B.D., et al., *Cryogenic Fluid Management Technology Development for Nuclear Thermal Propulsion.* 2015.
377. Poston, D.I., *Nuclear Thermal Propulsion: Benefits and Challenges.* 2021: p. 280-289.
378. Tatem, A.J., S.J. Goetz, and S.I. Hay, *Fifty Years of Earth-Observation Satellites.* American Scientist, 2008. **96**(5): p. 390.
379. Lee, W.-C., K.-S. Kim, and Y.H. Kwon, *Review of the History of Animals That Helped Human Life and Safety for Aerospace Medical Research and Space Exploration.* The Korean Journal of Aerospace and Environmental Medicine, 2020. **30**(1): p. 18-24.
380. Lettenmaier, D.P., et al., *Inroads of Remote Sensing Into Hydrologic Science During the WRR Era.* Water Resources Research, 2015. **51**(9): p. 7309-7342.
381. King, J.V., *Cospas-Sarsat Satellite System for Search and Rescue.* 2008: p. 69-87.
382. Kataoka, R., et al., *Unexpected Space Weather Causing the Reentry of 38 Starlink Satellites in February 2022.* Journal of Space Weather and Space Climate, 2022. **12**: p. 41.
383. Halferty, G., et al., *Photometric Characterization and Trajectory Accuracy of Starlink Satellites: Implications for Ground-Based Astronomical Surveys.* Monthly Notices of the Royal Astronomical Society, 2022. **516**(1): p. 1502-1508.
384. Curzi, G., D. Modenini, and P. Tortora, *Large Constellations of Small Satellites: A Survey of Near Future Challenges and Missions.* Aerospace, 2020. **7**(9): p. 133.
385. Tyson, J.A., et al., *Mitigation of LEO Satellite Brightness and Trail Effects on the Rubin Observatory LSST.* The Astronomical Journal, 2020. **160**(5): p. 226.
386. Burleigh, S., et al., *From Connectivity to Advanced Internet Services: A Comprehensive Review of Small Satellites Communications and Networks.* Wireless Communications and Mobile Computing, 2019. **2019**: p. 1-17.
387. Murphy, C.N. and J. Yates, *Afterword: The Globalizing Governance of International Communications: Market Creation and Voluntary Consensus Standard Setting.* Journal of Policy History, 2015. **27**(3): p. 550-558.
388. Sanctis, M.D., et al., *Satellite Communications Supporting Internet of Remote Things.* Ieee Internet of Things Journal, 2016. **3**(1): p. 113-123.
389. Hiatt, D. and Y.B. Choi, *Issues and Trends in Satellite Telecommunications.* International Journal of Advanced Computer Science and Applications, 2017. **8**(3).
390. Chen, J.Y., et al., *The Channel Capacity of Dual-Polarized MIMO Mobile Satellite System.* Applied Mechanics and Materials, 2014. **643**: p. 111-116.
391. Jena, J. and P.K. Sahu, *Rain Fade and Ka-Band Spot Beam Satellite Communication in India.* 2010: p. 304-306.
392. Chen, Y., X. Ma, and C. Wu, *The Concept, Technical Architecture, Applications and Impacts of Satellite Internet: A Systematic Literature Review.* Heliyon, 2024. **10**(13): p. e33793.

393. Kim, M. and O. Yang, *Precise Attitude Control System Design for the Tracking of Parabolic Satellite Antenna.* International Journal of Smart Home, 2013. **7**(5): p. 275-290.
394. Obiyemi, O. and K. Moloi, *Rainfall's Symphony: Understanding Its Influence on Communication Systems in Nigeria.* 2024.
395. Arapoglou, P.-D., et al., *MIMO Over Satellite: A Review.* Ieee Communications Surveys & Tutorials, 2011. **13**(1): p. 27-51.
396. Xu, Y., Q. Chang, and Z. Yu, *On New Measurement and Communication Techniques of GNSS Inter-Satellite Links.* Science China Technological Sciences, 2011. **55**(1): p. 285-294.
397. Slotten, H.R., *International Governance, Organizational Standards, and the First Global Satellite Communication System.* Journal of Policy History, 2015. **27**(3): p. 521-549.
398. Singh, J., *International Communication Regimes.* 2017.
399. Zhao, J., et al., *The First Result of Relative Positioning and Velocity Estimation Based on CAPS.* Sensors, 2018. **18**(5): p. 1528.
400. Li, B., et al., *Influence of Sweep Interference on Satellite Navigation Time-Domain Anti-Jamming.* Frontiers in Physics, 2023. **10**.
401. Asgari, J., T.H. Mohammadloo, and A.R. Amiri-Simkooei, *Geometrically Constrained Kinematic Global Navigation Satellite Systems Positioning: Implementation and Performance.* Advances in Space Research, 2015. **56**(6): p. 1067-1078.
402. Zhang, P., *Research on Satellite Selection Algorithm in Ship Positioning Based on Both Geometry and Geometric Dilution of Precision Contribution.* International Journal of Advanced Robotic Systems, 2019. **16**(1).
403. Li, M., et al., *Performance of Multi-GNSS in the Asia-Pacific Region: Signal Quality, Broadcast Ephemeris and Precise Point Positioning (PPP).* Remote Sensing, 2022. **14**(13): p. 3028.
404. Li, B., *Unmodeled Error Mitigation for Single-Frequency Multi-GNSS Precise Positioning Based on Multi-Epoch Partial Parameterization.* Measurement Science and Technology, 2019. **31**(2): p. 025008.
405. Rahim, N.A., *L-Band Amplitude Scintillations During Solar Maximum at a Low Latitude Station.* International Journal of Advanced Trends in Computer Science and Engineering, 2020. **9**(1.4): p. 465-470.
406. Shivani*, B. and S. Raghunath, *Low Latitude Ionosphere Error Correction Algorithms for Global Navigation Satellite System.* International Journal of Innovative Technology and Exploring Engineering, 2020. **9**(3): p. 3244-3248.
407. Musumeci, L., J. Samson, and F. Dovis, *Performance Assessment of Pulse Blanking Mitigation in Presence of Multiple Distance Measuring Equipment/Tactical Air Navigation Interference on Global Navigation Satellite Systems Signals.* Iet Radar Sonar & Navigation, 2014. **8**(6): p. 647-657.
408. Innac, A., et al., *The EGNOS Augmentation in Maritime Navigation.* Sensors, 2022. **22**(3): p. 775.
409. Fu, W., et al., *Multi-GNSS Combined Precise Point Positioning Using Additional Observations With Opposite Weight for Real-Time Quality Control.* Remote Sensing, 2019. **11**(3): p. 311.
410. Sapry, H.R.M., et al., *The Implementation of Global Position System (GPS) Among the Cement Transporters and Its Impact to Business Performance.* International Journal of Advanced Trends in Computer Science and Engineering, 2020. **9**(1.1 S I): p. 12-16.
411. Jin, S., T.v. Dam, and S. Wdowinski, *Observing and Understanding the Earth System Variations From Space Geodesy.* Journal of Geodynamics, 2013. **72**: p. 1-10.

412. Gu, S., et al., *Quasi-4-Dimension Ionospheric Modeling and Its Application in PPP.* Satellite Navigation, 2022. **3**(1).
413. Siejka, Z., *Verification of the Usefulness of the Trimble RTX Extended Satellite Technology With the Xfill Function in the Local Network Implementing RTK Measurements.* Artificial Satellites, 2014. **49**(4): p. 191-209.
414. Wang, Z., et al., *On-Orbit Calibration and Characterization of GOES-17 ABI IR Bands Under Dynamic Thermal Condition.* Journal of Applied Remote Sensing, 2020. **14**(03).
415. Xiong, X., et al., *VIIRS On-orbit Calibration Methodology and Performance.* Journal of Geophysical Research Atmospheres, 2014. **119**(9): p. 5065-5078.
416. Upadhyaya, S., et al., *On the Propagation of Satellite Precipitation Estimation Errors: From Passive Microwave to Infrared Estimates.* Journal of Hydrometeorology, 2020. **21**(6): p. 1367-1381.
417. Schmit, T.J., et al., *A Closer Look at the ABI on the GOES-R Series.* Bulletin of the American Meteorological Society, 2017. **98**(4): p. 681-698.
418. Singh, R., et al., *Evaluation and Assimilation of the COSMIC-2 Radio Occultation Constellation Observed Atmospheric Refractivity in the WRF Data Assimilation System.* Journal of Geophysical Research Atmospheres, 2021. **126**(18).
419. Stolle, C., et al., *Space Weather Opportunities From the Swarm Mission Including Near Real Time Applications.* Earth Planets and Space, 2013. **65**(11): p. 1375-1383.
420. Kress, B., et al., *Observations From NOAA's Newest Solar Proton Sensor.* Space Weather, 2021. **19**(12).
421. Li, L., et al., *Design and Implementation of Variable Coding and Modulation for LEO High-resolution Earth Observation Satellite.* International Journal of Satellite Communications and Networking, 2022. **41**(4): p. 303-314.
422. Zhang, S., et al., *An Effectiveness Evaluation Model for Satellite Observation and Data-Downlink Scheduling Considering Weather Uncertainties.* Remote Sensing, 2019. **11**(13): p. 1621.
423. Choudhary, K., M.S. Boori, and A. Kupriyanov, *Spatio-Temporal Analysis Through Remote Sensing and GIS in Moscow Region, Russia.* 2017: p. 42-46.
424. Ngcofe, L. and K. Gottschalk, *The Growth of Space Science in African Countries for Earth Observation in the 21st Century.* South African Journal of Science, 2013. **109**(1/2): p. 1-5.
425. He, C. and Y. Dong, *Multi-Satellite Observation-Relay Transmission-Downloading Coupling Scheduling Method.* Remote Sensing, 2023. **15**(24): p. 5639.
426. Harris, R. and I. Baumann, *Satellite Earth Observation and National Data Regulation.* Space Policy, 2021. **56**: p. 101422.
427. Hossain, F., et al., *Building User-Readiness for Satellite Earth Observing Missions: The Case of the Surface Water and Ocean Topography (SWOT) Mission.* Agu Advances, 2022. **3**(6).
428. Amin, E., et al., *Multi-Season Phenology Mapping of Nile Delta Croplands Using Time Series of Sentinel-2 and Landsat 8 Green LAI.* Remote Sensing, 2022. **14**(8): p. 1812.
429. Zhou, Q., et al., *Monitoring Landscape Dynamics in Central U.S. Grasslands With Harmonized Landsat-8 and Sentinel-2 Time Series Data.* Remote Sensing, 2019. **11**(3): p. 328.
430. Houborg, R. and M.F. McCabe, *High-Resolution NDVI From Planet's Constellation of Earth Observing Nano-Satellites: A New Data Source for Precision Agriculture.* Remote Sensing, 2016. **8**(9): p. 768.

431. Lu, F., et al., *System Demonstrations of Ka-band 5-Gbps Data Transmission for Satellite Applications.* International Journal of Satellite Communications and Networking, 2021. **40**(3): p. 204-217.
432. Oz, I., *Design Tradeoffs in Full Electric, Hybrid and Full Chemical Propulsion Communication Satellite.* Sakarya University Journal of Computer and Information Sciences, 2019. **2**(3): p. 124-133.
433. Choi, Y.-G., K.-B. Shin, and W.-H. Kim, *A Study on Size Optimization of Rocket Motor Case Using the Modified 2D Axisymmetric Finite Element Model.* International Journal of Precision Engineering and Manufacturing, 2010. **11**(6): p. 901-907.
434. Zhang, M., X. Qin, and Q. Zhang, *Aggregated Preference Value Analysis of Small Satellite Launch Opportunities.* Transactions of the Japan Society for Aeronautical and Space Sciences, 2018. **61**(2): p. 69-78.
435. Saleh, J.H., et al., *Electric Propulsion Reliability: Statistical Analysis of on-Orbit Anomalies and Comparative Analysis of Electric Versus Chemical Propulsion Failure Rates.* Acta Astronautica, 2017. **139**: p. 141-156.
436. Zheng, Y., *An Overview of Communication and Orbital Composition Technologies Based on Starlink LEO Satellite Constellation From a Technical Perspective.* Theoretical and Natural Science, 2023. **18**(1): p. 230-237.
437. Guyot, J., A. Rao, and S. Rouillon, *Oligopoly Competition Between Satellite Constellations Will Reduce Economic Welfare From Orbit Use.* Proceedings of the National Academy of Sciences, 2023. **120**(43).
438. Kang, K. and J. Kugler, *Assessment of Deterrence and Missile Defense in East Asia: A Power Transition Perspective.* International Area Studies Review, 2015. **18**(3): p. 280-296.
439. Fontana, S. and F.D. Lauro, *An Overview of Sensors for Long Range Missile Defense.* Sensors, 2022. **22**(24): p. 9871.
440. Dou, J., et al., *Evaluation Method of Formation Ship-to-Air Missile Air Defense Area Capability.* Scientific Journal of Technology, 2023. **4**(12): p. 7-12.
441. Skiba, R., *Mach 5 and Beyond: Hypersonic Warfare, Missile Technology and Global Strategic Implications*. 2024, Melbourne, Victoria, Australia: After Midnight Publishing.
442. Dawood, S.D.S. and M.Y. Harmin, *Structural Responses of a Conceptual Microsatellite Structure Incorporating Perforation Patterns to Dynamic Launch Loads.* Aerospace, 2022. **9**(8): p. 448.
443. Ryan, R.G., et al., *Impact of Rocket Launch and Space Debris Air Pollutant Emissions on Stratospheric Ozone and Global Climate.* 2022.
444. Slongo, A.G., et al., *Preliminary Study of Launch and Orbit of a CubeSat Using a Modified VSB-30 Launcher Vehicle.* Journal of Aerospace Technology and Management, 2020(12): p. 62-79.
445. Long, X., et al., *Mission Scheduling of Multi-Sensor Collaborative Observation for Space Surveillance Network.* Journal of Systems Engineering and Electronics, 2023. **34**(4): p. 906-923.
446. Chae, S.H., et al., *Performance Analysis of Dense Low Earth Orbit Satellite Communication Networks With Stochastic Geometry.* Journal of Communications and Networks, 2023. **25**(2): p. 208-221.
447. Sweeting, M., *Modern Small Satellites-Changing the Economics of Space.* Proceedings of the Ieee, 2018. **106**(3): p. 343-361.

448. Constantinou, V., et al., *Leveraging Deep Learning for High-Resolution Optical Satellite Imagery From Low-Cost Small Satellite Platforms.* Ieee Journal of Selected Topics in Applied Earth Observations and Remote Sensing, 2024. **17**: p. 6354-6365.
449. Nervold, A., et al., *A Pathway to Small Satellite Market Growth.* Advances in Aerospace Science and Technology, 2016. **01**(01): p. 14-20.
450. Zhang, Z. and X. Su, *Improvement of High Damping Structures Using a Photosensitive Resin Filled With Viscous Fluid.* Journal of the Brazilian Society of Mechanical Sciences and Engineering, 2020. **42**(3).
451. Hakia.com, *Space Tourism: Opening New Frontiers for Commercial Space Travel*. 2021, Hakia.com.
452. Sciti, D., et al., *Propulsion Tests on Ultra-High-Temperature Ceramic Matrix Composites for Reusable Rocket Nozzles.* Journal of Advanced Ceramics, 2023. **12**(7): p. 1345-1360.
453. Wang, C., et al., *Parameterized Design and Dynamic Analysis of a Reusable Launch Vehicle Landing System With Semi-Active Control.* Symmetry, 2020. **12**(9): p. 1572.
454. Edafetanure-Ibeh, F., *Mastering the Cosmos: Leveraging Machine Learning to Optimize Spacex Falcon 9 Launch Success Rates.* 2024.
455. Preclik, D., et al., *Reusability Aspects for Space Transportation Rocket Engines: Programmatic Status and Outlook.* Ceas Space Journal, 2011. **1**(1-4): p. 71-82.
456. Mian, A. and M.A. Mian, *Space Medicine: Inspiring a New Generation of Physicians.* Postgraduate Medical Journal, 2022. **99**(1173): p. 763-776.
457. Karukayil, J., H. Love, and N. None, *Optimal Leg Height of Landing Legs to Reduce Risk of Damage From Regolith Ejecta by Retrorocket Exhausts.* Hyperscience International Journals, 2023. **3**(3): p. 17-23.
458. Global Aerospace Editorial Team, *How Fully Reusable Rockets Are Transforming Spaceflight*. 2024, Global Aerospace.
459. Huang, M., *Analysis of Rocket Modelling Accuracy and Capsule Landing Safety.* International Journal of Aeronautical and Space Sciences, 2022. **23**(2): p. 392-405.
460. Dirloman, F.M., et al., *Novel Polyurethanes Based on Recycled Polyethylene Terephthalate: Synthesis, Characterization, and Formulation of Binders for Environmentally Responsible Rocket Propellants.* Polymers, 2021. **13**(21): p. 3828.
461. Okniński, A., et al., *Development of Green Storable Hybrid Rocket Propulsion Technology Using 98% Hydrogen Peroxide as Oxidizer.* Aerospace, 2021. **8**(9): p. 234.
462. Okniński, A., *On Use of Hybrid Rocket Propulsion for Suborbital Vehicles.* Acta Astronautica, 2018. **145**: p. 1-10.
463. Borowski, S.K., D.R. McCurdy, and T.W. Packard, *Conventional and Bimodal Nuclear Thermal Rocket (NTR) Artificial Gravity Mars Transfer Vehicle Concepts.* 2014.
464. Yam, C.H., et al., *Preliminary Design of Nuclear Electric Propulsion Missions to the Outer Planets.* 2004.
465. Casani, J.R., et al., *Enabling a New Generation of Outer Solar System Missions: Engineering Design Studies for Nuclear Electric Propulsion*. 2021. **53**(4).
466. Palaszewski, B., *Assessing Propulsion and Transportation Issues With Mars' Moons.* 2020.
467. McGuire, M.L., et al., *Use of High-Power Brayton Nuclear Electric Propulsion (NEP) for a 2033 Mars Round-Trip Mission*. 2006. **813**: p. 222-229.
468. Cassady, R.J., et al., *Recent Advances in Nuclear Powered Electric Propulsion for Space Exploration.* Energy Conversion and Management, 2008. **49**(3): p. 412-435.
469. Mason, L.S., *A Comparison of Energy Conversion Technologies for Space Nuclear Power Systems.* 2018.

470. Johnson, C.L., et al., *Status of Solar Sail Technology Within NASA*. Advances in Space Research, 2011. **48**(11): p. 1687-1694.
471. Guo, Y., et al., *The Earth-Mars Transfer Trajectory Optimization of Solar Sail Based On<i> Hp</I>-Adaptive Pseudospectral Method*. Discrete Dynamics in Nature and Society, 2018. **2018**: p. 1-14.
472. Bovesecchi, G., et al., *A Novel Self-Deployable Solar Sail System Activated by Shape Memory Alloys*. Aerospace, 2019. **6**(7): p. 78.
473. Boschetto, A., et al., *Shape Memory Activated Self-Deployable Solar Sails: Small-Scale Prototypes Manufacturing and Planarity Analysis by 3D Laser Scanner*. Actuators, 2019. **8**(2): p. 38.
474. Wilkie, W.K., et al., *Overview of the NASA Advanced Composite Solar Sail System (ACS3) Technology Demonstration Project*. 2021.
475. Quarta, A.A., et al., *Optimal Interplanetary Trajectories for Sun-Facing Ideal Diffractive Sails*. Astrodynamics, 2023. **7**(3): p. 285-299.
476. Kezerashvili, V.Y. and R.Y. Kezerashvili, *Solar Sail With Superconducting Circular Current-Carrying Wire*. 2021.
477. Costanza, G. and M.E. Tata, *Shape Memory Alloys for Aerospace, Recent Developments, and New Applications: A Short Review*. Materials, 2020. **13**(8): p. 1856.
478. Bassetto, M., et al., *Optimal Heliocentric Transfers of a Sun-Facing Heliogyro*. Aerospace Science and Technology, 2021. **119**: p. 107094.
479. Liu, J., et al., *Dynamics and Control of a Flexible Solar Sail*. Mathematical Problems in Engineering, 2014. **2014**: p. 1-25.
480. Gohardani, A.S., *A Historical Glance at Solar Sails*. 2014.
481. Heller, R. and M. Hippke, *Deceleration of High-Velocity Interstellar Photon Sails Into Bound Orbits at A Centauri*. The Astrophysical Journal, 2017. **835**(2): p. L32.
482. Matloff, G.L., *The Solar Photon Sail Comes of Age*. Astronomical Review, 2012. **7**(3): p. 5-12.
483. Kezerashvili, R.Y., et al., *Inflation Deployed Torus-Shaped Solar Sail Accelerated via Thermal Desorption of Coating*. 2019.
484. Campbell, M.F., et al., *Relativistic Light Sails Need to Billow*. 2021.
485. Kulkarni, N., P.M. Lubin, and Q. Zhang, *Relativistic Spacecraft Propelled by Directed Energy*. The Astronomical Journal, 2018. **155**(4): p. 155.
486. Holdman, G.R., et al., *Thermal Runaway of Silicon-Based Laser Sails*. Advanced Optical Materials, 2022. **10**(19).
487. She, H., W. Hettel, and P.M. Lubin, *Directed Energy Interception of Satellites*. Advances in Space Research, 2019. **63**(12): p. 3795-3815.
488. Daukantas, P., *Breakthrough Starshot*. Optics and Photonics News, 2017. **28**(5): p. 26.
489. Jin, W., et al., *Inverse Design of Lightweight Broadband Reflector for Relativistic Lightsail Propulsion*. Acs Photonics, 2020. **7**(9): p. 2350-2355.
490. Zhu, J.-P., B. Zhang, and Y.-P. Yang, *Relativistic Astronomy. II. In-Flight Solution of Motion and Test of Special Relativity Light Aberration*. The Astrophysical Journal, 2019. **877**(1): p. 14.
491. Kammash, T. and D.L. Galbraith, *Antimatter-Driven Fusion Propulsion Scheme for Solar System Exploration*. Journal of Propulsion and Power, 1992. **8**(3): p. 644-649.
492. Cassenti, B.N. and T. Kammash, *Future of Antiproton Triggered Fusion Propulsion*. 2009.
493. Cassenti, B.N., T. Kammash, and D.L. Galbraith, *Antiproton Catalyzed Fusion Propulsion for Interplanetary Missions*. 1996.

494. Lewis, R., et al., *Antiproton-Catalyzed Microfission/Fusion Propulsion Systems for Exploration of the Outer Solar System and Beyond.* 1998. **420**: p. 1365-1372.
495. Kezerashvili, R.Y., *Exploration of the Solar System and Beyond Using a Thermonuclear Fusion Drive.* 2021.
496. Emrich, W., *First Results of the Gasdynamic Mirror Fusion Propulsion Experiment.* 2003. **654**: p. 483-489.
497. Gabrielli, R., et al., *Effect of Nuclear Side Reactions on Magnetic Fusion Reactors in Space.* 2012.
498. Gajeri, M., P. Aime, and R.Y. Kezerashvili, *A Titan Mission Using the Direct Fusion Drive.* 2020.
499. Romanelli, F. and C. Bruno, *Assessment of Open Magnetic Fusion for Space Propulsion.* 2006.
500. Emrich, W. and C.W. Hawk, *Magnetohydrodynamic Instabilities in a Simple Gasdynamic Mirror Propulsion System.* Journal of Propulsion and Power, 2005. **21**(3): p. 401-407.
501. Gnesi, S. and T. Margaria, *Requirements of an Integrated Formal Method for Intelligent Swarms.* 2012: p. 33-59.
502. Ye, Z. and Q. Zhou, *Performance Evaluation Indicators of Space Dynamic Networks Under Broadcast Mechanism.* Space Science & Technology, 2021. **2021**.
503. Mejía–Kaiser, M., *Space Law and Hazardous Space Debris.* 2020.
504. Paliouras, Z.A., *The Non-Appropriation Principle: The <i>Grundnorm</I> of International Space Law.* Leiden Journal of International Law, 2014. **27**(1): p. 37-54.
505. Gupta, B. and R. Kd, *Understanding International Space Law and the Liability Mechanism for Commercial Outer Space Activities—Unravelling the Sources.* India Quarterly a Journal of International Affairs, 2019. **75**(4): p. 555-578.
506. Popova, R. and V. Schaus, *The Legal Framework for Space Debris Remediation as a Tool for Sustainability in Outer Space.* Aerospace, 2018. **5**(2): p. 55.
507. Nucera, G.G., *International Geopolitics and Space Regulation.* 2019.
508. Marinich, V.K. and M.I. Myklush, *Space Law, Subjects and Jurisdictions: Pre-1963 Period.* Analytical and Comparative Jurisprudence, 2023(4): p. 569-581.
509. Ross, M.N. and P. Sheaffer, *Radiative Forcing Caused by Rocket Engine Emissions.* Earth S Future, 2014. **2**(4): p. 177-196.
510. Dallas, J., et al., *The Environmental Impact of Emissions From Space Launches: A Comprehensive Review.* Journal of Cleaner Production, 2020. **255**: p. 120209.
511. Voigt, C., et al., *Impact of Rocket Exhaust Plumes on Atmospheric Composition and Climate — An Overview.* 2013.
512. Ross, M.N., M.J. Mills, and D.W. Toohey, *Potential Climate Impact of Black Carbon Emitted by Rockets.* Geophysical Research Letters, 2010. **37**(24).
513. Larsson, A. and N. Wingborg, *Green Propellants Based on Ammonium Dinitramide (ADN).* 2011.
514. Lock, A.C. and A. Sóbester, *Suborbital Air-Launch of Very Light Payloads From a Fixed Wing Platform.* 2016.
515. Orgeira-Crespo, P., et al., *Optimization of the Conceptual Design of a Multistage Rocket Launcher.* Aerospace, 2022. **9**(6): p. 286.
516. Fukunari, M., et al., *Replacement of Chemical Rocket Launchers by Beamed Energy Propulsion.* Applied Optics, 2014. **53**(31): p. I16.

517. Chen, D., et al., *Aerodynamic and Static Aeroelastic Computations of a Slender Rocket With All-Movable Canard Surface.* Proceedings of the Institution of Mechanical Engineers Part G Journal of Aerospace Engineering, 2017. **232**(6): p. 1103-1119.
518. Okniński, A., J. Kindracki, and P. Wolański, *Rocket Rotating Detonation Engine Flight Demonstrator.* Aircraft Engineering and Aerospace Technology, 2016. **88**(4): p. 480-491.
519. Cichocki, M. and D. Sokolowski, *Lessons Learned and the Recent Achievements of a Three-Stage Suborbital Rocket Production.* Safety & Defense, 2023. **9**(1): p. 47-57.
520. Strunz, R. and J.W. Herrmann, *Reliability as an Independent Variable Applied to Liquid Rocket Engine Hot Fire Test Plans.* Journal of Propulsion and Power, 2011. **27**(5): p. 1032-1044.
521. Kobald, M., et al., *The HyEnD Stern Hybrid Sounding Rocket Project.* 2019: p. 25-64.
522. Nomura, K., et al., *Tipping Points of Space Debris in Low Earth Orbit.* International Journal of the Commons, 2024. **18**(1).
523. Yang, W., et al., *Target Selection for a Space-Energy Driven Laser-Ablation Debris Removal System Based on Ant Colony Optimization.* Sustainability, 2023. **15**(13): p. 10380.
524. Costigliola, D. and L. Casalino, *Simplified Maneuvering Strategies for Rendezvous in Near-Circular Earth Orbits.* Aerospace, 2023. **10**(12): p. 1027.
525. Blaise, J. and M.C.F. Bazzocchi, *Space Manipulator Collision Avoidance Using a Deep Reinforcement Learning Control.* Aerospace, 2023. **10**(9): p. 778.
526. Pesaresi, M., V. Syrris, and A. Julea, *A New Method for Earth Observation Data Analytics Based on Symbolic Machine Learning.* Remote Sensing, 2016. **8**(5): p. 399.
527. Kanazaki, M., Y. Yamada, and M. Nakamiya, *Multi-Objective Path Optimization of a Satellite for Multiple Active Space Debris Removal Based on a Method for the Travelling Serviceman Problem.* Advances in Science Technology and Engineering Systems Journal, 2018. **3**(6): p. 479-488.
528. Ma, B., et al., *Advances in Space Robots for On-Orbit Servicing: A Comprehensive Review.* Advanced Intelligent Systems, 2023. **5**(8).
529. Rybus, T., et al., *Application of the Obstacle Vector Field Method for Trajectory Planning of a Planar Manipulator in Simulated Microgravity.* Artificial Satellites, 2023. **58**(s1): p. 171-187.
530. Wilde, M., J. Harder, and E. Stoll, *Editorial: On-Orbit Servicing and Active Debris Removal: Enabling a Paradigm Shift in Spaceflight.* Frontiers in Robotics and Ai, 2019. **6**.
531. Seddaoui, A., C.M. Saaj, and M.H. Nair, *Modeling a Controlled-Floating Space Robot for in-Space Services: A Beginner's Tutorial.* Frontiers in Robotics and Ai, 2021. **8**.
532. Albee, K., et al., *A Robust Observation, Planning, and Control Pipeline for Autonomous Rendezvous With Tumbling Targets.* Frontiers in Robotics and Ai, 2021. **8**.
533. Kempler, S. and T.J. Mathews, *Earth Science Data Analytics: Definitions, Techniques and Skills.* Data Science Journal, 2017. **16**.
534. Rao, A., M.G. Burgess, and D.T. Kaffine, *Orbital-Use Fees Could More Than Quadruple the Value of the Space Industry.* Proceedings of the National Academy of Sciences, 2020. **117**(23): p. 12756-12762.
535. Bhattarai, S. and J.-R. Shang, *Space Debris Removal Mechanism Using CubeSat With Gun Shot Facilities.* American Journal of Applied Sciences, 2018. **15**(9): p. 456-463.
536. Zhao, S., et al., *Target Sequence Optimization for Multiple Debris Rendezvous Using Low Thrust Based on Characteristics of SSO.* Astrodynamics, 2017. **1**(1): p. 85-99.
537. Papadopoulos, E., et al., *Robotic Manipulation and Capture in Space: A Survey.* Frontiers in Robotics and Ai, 2021. **8**.

538. Murtaza, A., et al., *Orbital Debris Threat for Space Sustainability and Way Forward (Review Article).* Ieee Access, 2020. **8**: p. 61000-61019.
539. Li, W., et al., *On-Orbit Service (OOS) of Spacecraft: A Review of Engineering Developments.* Progress in Aerospace Sciences, 2019. **108**: p. 32-120.
540. Adamopoulos, G., *Optimizing the Space Debris Removal Process: An in-Depth Analysis of Current Debris Removal Technologies.* Journal of Student Research, 2023. **12**(4).
541. Becker, M., I. Retat, and E. Stoll, *Influence of Orbital Perturbations on Tethered Space Systems for Active Debris Removal Missions.* 2015.
542. Mark, C.P. and S. Kamath, *Review of Active Space Debris Removal Methods.* Space Policy, 2019. **47**: p. 194-206.
543. Алпатов, А.П. and Y.M. Holdshtein, *Assessment Perspectives for the Orbital Utilization of Space Debris.* Kosmìčna Nauka Ì Tehnologìâ, 2021. **27**(3): p. 3-12.
544. Rivière, A., *Potential Export Control Challenges and Constraints for Emerging Space Debris Detection and Removal Technologies: The Case of on-Orbit Collision.* Advances in Astronautics Science and Technology, 2020. **3**(2): p. 105-114.
545. Yang, C., et al., *Big Earth Data Analytics: A Survey.* Big Earth Data, 2019. **3**(2): p. 83-107.

Index